AF606488

NEURAL ADAPTIVE CONTROL TECHNOLOGY

WORLD SCIENTIFIC SERIES IN ROBOTICS AND INTELLIGENT SYSTEMS

Editor-in-Charge: C J Harris (*University of Southampton*)
Advisor: T M Husband (*University of Salford*)

Vol. 1: Genetic Algorithms and Robotics — A Heuristic Strategy for Optimization (*Y Davidor*)

Vol. 2: Parallel Computation Systems for Robotics: Algorithms and Architectures (*Eds. A Bejczy and A Fijany*)

Vol. 3: Intelligent Robotic Planning Systems (*P C-Y Sheu and Q Xue*)

Vol. 4: Computer Vision, Models and Inspection (*A D Marshall and R R Martin*)

Vol. 5: Advanced Tactile Sensing for Robotics (*Ed. H R Nicholls*)

Vol. 6: Intelligent Control: Aspects of Fuzzy Logic and Neural Nets (*C J Harris, C G Moore and M Brown*)

Vol. 7: Visual Servoing: Real-Time Control of Robot Manipulators Based on Visual Sensory Feedback (*Ed. K Hashimoto*)

Vol. 8: Modelling and Simulation of Robot Manipulators: A Parallel Processing Approach (*A Y Zomaya*)

Vol. 9: Advanced Guided Vehicles – Aspects of the Oxford AGV Project (*S Cameron and P Probert*)

Vol. 10: Cellular Robotics and Micro Robotic Systems (*T Fukuda and T Ueyama*)

Vol. 11: Recent Trends in Mobile Robots (*Ed. Y F Zheng*)

Vol. 12: Intelligent Assembly Systems (*Eds. M Lee and J J Rowland*)

Vol. 15: Neural Adaptive Control Technology (*Eds. R Żbikowski and K J Hunt*)

Forthcoming:

Vol. 13: A Methodology of Systems Integration for Intelligent Autonomous Systems (*R J C Fraser and C J Harris*)

Vol. 14: Intelligent Supervisory Control Based on Qualitative Bond Graph Reasoning (*D A Linkens and H Wang*)

Vol. 16: Advances in Robotics & Automation for Hazardous Environment (*Eds. P Lever and F Y Wang*)

World Scientific Series in Robotics and Intelligent Systems – Vol.15

NEURAL ADAPTIVE CONTROL TECHNOLOGY

Editors

RAFAŁ ŻBIKOWSKI
Glasgow University

KENNETH J HUNT
Daimler-Benz AG

Published by

World Scientific Publishing Co Pte Ltd
P O Box 128, Farrer Road, Singapore 912805
USA office: Suite 1B, 1060 Main Street, River Edge, NJ 07661
UK office: 57 Shelton Street, Covent Garden, London WC2H 9HE

Library of Congress Cataloging-in-Publication Data
Neural Adaptive Control Technology / editors. R. Żbikowski, K. J. Hunt.
p. cm. -- (World Scientific series in robotics and intelligent systems ; vol. 15)
Includes bibliographical references and index.
ISBN 9810225571
1. Adaptive control systems. 2. Neural networks (Computer science)
3. Nonlinear control theory. I. Żbikowski, R. (Rafał)
II. Hunt, K. J. (Kenneth J.), 1963– . III. Series.
TJ217.N47 1996
629.8'9--dc20 96-951
CIP

British Library Cataloguing-in-Publication Data
A catalogue record for this book is available from the British Library.

This book is printed on acid-free paper.

Printed in Singapore by Uto-Print

PREFACE

This book is an outgrowth of the workshop on Neural Adaptive Control Technology, *NACT I*, held on May 18–19, 1995 in Glasgow. However, this book is not simply the conference proceedings. Instead, selected workshop participants were asked to substantially expand and revise their contributions to make them into full papers.

Before the contents of the book is discussed, it seems in order to briefly sketch the background and purpose of the workshop. The event was organised in connection with a three-year European Union funded Basic Research Project in the ESPRIT framework, called NACT, a collaboration between Daimler-Benz (Germany) and the University of Glasgow (Scotland).

The NACT project, which began on 1 April 1994, is a study of the fundamental properties of neural network based adaptive control systems. Where possible, links with traditional adaptive control systems are exploited. A major aim is to develop a systematic engineering procedure for designing neural controllers for non-linear dynamic systems. The techniques developed are being evaluated on concrete industrial problems from within the Daimler-Benz group of companies.

This context dictated the focus of the workshop and guided the editors in the choice of the papers and their subsequent reshaping into substantive book chapters. Thus, emphasis is put on development of a sound theory of neural adaptive control for non-linear control systems, but firmly anchored in the engineering context of industrial practice. Therefore, the contributors are both renowned academics and practitioners from major industrial users of neurocontrol.

The book naturally divides into three parts.

Part I is devoted to the theoretical and practical results on neural adaptive control technology resulting from the NACT project. Chapter 1 by J. C. Kalkkuhl and K. J. Hunt analyses several important fundamental issues so far largely ignored in the neurocontrol context. The issue of prime importance, and thus treated first, is that of the discretisation of continuous-time models. The physical plants are continuous-time, but the practicality of digital implementations require discrete-time representations. A careful discussion is presented exposing the limitations of NARMAX models, widely used in neurocontrol. This sets the stage for the finite-element method approach to approximation of NARMAX models. Chapter 2 is written by Peter J. Gawthrop, a well-known contributor to adaptive control, in particular, continuous-time self-tuning. Following this line, the continuous-time version of Local Model Networks/Controllers is introduced. The structure has some surprising connections to the state observation problem. This leads to the important distinction between local and global states of the model. The exposition is accompanied by simulations of essentially non-linear systems. Chapter 3 by R. Żbikowski and A. Dzieliński introduces and describes in some detail the nonuniform multi-dimensional sampling approach to neurocontrol with feedforward neural networks. The authors argue that this is a natural theoretical framework for practical control engineering problems, because the measured data representing the NARMA model come as multidimensional samples. The dynamics of the underlying system manifest themselves by nonuniformity of the data and thus the irregular spread of the samples is an essential feature of the representation. A novel method of neural modelling of NARMA systems is given. Important practical issues of distortions caused by approximation of the Fourier transform are addressed. A tutorial survey of the theory of Paley-Wiener functions, with emphasis on the neural modelling aspects, completes the presentation.

Part II is devoted to results of non-linear control, relevant to theory of neurocontrol. It opens with Chapter 4 by Witold Respondek, whose pioneering work in the beginning of the 1980s resulted in explosive development (lasting to this day) of geometric methods in non-linear control. In fact, 'geometric control' has practically become synonymous with non-linear control. Recently, the rich and consistent theory has been used more often in the context of neurocontrol, because of the need for a control framework for (non-linear) neural models. Being an important and active participant in the development of the geometric approach, Respondek offers valuable insights into the underlying mathematics, while never compromising on rigour. His lucid style is supported by numerous illustrations and examples making the chapter a readable and informative introduction. It is a welcome feature for a subject dominated by presentations often deprived of geometric feeling and overloaded with distracting technicalities. Chapter 5 gives another theoretical perspective, this time from Tadeusz Kaczorek, a major contributor to the theory of 2-D control systems. Chapter 6 by T. A. Johansen and M. M. Polycarpou deals with the important, yet often neglected, issue of stability of adaptive control of non-linear systems.

Part III presents various aspects of neural control and its applications. It starts with Chapter 7 by J-M. Renders and M. Saerens. These authors develop local stability results for an adaptive neurocontrol strategy. Weight adaptation is based upon Lyapunov stability theory. The next chapter, Chapter 8, is co-authored by the well-known neural networks researchers I. Rivals and L. Personnaz who consider state-space models for neurocontrol as an alternative to the predominant input-output approach. Chapter 9 brings the intelligent control perspective on neural issues by one of the major players in the field, Donald A. Sofge. Chapter 10 by W. S. Mischo (from Henning Tolle's CMAC school) presents theory and applications of CMAC-type memories for learning control. Finally, Chapter 11 by G. Lines and T. Kavli presents the results of applying spline-based adaptive methods for the development of dynamics models. The interpretation of the spline models as fuzzy systems is also examined.

Rafał Żbikowski, Kenneth Hunt
Glasgow, Berlin: December, 1995

CONTENTS

Part I

Neural Adaptive Control Technology

DISCRETE-TIME NEURAL MODEL STRUCTURES FOR CONTINUOUS-TIME NONLINEAR SYSTEMS: FUNDAMENTAL PROPERTIES AND CONTROL ASPECTS

J. Kalkkuhl and K. J. Hunt
Daimler-Benz AG, Alt Moabit 96A,
10559 Berlin, Germany
E-mail: kalkkuhl@DBresearch-berlin.de, hunt@DBresearch-berlin.de

ABSTRACT

A unified framework for discrete-time control of continuous-time nonlinear systems based on a neural plant model is presented. The exploitation of *a-priori* structural information for setting up neural model structures is combined with a special type of basis function for modelling and a two-degrees of freedom nonlinear internal model control strategy.

1. Introduction

Any control design is based on a model of the plant to be controlled. Neural network models of the plant have been widely used as a basis for control design in the following cases:

1. the plant is a continuous-time nonlinear dynamical system
2. with only partially known, uncertain or unknown structure and/or parameters

The identification of a discrete-time plant model from input output data is of particular interest for applications and a number of neural network structures for representation of the dynamic behaviour of the plant have been suggested [33,37,8,16].

From a system theoretic as well as from a practical point of view it is highly unsatisfactory to treat a nonlinear plant as a black box and to identify it using general neural network structures that do not exploit structural information. This is due to the following reasons:

- The fact that a flexible neural network structure is able to reproduce a limited set of input output sequences does not necessarily mean that the dynamic behaviour of the plant is represented properly.
- The number of parameters in black box neural networks tends to be very large.
- System properties that may be essential for control design cannot be easily analysed for a complex neural network model.

One of the main obstacles to using available structural information is that for nonlinear systems the relation between the original system structure and the structure of the discrete-time input-output model to be identified has not been well understood. However, parallel to the development of neural network based identification, the theory of sampling of certain classes of nonlinear systems has been developed by Monaco and Normand-Cyrot [27,30] and Kotta [20]. Thus, it has become possible to analyse more deeply the correspondence between the original system and its discrete-time representations and to incorporate structural information into a neural network model of the system. Of particular interest is the result that any sampled representation of a continuous-time nonlinear system will always be nonlinear in the control variable even if the original system is linear in the control. This result has often been ignored leading to neural network model structures that are not meaningful for controlling nonlinear continuous-time systems (e.g. in [9]). A paramount role for nonlinear systems identification is played by the external or NARX (Nonlinear Autoregressive with eXogenous input) representation [10,34,24] which relates the output $y[k]$ at the discrete time instant k to past inputs and outputs:

$$y[k] = \phi(y[k-1], \ldots, y[k-n], u[k-d], \ldots, u[k-n]).$$

A key problem in this respect is the question of existence of this type of model for a given discrete-time state space system. This problem (closely related to the observability problem) has been investigated in [23,10,34,1,32,24].

Recently, a number of adaptive and nonadaptive control approaches based on discrete-time neural plant models have been suggested one of the most popular being nonlinear model predictive control techniques of which nonlinear internal-model control is a special case [11]. In contrast to the linear case, however, in the nonlinear case it is not easy to relate the parameters of the NMPC cost index to the desired performance of the control system. A first theoretical analysis of the properties of nonlinear internal model control was attempted by Henson and Seborg [15].

Meanwhile, the geometric approach to nonlinear control theory has provided a variety of tools for the analysis and design of control systems[31,12,29,28,17,35,21]. Based on this approach, a two-degrees of freedom internal model controller based on geometric design techniques was proposed in [3,25]. For this type of controller no cost index needs to be specified and no optimisation procedure is required.

The purpose of this chapter is to give an overview of issues of neural modelling and control for a class of nonlinear continuous-time systems. Our main objective is to incorporate plant modelling and control design aspects within a unified framework that allows for a straightforward practical implementation. As depicted in Fig. 1 the focus of our analysis will be the incorporation of both *a priori* information on the plant as well as the requirements of the control design in the neural plant model.

It is our belief that model structures and basis function should not only be chosen to fit the available input output data with high accuracy. They also should reflect

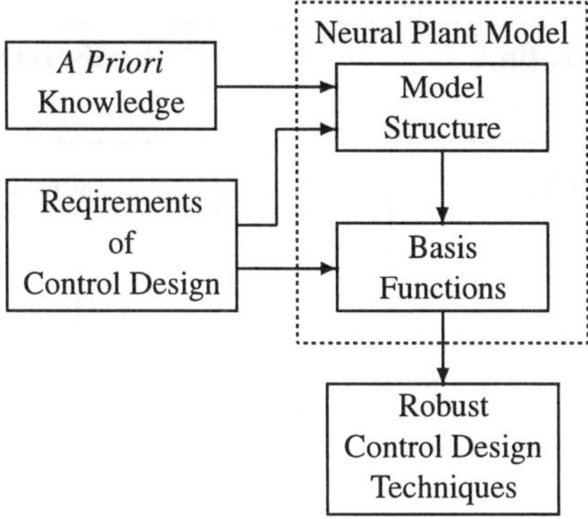

Figure 1: Neural Network Plant Model and Control

important structural properties of the original plant and allow for an efficient control design.

Since there will always be modelling errors, the robustness of the control design techniques used is also an important issue.

Section 2 of this chapter will be dedicated to structural questions of neural modelling while in Section 3 the finite element method (FEM) to the NARX identification problem will be presented. In Section 4 we will discuss a robust two-degrees of freedom internal model control strategy that can be based on neural process models.

2. Sampling of continuous-time nonlinear systems

In this section the problem of modelling a nonlinear process from discrete time input-output data using neural network techniques is considered. The main objective is to use the resulting model for control purposes. It will be analysed how the dynamical properties and structure of the continuous-time plant are reflected in the properties and structure of the discrete-time input-output model. Such an analysis will be carried out here for the case that the plant can be represented in continuous-time by a linear analytic SISO system.

Fig. 2 indicates that a NARX model can be derived from the continuous state equation by discretisation and state elimination. Note that while the original continuous system is linear in the control variable, the discrete representation is not.

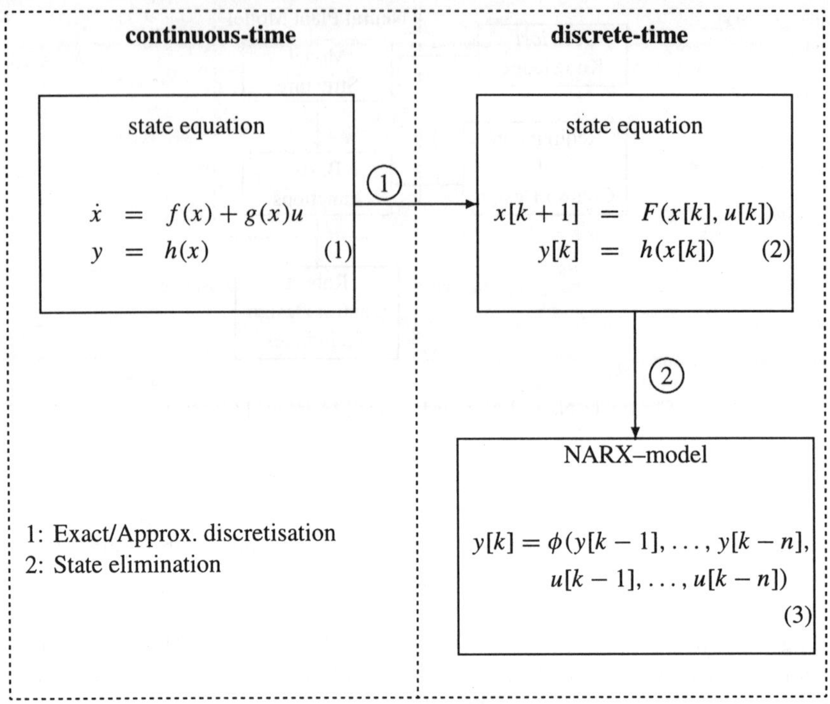

Figure 2: Relation between continuous-time state space system and NARX representation

2.1. *Exact Discretisation*

The *exact discretisation* (ED) of system (1) in Fig. 2 is the discrete-time state equation (2) that reproduces the continuous-time solution at the sampling instants, i.e. with the initial states $x(0) = x[0] = x_0$ being equal, the solution $x[k]$ of (2) will coincide with the solution of (1) at the sampling instants:

$$x[k] = x(t) \quad \text{for} \quad t = kT.$$

It is shown [27] that the exact discretisation is given by the Lie-series

$$\begin{aligned} x(t+T) &= e^{T(L_f + u(t)L_g)}(x(t)) \\ &= (I + TL_f + Tu(t)L_g + \frac{1}{2}T^2[L_f + u(t)L_g]^{\circ 2} + \ldots)(x(t)) \\ &= x(t) + Tf(x(t)) + Tu(t)g(x(t)) + \frac{1}{2}T^2 L_f f(x(t)) \end{aligned}$$

$$+\frac{1}{2}T^2u(t)[L_fg(x(t)) + L_gf(x(t))] + \frac{1}{2}T^2u^2(t)L_gg(x(t)) + \ldots$$

which leads to

$$\begin{aligned} x[k+1] &= F(x[k], u[k]) = F_0(x[k]) + \sum_{i=1}^{\infty} F_i(x[k]) \cdot u^i[k], \\ \bar{y}[k] &= h(x[k]). \end{aligned}$$

The ED is an infinite power series in u and T that will converge if the input signal u is bounded and the sampling period T is sufficiently small.

A pth order *approximate discretisation* of (1) can be obtained by considering terms up to the order of T^p only. For the first order approximation $p = 1$ we obtain the well known Euler-discretisation

$$\begin{aligned} x[k+1] &= \left\{I + T(L_f + u[k]L_g)\right\}(\bar{x}[k]) + \mathcal{O}(T^2) \\ &= \bar{x}[k] + T(f(\bar{x}[k]) + u[k]g(\bar{x}[k]) + \mathcal{O}(T^2). \end{aligned} \tag{4}$$

Euler discretisations in general will require very small T. An extension of the ED problem to a more general class of nonlinear systems was given in a paper by Kotta [20] where a recursive scheme for calculating the elements of the Lie series can also be found.

System (2) is said to have relative degree d at x^0 if

$$\begin{aligned} \frac{\partial(h \circ F_0^{\circ k} \circ F(x, u))}{\partial u} &\equiv 0 \quad \text{for all } x \quad \text{in a neighbourhood of } x^0 \\ & \qquad\qquad \text{and} \quad 0 \le k < d-1 \\ \frac{\partial(h \circ F_0^{\circ d-1} \circ F(x, u))}{\partial u} &\neq 0 \quad \text{at } x = x^0 \end{aligned} \tag{5}$$

holds where $F_0(\bullet) = F(\bullet, 0)$ represents the undriven state dynamics and $F_0^{\circ k}$ the k-times iterated composition of F_0 [28]. Equivalent to the notion of relative degree for continuous time systems [17], $y[d]$ will be the first output affected by $u[0]$. Consequently, there is a delay of d time steps between input and output of the system. For the relative degree we will either obtain $0 < d \le n$ or $d = \infty$ [28]. In the latter case the output is not affected by the input at all. It can easily be shown [30] that any exact discretisation of a continuous-time system will have relative degree $d = 1$, while the relative degree of the Euler discretisation equals that of the original system $d = r$.

If the system (2) has relative degree d at $x^0 = 0$ then there exists a smooth local coordinate transform

$$\begin{pmatrix} z \\ \eta \end{pmatrix} = \Phi(x), \quad \text{with} \quad \Phi(0) = 0 \quad \text{and} \quad \left.\frac{\partial \Phi}{\partial x}\right|_{x=0} \neq 0$$

such that in the new coordinates the system can be represented in *normal form* [28]

$$
\begin{aligned}
z_1[k+1] &= z_2[k] \\
z_2[k+1] &= z_3[k] \\
&\vdots \\
z_{d-1}[k+1] &= z_d[k] \\
z_d[k+1] &= \alpha_1(z[k], \eta[k], u[k]) \qquad (6)\\
\eta[k+1] &= \alpha_2(z[k], \eta[k], u[k]) \qquad (7)\\
y[k] &= z_1[k] \qquad (8)
\end{aligned}
$$

with

$$
z[k] = \begin{pmatrix} z_1[k] \\ \vdots \\ z_d[k] \end{pmatrix}, \qquad \text{and} \qquad \frac{\partial \alpha_1(z, \eta, u)}{\partial u} \neq 0.
$$

In Figure 3 the structure of a discrete-time state space representation in normal form is shown. It is easily seen that a model of the system can be realised using a neural

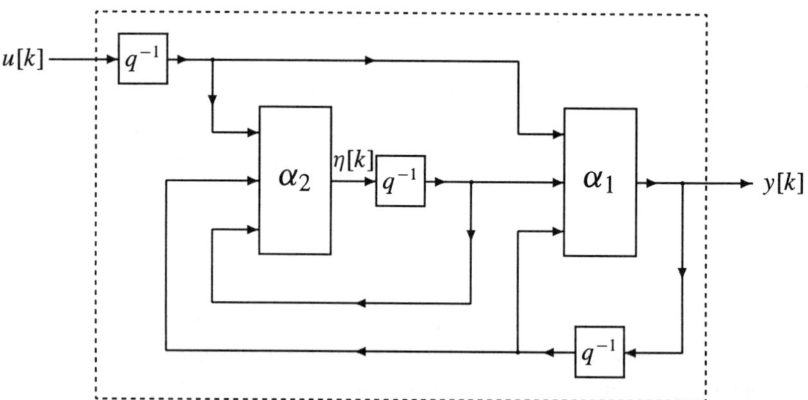

Figure 3: Structure of a discrete-time state space system in normal form.

network with at least one internal feedback loop where the nonlinear functions α_1 and α_2 can be parameterised appropriately. The identification of such a state space model from input output data always requires the solution of a nonlinear parameter estimation problem, since the state vector η cannot be measured. As pointed out by Narendra et al.[33] in this case identification can be quite complicated and convergence of the parameters is not guaranteed.

2.2. *NARX model*

The states z_i of the normal form are equal to the output shifted $i-1$ steps forward in time:

$$y[k+i-1] = z_i[k], \quad i = 1, \ldots, d. \tag{9}$$

The problem of expressing the remaining part η of the state vector in terms of input and output values is considered now.

The state vector η has the dimension $n-d$. From Eqs. (6)–(9) the following system of $n-d+1$ nonlinear equations can be obtained:

$$\begin{aligned}
y[k+d] &= \alpha_1(z[k], \eta[k], u[k]) \\
&= \Psi_1(y[k+d-1], \ldots, y[k], \eta[k], u[k]) \\
y[k+d+1] &= \alpha_1(z[k+1], \eta[k+1], u[k+1]) \\
&= \alpha_1(z[k+1], \alpha_2(z[k], \eta[k], u[k]), u[k+1]) \\
&= \Psi_2(y[k+d], \ldots, y[k], \eta[k], u[k+1], u[k]) \\
&\vdots \\
y[k+n-1] &= \Psi_{n-d}(y[k+n-2], \ldots, y[k], \eta[k], u[k+n-d-1], \ldots, u[k]) \\
y[k+n] &= \Psi_{n-d+1}(y[k+n-1], \ldots, y[k], \eta[k], u[k+n-d], \ldots, u[k])
\end{aligned} \tag{10}$$

Let P^0 denote the abbreviation for the equilibrium point $(z^0, \eta^0) = (0, 0)$ of the system where

$$y[k+i] = 0,\ i \geq 0, \qquad u[k+j] = 0,\ j \geq 0$$

holds. Consider the $n-d$-dimensional vector function

$$\Psi(y[k+n-2], \ldots, y[k], \eta[k], u[k+n-d-1], \ldots, u[k]) = \begin{bmatrix} \Psi_1 \\ \vdots \\ \Psi_{n-d} \end{bmatrix}.$$

If the linearisation of system (6)–(7) about P^0 is observable then Ψ will have a nonsingular Jacobian

$$\left.\frac{\partial \Psi}{\partial \eta}\right|_{P^0} \neq 0$$

at P^0 [23,34]. Then according to the implicit function theorem there exists a function Θ such that

$$\eta[k] = \Theta(y[k+n-1], \ldots, y[k], u[k+n-d-1], \ldots, u[k]) \tag{11}$$

in a neighbourhood of P^0. Thus, Eq. (11) can be used to replace the remaining state variables in Eq. (10) by input and output variables and an *external representation*

$$\begin{aligned} y[k+n] &= \Psi_{n-d+1}(y[k+n-1], \ldots, y[k], \\ &\qquad \Theta(y[k+n-1], \ldots, y[k], u[k+n-d-1], \ldots, u[k]), \\ &\qquad u[k+n-d], \ldots, u[k]) \\ &= \phi(y[k+n-1], \ldots, y[k], u[k+n-d], \ldots, u[k]) \end{aligned}$$

can be obtained which is at least locally valid. By shifting backwards $n-d$ steps we finally obtain

$$y[k+d] = \phi(y[k+d-1], \ldots, y[k-n+d], u[k], \ldots, u[k-n+d])$$

where the output value at time $y[k+d]$ is related to output and input values at previous time instants.

It has been shown in [23,34] that the external representation of system (2) exists at P^0 if the linearisation of (2) at this point is observable. This can be veryfied in the following way. The linearisation of (6)–(7) around P^0 is observable if and only if the pair

$$(C_0, A_0) = \left(\left.\frac{\partial \alpha}{\partial \eta}\right|_{P^0}, \left.\frac{\partial q}{\partial \eta}\right|_{P^0} \right)$$

is observable. However, the Jacobian of Ψ at P^0 is given by

$$\left.\frac{\partial \Psi}{\partial \eta}\right|_{P^0} = \begin{bmatrix} C_0 \\ C_0 A_0 \\ \vdots \\ C_0 A^{n-d-1} \end{bmatrix}$$

since

$$\begin{aligned} \left.\frac{\partial \Psi_i}{\partial \eta[k]}\right|_{P^0} &= \left.\frac{\partial \alpha_1(z[k+i-1], \eta[k+i-1], u[k+i-1])}{\partial \eta[k+i-1]}\right|_{P^0} \cdot \left.\frac{\partial \eta[k+i-1]}{\partial \eta[k]}\right|_{P^0} \\ &= C_0 A_0^{i-1}. \end{aligned}$$

Therefore, if the observability condition holds then the Jacobian of Ψ will be non-singular. The type of observability required here is referred to as strong observability.

It is well known that strong observability for nonlinear systems is only a local property since in general observability will depend on the input sequence. This means that there will be input sequences $u[k], \ldots, u[k-n+d]$ for which the external representation (3) does not exist. This problem can be solved by increasing the number of delayed inputs and outputs in the external representation in which case the observability condition can be weakened. It has been shown in [1,32,24] that any generically observable nonlinear system of order n can be represented by an external representation of the form

$$y[k+d] = \phi(y[k+d-1], \ldots, y[k-2n+d], u[k], \ldots, u[k-2n+d]).$$

The discrete-time external representation (3) of a nonlinear system is often referred to as a NARX representation (Nonlinear AutoRegressive with eXogenous input) in analogy to the linear ARX representation [23,10]. It plays a paramount role in nonlinear systems identification. The structure of a NARX model is depicted in Figure 4. The delay elements within the structure can be interpreted as states. Since

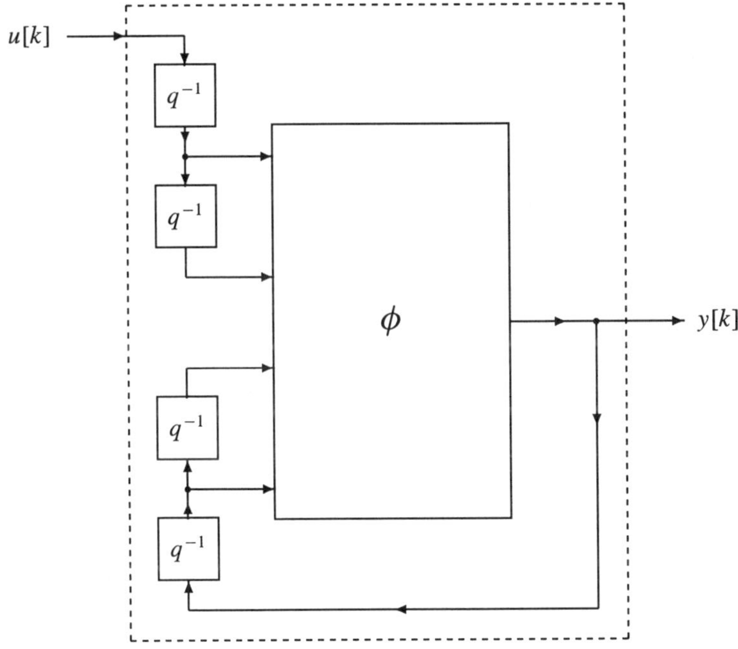

Figure 4: Structure of a NARX model

an external representation has at least $2n$ states it can be considered a non-minimal realisation of the state space system. All state variables are measurable and thus, from the identification viewpoint, there will be no internal feedback loop in the network structure. Consequently, as a difference to the state space model shown in Fig. 3

a NARX model can be realised as a feedforward network which can be *linearly parameterised* and requires only standard identification techniques. This advantage is traded off against an increase in the number of parameters in comparison to the state space representation, due to the increased number of states.

Using the techniques introduced above it can be shown that certain classes of systems, for instance polynomial systems or bilinear systems will result in special structures of the corresponding NARX representations. Thus, structural information about the continuous-time system can be used to generate specialised types of neural network models that will be more parsimonious in their parameterisation.

If input and initial conditions of system (2) have been chosen in such a way that the output remains constant for all times , the dynamics governing the internal behaviour of the system are called the zero dynamics [17]. For the linear case the zero dynamics are equivalent to the zeros of the transfer function whereas for a nonlinear system the zero dynamics can be obtained from the normal form (6)–(7): setting $y[k] = y[k+1] = \ldots = y[k+d] = y^0$ will result in an equation for the corresponding input value $u[k]$.

$$y^0 = \alpha_1(z^0, \eta, u[k]) \quad \text{where} \quad z^0 = \begin{bmatrix} z_1^0 \\ \vdots \\ z_d^0 \end{bmatrix}, \quad z_i^0 = y^0, \quad i = 1, \ldots, d. \tag{12}$$

If $u[k] = u^0$ is a solution to (12) then we obtain from (7)

$$\eta[k+1] = \alpha_2(z^0, \eta[k], u^0).$$

This is an $n-d$-dimensional autonomous state equation governing η and describing the zero dynamics of the system. The equivalent representation of the zero dynamics in terms of the NARX model is the autonomous difference equation in u

$$\phi(y^0, \ldots, y^0, u[k], \ldots, u[k-n+d]) = y^0 \tag{13}$$

obtained by setting $y[k+d], \ldots, y[k-n+d]$ in (3) to y^0. The zero-dynamics can be directly obtained from the external representation. Thus, in regions where it exists, the external representation is equivalent to the normal form of the system.

For control design it is essential to know whether the zero dynamics are stable or not. As in the linear case a nonlinear system is called

- minimum phase if its zero dynamics are stable and
- non-minimum phase otherwise.

Often, somewhat stronger assumptions about the zero-dynamics will be necessary:

- Considering the zero dynamics for the equilibrium point P^0 a system is said to be *hyperbolically minimum phase* if the zero dynamics are asymptotically stable in its linearisation about $\eta^0 = 0$ and therefore

$$\xi[k+1] = \left.\frac{\partial \alpha_2(0, \eta, 0)}{\partial \eta}\right|_{\eta=\eta^0} \xi[k]$$

  and

$$\sum_{i=0}^{n-d} B_i u[k-i] = 0, \qquad B_i = \left.\frac{\partial f(0, \ldots, 0, w_0, \ldots, w_{n-d})}{\partial w_i}\right|_{w_0=w_1=\ldots=w_{n-d}=0}$$

  are asymptotically stable linear systems.

- On the other hand, independently from a particular equilibrium point the zero-dynamics are said to be *bounded-input bounded-state stable* (BIBS) if
  - in Eq. (7) the state η will remain bounded for any bounded $z[k]$ or respectively
  - for Eq. (3) the time-varying difference equation

$$y_t[k+d] = \phi(y_t[k+d-1], \ldots, y_t[k-n+d], u[k], \ldots, u[k-n+d])$$

    is bounded in u for any prescribed bounded output signal y_t.

Since the relative degree of an exact discretisation is always $d = 1$ it is obvious that sampling of a system with $r > 1$ introduces additional zero dynamics whose stability will in general depend on r and the sampling period T:

- Continuous-time *linear* systems with $r > 2$ will result in sampled systems with unstable zero dynamics if the sampling period is sufficiently small [6].

- The latter result carries over to nonlinear systems since the operations of time-discretisation and linearisation commute [30] and consequently

- the exact discretisation of a nonlinear system with relative degree $r > 2$ will be non-minimum phase if T is chosen too small.

On the other hand, for a system with $r = 1$ the property of being hyperbolically minimum phase will not be affected by sampling [30].

Unstable zero dynamics may also be introduced if a NARX model is *identified* from input-output data samples with the sampling period being small and the relative degree of the plant being $r > 2$.

2.3. Examples

Bilinear System

For the bilinear system

$$\begin{aligned}\dot{x} &= Ax + (b + Dx)u \\ y &= Cx\end{aligned}$$

the p-th order discretisation assumes the form:

$$\begin{aligned}\bar{x}[k+1] &= \bar{A}_0\bar{x}[k] + (\bar{A}_1\bar{x}[k] + \bar{b}_1)u[k] + (\bar{A}_2\bar{x}[k] + \bar{b}_2)u^2[k] + \ldots \\ &\quad +(\bar{A}_p\bar{x}[k] + \bar{b}_p)u^p[k] \\ \bar{y}[k] &= C\bar{x}[k]\end{aligned}$$

which can be rewritten as

$$\begin{aligned}\bar{x}[k+1] &= \left\{\bar{A}_0 + u[k]\bar{A}_1 + u^2[k]\bar{A}_2 + \ldots + u^p[k]A_p\right\}\bar{x}[k] \\ &\quad +\bar{b}_1u[k] + \ldots + \bar{b}_pu^p[k] \\ \bar{y}[k] &= C\bar{x}[k]\end{aligned}$$

This can be abbreviated to

$$\begin{aligned}\bar{x}[k+1] &= P(u[k])\bar{x}[k] + Q(u[k]) \\ \bar{y}[k] &= C\bar{x}[k]\end{aligned} \tag{14}$$

with $P(u[k])$ and $Q(u[k])$ being polynomial matrices in $u[k]$.

To calculate the corresponding NARX representation we set

$$\begin{aligned}y[k] &= Cx[k] \\ y[k+1] &= Cx[k+1] = CP(u[k])x[k] + CQ(u[k]) \\ y[k+2] &= CP(u[k+1])P(u[k])x[k] \\ &\quad +CP(u[k+1])Q(u[k]) + CQ(u[k+1]) \\ y[k+3] &= CP(u[k+2])P(u[k+1])P(u[k])x[k] \\ &\quad +CP(u[k+2])P(u[k+1])Q(u[k]) + \\ &\quad +CP(u[k+2])Q(u[k+1]) + CQ(u[k+2])\end{aligned}$$

$$\begin{aligned}
&\vdots \\
y[k+n-1] &= CP(u[k+n-2])\cdots P(u[k])x[k] \\
&\quad +CP(u[k+n-2])\cdots P(u[k+1])Q(u[k]) + \ldots \\
&\quad \ldots + CQ(u[k+n-2])
\end{aligned}$$

This system can be solved for $x[k]$ if and only if the matrix

$$\Phi = C \cdot \begin{bmatrix} I \\ P(u[k]) \\ P(u[k+1])P(u[k]) \\ \vdots \\ P(u[k+n-2])\cdots P(u[k]) \end{bmatrix}$$

is nonsingular. For arguments $(u[k], \ldots, u[k+n-2])$ for which Φ is singular there does not exist a NARX representation. At such points the states of the bilinear system are not strongly observable. In general, at points where it exists, the NARX representation of a bilinear system has the structure

$$y[k] = p_0 + \sum_{i=1}^{n} p_i \cdot y[k-i] \tag{15}$$

where

$$p_i = p_i(u[k-1], u[k-2], \ldots, u[k-n]), \qquad i = 0, 1, \ldots, n$$

are rational functions in u. Since this representation is linear with respect to the output it becomes obvious that the corresponding structure will be much more parsimonious with respect to the number of parameters than a general neural network structure could be.

Forced Duffing Oscillator

The equation of a forced Duffing oscillator is

$$\begin{aligned}
\dot{x}_1 &= x_2 \\
\dot{x}_2 &= x_1 - x_1^3 - \delta x_2 + u \\
y &= x_1.
\end{aligned}$$

resulting in

$$f(x) = \begin{pmatrix} x_2 \\ x_1 - x_1^3 - \delta x_2 \end{pmatrix}, \qquad g(x) = \begin{pmatrix} 0 \\ 1 \end{pmatrix}.$$

For these vector fields the Lie derivatives up to third order are

$$
\begin{aligned}
L_g^{\circ k} &= \frac{\partial^k}{\partial x_2^k} \\
L_g(x) &= \frac{\partial}{\partial x_2}\begin{pmatrix} x_1 \\ x_2 \end{pmatrix} = \begin{pmatrix} 0 \\ 1 \end{pmatrix}, \qquad L_g^{\circ k}(x) = 0 \quad \text{for} \quad k > 1 \\
L_g \circ L_f(x) &= \frac{\partial}{\partial x_2} f(x) = \begin{pmatrix} 1 \\ -\delta \end{pmatrix}, \qquad L_g^{\circ k} L_f(x) = 0 \quad \text{for} \quad k > 1 \\
L_f(x) &= f(x) \\
L_f^{\circ 2}(x) &= \left(x_2 \frac{\partial}{\partial x_1} + (x_1 - x_1^3 - \delta x_2) \frac{\partial}{\partial x_2} \right) (f(x)) \\
&= \begin{pmatrix} x_1 - x_1^3 - \delta x_2 \\ x_2(1 - 3x_1^2) - \delta(x_1 - x_1^3 - \delta x_2) \end{pmatrix} \\
L_f^{\circ 3}(x) &= \left(x_2 \frac{\partial}{\partial x_1} + (x_1 - x_1^3 - \delta x_2) \frac{\partial}{\partial x_2} \right) (L_f^{\circ 2}(x)) \\
&= \begin{pmatrix} x_2(1 - 3x_1^2) - \delta(x_1 - x_1^3 - \delta x_2) \\ -x_2(6x_1 + \delta) + (x_1 - x_1^3 - \delta x_2)(1 + \delta^2 - 3x_1^2) \end{pmatrix} \\
L_g \circ L_f^{\circ 2}(x) &= \begin{pmatrix} -\delta \\ 1 + \delta^2 - 3x_1^2 \end{pmatrix}, \qquad L_f^{\circ k} \circ L_g(x) = 0
\end{aligned}
$$

The Euler discretisation of the Duffing oscillator is

$$
\begin{aligned}
\bar{x}_1[k+1] &= \bar{x}_1[k] + T\bar{x}_2[k] &(16) \\
\bar{x}_2[k+1] &= (1 - T\delta)\bar{x}_2[k] + T(\bar{x}_1[k] - \bar{x}_1^3[k]) + Tu[k] &(17)
\end{aligned}
$$

and for the second order approximation we obtain

$$
\begin{aligned}
\bar{x}_1[k+1] &= (1 + \frac{T^2}{2})\bar{x}_1[k] + (T - \frac{T^2}{2}\delta)\bar{x}_2[k] - \frac{T^2}{2}\bar{x}_1^3[k] + \frac{T^2}{2}u[k] \quad (18) \\
\bar{x}_2[k+1] &= (1 - T\delta + \frac{T^2}{2}(1 + \delta^2))\bar{x}_2[k] + \\
&\quad +(T - \frac{T^2}{2}\delta)(\bar{x}_1[k] - T\bar{x}_1^3[k]) \\
&\quad -3\frac{T^2}{2}\bar{x}_1^2[k]\bar{x}_2[k] + (T - \frac{T^2}{2}\delta)u[k].
\end{aligned}
$$

From the Euler discretisation of the Duffing equation (16) using $y[k] = \bar{x}_1[k]$ we can easily obtain the NARX model:

$$
y[k+2] = (2 - T\delta)y[k+1] + (T + T\delta - 1)y[k] - Ty^3[k] + Tu[k] \qquad (19)
$$

The Euler discretisation preserves the equilibrium points of the system $y_1 = 0$, $y_2 = 1$, $y_3 = -1$. The stable inverse of the system is given by

$$u[k] = \frac{1}{T} \cdot \left\{-(2 - T\delta)y[k+1] - (T + T\delta - 1)y[k] + Ty^3[k] + v[k]\right\}$$

Now, the second order discretisation is considered. The corresponding NARX model is

$$\begin{aligned} y[k+1] &= (k_1 + k_3)y[k] + (k_2^2 - k_1 k_3)y[k-1] - \frac{T^2}{2}y^3[k] + \\ &+(\frac{T^2}{2} - k_2^2 T + 3\frac{T^2}{2}k_1)y^3[k-1] - \\ &-3\frac{T^4}{4}y^5[k-1] - 3\frac{T^2}{2}y^2[k-1]y[k] + \\ &+\frac{T^2}{2}u[k] - \frac{T^2}{2}u[k-1] + k_2 u[k-1] + 3\frac{T^4}{4}y^2[k-1]u[k] \end{aligned} \tag{20}$$

with

$$k_1 = \frac{T^2}{2} + 1, \quad k_2 = T - \frac{T^2}{2}\delta, \quad k_3 = 1 - T\delta + \frac{T^2}{2}(1 + \delta^2).$$

The relative degree of the second order NARX model is $d = 1$ with the linearised zero dynamics about the equilibrium point $(y, u) = (0, 0)$ given by

$$\frac{T^2}{2}u[k+1] + (k_2 - \frac{T^2}{2})u[k] = 0. \tag{21}$$

This means that the linearisation of (20) has a zero which will depend on the sampling period T and the damping constant δ while the original system does not have zero dynamics at all. The zero is inside the unit circle if

$$\frac{2}{2+\delta} < T < \frac{2}{\delta} \tag{22}$$

holds. Only in this case the system will be asymptotically minimum phase and its inverse will be asymptotically stable. In particular, there is a minimum sampling time for obtaining a minimum phase model.

A polynomial system

It is easy to calculate the exact discretisation of polynomial systems i.e. systems that can be represented by a finite Volterra series. This is due to the fact that in this case there exists a finite m such that

$$L_g^{\circ i} \circ L_f^{\circ k}(x) = 0 \quad \text{for} \quad i > m$$

thus the corresponding discrete time state transition function $F(\bar{x}[k], u[k])$ will be polynomial in $u[k]$. Consider for instance the polynomial system

$$
\begin{aligned}
\dot{x}_1 &= -x_1 + bu \\
\dot{x}_2 &= -x_2 + ax_1 + x_1^2 + u \\
\\
y &= x_2 + x_1^2.
\end{aligned} \tag{23}
$$

with the parameter values being

$$a = 2, \qquad b = 0.25.$$

This system has the exact discretisation

$$
\begin{aligned}
\bar{x}_1[k+1] &= \alpha\bar{x}_1[k] + b(1-\alpha)u[k] \\
\bar{x}_2[k+1] &= \alpha\bar{x}_2[k] + \beta\bar{x}_1[k] + \alpha(1-\alpha)\bar{x}_1^2[k] + \\
&\quad + \left\{\gamma + (1-\alpha)^2 b\bar{x}_1[k]\right\} u[k] + \varrho u^2[k]
\end{aligned} \tag{24}
$$

$$y[k] = \bar{x}_2[k] + \bar{x}_1^2[k] \tag{25}$$

where T is the sampling time and

$$
\begin{aligned}
\alpha &= e^{-T} \\
\beta &= aT\alpha \\
\gamma &= (1+ab)(1-\alpha) - \beta b \\
\varrho &= \left[1 - \alpha^2 - 4(1-\alpha) + 2T\right] \cdot \frac{b^2}{2}.
\end{aligned}
$$

The normal form of the sampled polynomial system is

$$
\begin{aligned}
y[k+1] &= \alpha y[k] + \beta x_1[k] + \{\gamma + \lambda x_1[k]\}\, u[k] + \varrho_1 u^2[k] \\
\\
x_1[k+1] &= \alpha x_1[k] + \mu u[k].
\end{aligned} \tag{26}
$$

The corresponding NARX representation is given by

$$
\begin{aligned}
y[k+1] &= \alpha y[k] + \alpha\frac{\beta + \lambda u[k]}{\beta + \lambda u[k-1]} \cdot \left\{y[k] - \alpha y[k-1] - \gamma u[k-1] - \varrho_1 u^2[k-1]\right\} + \\
&\quad + \beta\mu u[k-1] + \{\gamma + \lambda\mu u[k-1]\}\, u[k] + \varrho_1 u^2[k]
\end{aligned} \tag{27}
$$

where

$$\begin{aligned}\lambda &= b(1-\alpha^2)\\ \mu &= b(1-\alpha)\\ \varrho_1 &= \varrho + b^2(1-\alpha)^2 = (\alpha^2 - 1 + 2T)\frac{b^2}{2}.\end{aligned}$$

Note that this representation is valid everywhere except for the input value

$$u^* = -\frac{\beta}{\lambda}.$$

The zero dynamics corresponding to (26) are

$$\begin{aligned}x_1[k+1] &= f_0(x[k]) = \alpha x_1[k] + \\ &+\frac{\mu}{2\varrho_1}\left\{-\gamma - \lambda x_1[k] \pm \sqrt{(\gamma + \lambda x_1[k])^2 + 4\varrho_1((1-\alpha)y^0 - \beta x_1[k])}\right\}\end{aligned}$$

For the given parameter values and sufficiently large values of x_1 the slope of $f_0(x)$ approaches

$$\lim_{x_1\to\infty}\left|\frac{f_0(x_1)}{x_1}\right| = \left|\alpha - \frac{\mu}{2\varrho_1}(\lambda \pm \lambda)\right| < 1 \quad \text{for} \quad T = 0.25.$$

Thus, the zero dynamics will be globally stable.

3. The finite element approach to NARX modelling

It has been pointed out in Section 2 that if a continuous-time representation of the system is known (for instance the state space description), for certain classes of systems it is possible to calculate a NARX representation directly. However, if the continuous-time model is unknown or the direct calculation of (39) is too difficult then the only way to obtain the NARX model is by identification from a set of input-output data whereby structural information can still be used.

For the purposes of identification, the nonlinear function ϕ of the NARX model has first to be parameterised. The most straightforward way is to facilitate a linear parameterisation by using a set of n basis functions

$$\rho_i(\psi[k-1]), \quad i = 1, \dots, n$$

over an N-dimensional subdomain of the N-dimensional space with ϕ being approximated by

$$y[k] = \phi(\psi[k-1]) \approx \hat{y}[k] = \sum_{i=1}^{n} \theta_i \rho_i(\psi[k-1]), \tag{28}$$

where

$$\psi[k-1] = (y[k-1], \ldots, y[k-n], u[k-d-1], \ldots, u[k-d-n])^T$$

is the data vector.

The choice of basis functions is a frequently discussed subject with the main concern usually being that the set of basis functions has the best capability of approximating ϕ. However, if the identified model is used for control design, good approximation properties, and consequently high accuracy of the model in itself, is not the only point that should influence the decision. The following criteria are equally important and should also be taken into account:

- Is the model useful for control design? Can properties of the model that are important for control design (such as minimum-phase property or relative degree) be detected? Is the calculation of a control law numerically easy or difficult?

- What are the numerical costs of identification?

In this section we suggest the use of finite element methods (FEM) for approximation of NARX models since the criteria mentioned above can apparently be met more easily using FEM in comparison to the use of RBF's, especially as far as control applications are concerned.

3.1. Finite elements

The FEM approximation techniqe will be explained using the example of N-linear interpolation functions over rectangular elements. The two dimensional case $N = 2$ will be considered first. The results presented here carry easily over into any dimension and the numerical implementation is straightforward. Note however, that this is not true for higher order interpolation functions (quadratic or cubic) where the geometry becomes quite involved for $N > 3$.

Consider the NARX model

$$y[k] = f(y[k-1], u[k-1])$$

and suppose the input-output signals are bounded in the intervals

$$u \in [u_{min}, u_{max}], \qquad y \in [y_{min}, y_{max}]$$

for all time. Without loss of generality, it will be assumed that

$$u \in [0, 1], \qquad y \in [0, 1],$$

since this can always be achieved by normalisation.

We partition the corresponding two-dimensional domain spanned by the variables $y[k-1]$ and $u[k-1]$ into $(M-1)$ rectangular subdomains and obtain a grid as shown in Fig. 5 for $M = 4$.

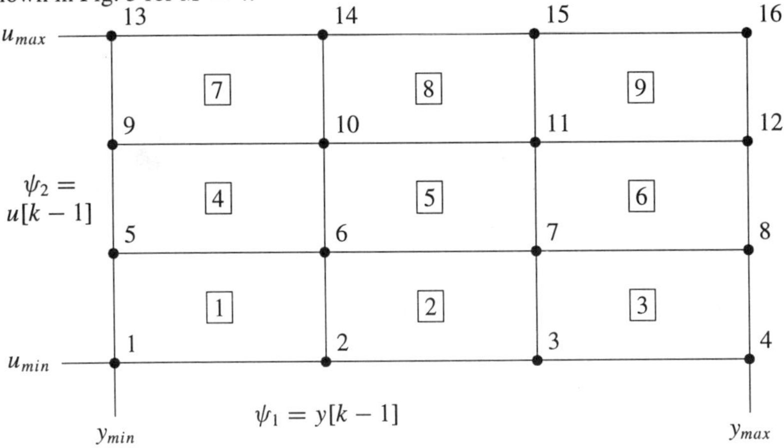

Figure 5: Two-dimensional rectangular grid

The grid consists of $n = M^2 = 16$ meshpoints and $n_{el} = (M-1)^2 = 9$ meshes, the so-called elements. The elements of the grid are formed by groups of $2^N = 4$ adjacent meshpoints. An N-linear C^0-interpolation element is depicted in Fig 6 for the two-dimensional case.

Such an element is called a C^0-interpolation element since the interpolated function will be continuous. Meshpoints and elements are numbered consecutively in an arbitrary way to get a global frame of reference.

3.2. *Coordinates*

Each element is uniquely determined by the *global coordinates* $x_1, \ldots, x_N$ of its reference point P and its geometric dimensions. As reference point we chose the meshpoint of the element with the smallest coordinate values and as geometric description the lengths l_i, $i = 1, \ldots, N$ of the edges adjacent to P.

Each meshpoint of an element is given an index $j = 1, \ldots, 2^N$. For the meshpoints of the element we introduce a local coordinate system $\xi_1, \ldots, \xi_N$ such that the reference point has the *local coordinate* $\xi_P = (0, \ldots, 0)$ and forms with the other meshpoints a unit-hypercube in local coordinates. Thus, the local coordinates are independent of the size of the element and of its position within the grid.

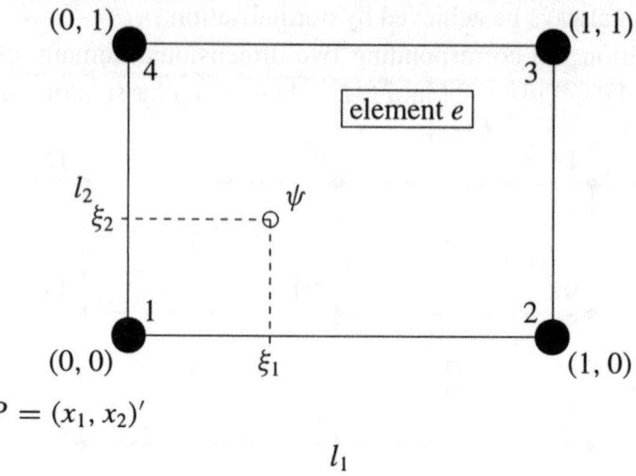

Figure 6: bilinear C^0-element

Each N-dimensional data vector $\psi[k-1]$ is contained in one of the elements of the grid and can therefore be expressed in the local coordinate vector ξ of this element using the equation

$$\xi_i = \frac{\psi_i - x_i}{l_i} \qquad i = 1, \ldots, N. \tag{29}$$

We define a 2^N-vector λ of *barycentric coordinates* (relating the point ξ to the meshpoints of the element) with the components given for a bilinear C^0-element by

$$\begin{aligned} \lambda_1 &= (1-\xi_1)(1-\xi_2) \\ \lambda_2 &= \xi_1(1-\xi_2) \\ \lambda_3 &= \xi_1\xi_2 \\ \lambda_4 &= (1-\xi_1)\xi_2 \end{aligned} \tag{30}$$

The scheme of generating the λ-coordinates can be easily carried over to higher dimensions, since there is a unique relation between the local coordinates of a meshpoint and the pattern of the N-linear term expressing the corresponding λ-coordinate. At a meshpoint j ($j = 1, \ldots, 2^N$) only λ_j assumes the value 1 whereas all the other coordinates assume the value zero. Note that

$$\lambda_j \geq 0, \quad \sum_{j=1}^{2^N} \lambda_j = 1. \tag{31}$$

Using tensor product notation the vector λ of barycentric coordinates can be written as

$$\lambda = (\lambda_1, \ldots, \lambda_{2^N})^T = \begin{bmatrix} 1-\xi_1 \\ \xi_1 \end{bmatrix} \otimes \begin{bmatrix} 1-\xi_2 \\ \xi_2 \end{bmatrix} \otimes \ldots \otimes \begin{bmatrix} 1-\xi_N \\ \xi_N \end{bmatrix} \tag{32}$$

3.3. *Interpolation within an element*

Now, if a data vector $\psi[k-1]$ is within an element e and has the local coordinates ξ with respect to this element then we interpolate the corresponding output value $\hat{y}[k]$ of the FEM NARX model by

$$\hat{y}[k] = \sum_{j=1}^{2^N} \lambda_j \theta_j^e \tag{33}$$

with the λ_j given by (30) in the two-dimensional case, and by (32) in the general case. The θ_j^e is the parameter value of the FEM NARX model corresponding to the meshpoint j of the element e.

Note that if the output $\hat{y}[k]$ matches exactly a meshpoint j then $\hat{y}[k]$ will assume exactly the value θ_j^e. Otherwise the output y will be interpolated between the parameter values of 2^N adjacent meshpoints. For the sake of convenience in control design we will rename the local coordinates according to their associated input and output variables:

$$\begin{aligned} \xi_1 &= \tilde{y}[k-1], \quad \ldots, \quad \xi_n = \tilde{y}[k-n] \\ \xi_{n+1} &= \tilde{u}[k-d], \quad \ldots, \quad \xi_{2n-d+1} = \tilde{u}[k-n] \end{aligned}$$

Furthermore, we introduce the following tensor products:

$$Y_{k-1}^{k-n} = \begin{bmatrix} 1-\tilde{y}[k-1] \\ \tilde{y}[k-1] \end{bmatrix} \otimes \ldots \otimes \begin{bmatrix} 1-\tilde{y}[k-n] \\ \tilde{y}[k-n] \end{bmatrix} \tag{34}$$

$$U_{k-d}^{k-n} = \begin{bmatrix} 1-\tilde{u}[k-d] \\ \tilde{u}[k-d] \end{bmatrix} \otimes \ldots \otimes \begin{bmatrix} 1-\tilde{u}[k-n] \\ \tilde{u}[k-n] \end{bmatrix} \tag{35}$$

$$U_{k-d-1}^{k-n} = \begin{bmatrix} 1-\tilde{u}[k-d-1] \\ \tilde{u}[k-d-1] \end{bmatrix} \otimes \ldots \otimes \begin{bmatrix} 1-\tilde{u}[k-n] \\ \tilde{u}[k-n] \end{bmatrix} \tag{36}$$

Using this notation equation (33) can be given the form

$$\hat{y}[k] = \theta^T \cdot \left[Y_{k-1}^{k-n} \otimes U_{k-d}^{k-n}\right] \tag{37}$$

where the index j denoting the element number has been omitted. Model (37) is multilinear since it is linear with respect to any of the variables $\tilde{y}[k-1], \ldots, \tilde{y}[k-$

$n], \tilde{u}[k-d], \ldots, \tilde{u}[k-n]$. It can be considered as a special case in the class of polynomial NARX models. The property of multilinearity arising from the use of N-linear C^0 elements simplifies controller design considerably. Rewriting (37) in predictor form

$$\begin{aligned}
\hat{y}[k+d] &= \theta^T \cdot \left[Y_{k+d-1}^{k+d-n} \otimes U_k^{k-n+d}\right] \\
&= \theta^T \cdot \left[Y_{k+d-1}^{k+d-n} \otimes \left(\begin{bmatrix} 1 \\ 0 \end{bmatrix} + \begin{bmatrix} -1 \\ 1 \end{bmatrix} \cdot \tilde{u}[k]\right) \otimes U_{k-1}^{k-n}\right] \\
&= \theta^T \cdot \left[Y_{k+d-1}^{k+d-n} \otimes \begin{bmatrix} 1 \\ 0 \end{bmatrix} \otimes U_{k-1}^{k-n+d}\right] + \\
&\quad +\theta^T \cdot \left[Y_{k+d-1}^{k+d-n} \otimes \begin{bmatrix} -1 \\ 1 \end{bmatrix} \otimes U_{k-1}^{k-n+d}\right] \cdot \tilde{u}[k] \\
&= P_{k+d} + Q_{k+d} \cdot \tilde{u}[k]
\end{aligned} \tag{38}$$

we obtain a model that is locally linear in the control variable $\tilde{u}[k]$.

Another important advantage of the finite element techniques is that due to the linear parameterisation standard least squares methods can be used to solve the identification problem. The sparsity of the associated observation matrices can be exploited to make the algorithms more efficient.

An interesting option is to use finite elements with triangular geometry in which case linear interpolation functions can be applied.

Finite element methods have been used for a long time for the numerical solution of partial differential equations. Therefore, their approximation properties are well known [14] and standard numerical packages are readily available. An application of FEM techniques in neural network modelling is reported in [7].

4. Control aspects

In this section some geometric aspects of nonlinear discrete-time output feedback control based on a NARX type plant description will be briefly discussed. We review the concepts of input-output linearisation by static state feedback, model inversion and asymptotic output tracking and use these techniques afterwards to anlyse a nonlinear two-degrees of freedom output feedback control structure [3,18].

4.1. Input-Output linearisation and asymptotic output tracking

Consider the NARX representation

$$y[k+d] = \phi(y[k-1+d], \ldots, y[k-n+d], u[k], \ldots, u[k-n+d]) \tag{39}$$

where ϕ is a nonlinear function in $N = 2n - d + 1$ variables, with n being the order of the system and d the relative degree. It is assumed that the system has uniform relative degree in a neighbourhood of the equilibrium point P^0.

The problem of *input-output linearisation* [21] of system (39) can be solved by finding a feedback law

$$u[k] = \gamma(y[k-1+d], \ldots, y[k-n+d], u[k-1], \ldots, u[k-n+d], v[k]) \quad (40)$$

which renders the closed loop system:

$$y[k+d] = v[k].$$

This means that the signal $v[k]$ is reproduced by the system output with a delay of d steps. Conditions for the existence of such a linearising feedback can be found in [21,28,22,35,5]. The problem can be solved if for any choice of initial conditions $y_1, \ldots, y_{n-d}$ and $u_1, \ldots, u_{n-d}$ and for a prescribed output value y_t there exist a control u with

$$y_t = \phi(y_1, \ldots, y_n, u, u_1 \ldots, u_{n-d}) \quad (41)$$

which means that from an arbitrary initial state the system can be controlled to reach a certain output value *within one time step*. According to the implicit function theorem, the solution is guaranteed to exist at least in some neighbourhood of P^0.

Replacing $v[k]$ by a reference output $y_t[k+d]$ we obtain:

$$y[k] = y_t[k].$$

The closed loop system is linear and after d steps the reference output is tracked exactly. Consider the case of a constant reference output

$$y[k] = y_t[k] = y^0 = \text{const.}, \forall k.$$

Then the internal variable $u[k]$ of the closed loop system will be governed exactly by the zero dynamics (13).

- Asymptotic stability of the zero dynamics is necessary for stability of the feedback linearised closed loop system.

- For unstable zero dynamics in particular the control variable u will not be bounded.

- Note that feedback linearisation renders the zero dynamics unobservable.

- Most of the control strategies for nonlinear systems are based on the concept of feedback linearisation and therefore require asymptotically stable zero dynamics of the plant model used.

Control law (40) represents a dead-beat controller and in connection with the system equation it allows us to find an expression for the *system inverse*, namely:

$$\begin{aligned} y[k+d] &= \phi(y[k-1+d], \ldots, y[k-n+d], u[k], \ldots, u[k-n+d]) \\ u[k] &= \gamma(y[k-1+d], \ldots, y[k-n+d], u[k-1], \ldots, u[k-n+d], v[k]) \end{aligned} \tag{42}$$

with $v[k]$ taken as the input and $u[k]$ as the output. This type of inverse has been called forward time-shift right inverse in [19,21]. The inverse system will be stable if the zero dynamics of the original system are stable.

In most cases the linearising feedback will have to be calculated numerically. This requires the solution of a nonlinear equation in the variable $u[k]$ for which a number of algorithms is available. Analytical expressions for approximate solutions can be obtained using the Lie series formalism [13,21]. For the proposed FEM NARX model, however, the calculation of the solution of the inversion problem is straightforward since within the element such a model is linear in the control.

Asymptotic output tracking can be achieved using feedback linearisation together with the control law

$$v[k] = y_t[k+d] - \sum_{i=1}^{n} c_{n-i}(y[k+d-i] - y_t[k+d-i]) \tag{43}$$

with the coefficients of the polynomial

$$z^n + c_{n-1}z^{n-1} + c_{n-2}z^{n-2} + \ldots + c_1 z + c_0 = 0$$

chosen such that its roots are inside the unit circle. If the tracking error is denoted by

$$e_t[k] = y[k] - y_t[k]$$

then the closed loop system will result in a linear autonomous difference equation for e_t:

$$e_t[k] + c_{n-1}e_t[k-1] + \ldots + a_0 e_t[k-n] = 0$$

with the tracking error asymptotically reaching zero.

The control strategies suggested so far are only applicable if there is *little uncertainty* in the plant structure and parameters since

- modelling errors are not taken into account and
- the required future values of the plant output are not available for feedback necessitating the use of an observer.

A more robust approach that only requires measurement of the current plant output without using an observer is offered by two degrees of freedom model reference controllers discussed in the next subsection.

4.2. A two degrees of freedom nonlinear controller (2DFNC)

This type of strategy was first proposed in its continuous-time version in [4] and in its discrete-time version in [3]. An analysis of the resulting control system structure and an extension to the control of unstable systems can be found in [25,18]. A robustness analysis of the control system has been carried out in [2].

4.2.1. Control system structure

Consider the following set-up for a control system [3]:
Let the *plant* be described by a NARX model:

$$y[k+d] = \phi(y[k+d-1], \ldots, y[k-n+d], u[k], \ldots, u[k-n+d]) \quad (44)$$

with the true parameters not exactly known.

A *nominal NARX model* of the plant is assumed to be given by

$$y_m[k+d] = \phi_m(y_m[k+d-1], \ldots, y_m[k-n+d], u[k], \ldots, u[k-n+d]) \quad (45)$$

The performance of the closed loop system is determined by a discrete-time linear *tracking filter* with unity gain generating a reference output y_t from a tracking input signal z:

$$y_t[k] = A_t(q^{-1})y_t[k] + q^{-d_t}B_t(q^{-1})z[k], \quad (46)$$

where

$$\begin{aligned} A_t(q^{-1}) &= -a_{t,n_t-1}q^{-1} - a_{t,n_t-2}q^{-2} - \ldots - a_{t,1}q^{-n_t+1} - a_{t,0}q^{-n_t} \\ B_t(q^{-1}) &= b_{t,n_t-d_t} + b_{t,n_t-d_t-1}q^{-1} + \ldots + b_{t,1}q^{-n_t+d_t+1} + b_{t,0}q^{-n_t+d_t} \end{aligned}$$

denote polynomials in q^{-1}.

The modelling error

$$\varepsilon[k] = y[k] - y_m[k]$$

is fed into a unity gain *regulating filter*

$$y_r[k] = A_r(q^{-1})y_r[k] + q^{-d_r}B_r(q^{-1})\varepsilon[k] \quad (47)$$

in order to obtain estimates of the modelling error dynamics with

$$\begin{aligned} A_r(q^{-1}) &= -a_{r,n_r-1}q^{-1} - a_{r,n_r-2}q^{-2} - \ldots - a_{r,1}q^{-n_r+1} - a_{r,0}q^{-n_r} \\ B_r(q^{-1}) &= b_{r,n_r-d_r} + b_{r,n_r-d_r-1}q^{-1} + \ldots + b_{r,1}q^{-n_r+d_r+1} + b_{r,0}q^{-n_r+d_r}. \end{aligned}$$

The following assumptions are made for the control system

A1: Nominal model (45) and discrete-time external representation (44) of the plant have the same relative degree d, whereas tracking filter and regulating filter have at least relative degree $d_t, d_r \geq d$.

A2: The modelling error ε is bounded.

A3: The nominal model has BIBS stable zero dynamics.

A4: The nominal model satisfies condition (41) (i.e. it is invertible).

The structure formed by nominal model, tracking and regulating filter will be called the *extended system* with the extended system output defined by the *extended error*

$$y_E[k] = y_t[k] - y_m[k] - y_r[k].$$

For the extended system a *nonlinear feedback law* will be designed which

- drives the extended error y_E to zero while
- decoupling y_E from external disturbances represented by the plant output y and the tracking input z

For this purpose only *internal states of the extended system* are used, represented by time-shifted output values of nominal model, regulating and tracking filter

$$y_m[k+i],\ y_r[k+i],\ y_t[k+i], \quad \text{for} \quad i = 0, \ldots, d-1.$$

Note that those values are available at time k.

The resulting structure of the control system consisting of

- extended system,
- feedback law,
- process with sample and hold elements,

is shown in Fig. 7.

Figure 7: Structure of the control system

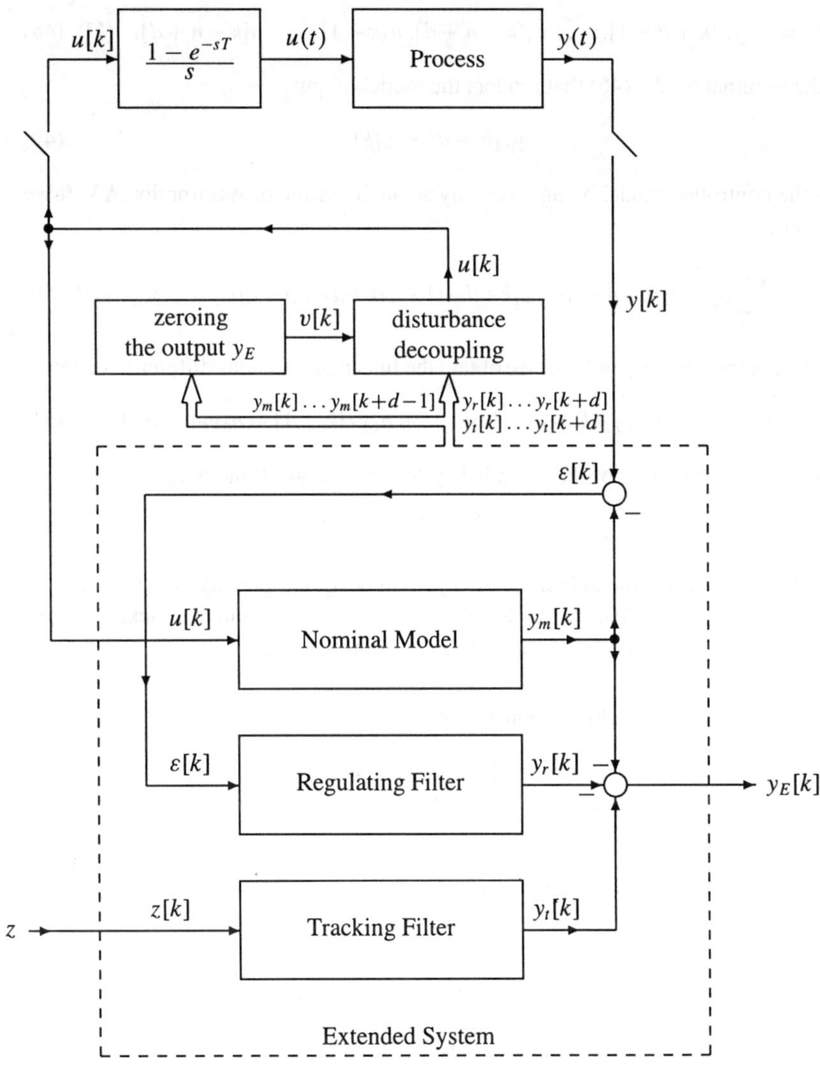

4.2.2. Control Law

The structure of the feedback law necessary for zeroing and disturbance decoupling y_E will now be discussed.

Due to Assumption **A4** there exists a feedback law

$$u[k] = \gamma(y_m[k+d-1], \ldots, y_m[k-n+d], u[k-1], \ldots, u[k-n+d], v[k]) \quad (48)$$

for the nominal model (45) that renders the model output

$$y_m[k+d] = v[k] \quad (49)$$

with the controlled model being internally stable by virtue of Assumption **A3**. If we now set

$$v[k] = \sum_{i=1}^{d} a_{d-i}(y_t[k+d-i] - y_m[k+d-i] - y_r[k+d-i]) + y_t[k+d] - y_r[k+d] \quad (50)$$

then by combining (49) and (50) we obtain the linear autonomous difference equation

$$y_E[k+d] + a_{d-1}y_E[k+d-1] + \ldots + a_1 y_E[k+1] + a_0 y_E[k] = 0 \quad (51)$$

governing the dynamics of the extended system output y_E. If the polynomial

$$p(z) = z^d + a_{d-1}z^{d-1} + \ldots + a_1 z + a_0 \quad (52)$$

has all its roots inside the unit circle then y_E will be asymptotically driven to zero. In fact, the control law (48) and (50) decouples the variable y_E from the tracking input z and the plant output y with (51) describing a sliding surface.

4.2.3. Analysis of the closed loop system

The properties of the controller described above have been discussed in [25]. From the extended error dynamics and using $y_m[k] = y[k] - \varepsilon[k]$ we obtain the following asymptotic result.

$$y[k] \to (\varepsilon[k] - y_r[k]) + y_t[k] \qquad \text{as} \quad k \to \infty.$$

Therefore, if

- the modelling error ε is bounded (as assumed in **A2**) and
- the bandwidth of the regulating filter is high enough to guarantee that

$$y_r[k] \to \varepsilon[k] \qquad \text{as} \quad k \to \infty$$

then the plant output y will converge asymptotically to y_t. Boundedness of the modelling error has to be shown for every particular case.

It is very useful to analyse the closed loop control system in terms of an input-output description. For this purpose we introduce operator representations for all elements of the control system relating any input signal to the corresponding output signal *for fixed initial states*. If the element is linear the corresponding z-transfer functions will also be used:

Element		Operator	Transfer Function
Tracking Filter	linear	$y_t[k] = G_t[z[k]]$	$G_t(z^{-1}) = \frac{z^{-d}B_t(z^{-1})}{A_t(z^{-1})}$
Regulating Filter	linear	$y_r[k] = G_r[\varepsilon[k]]$	$G_r(z^{-1}) = \frac{z^{-d}B_r(z^{-1})}{A_r(z^{-1})}$
Error Polynomial (52)	linear	$0 = D[y_E[k]]$	$p(z)$
Nominal Model	nonlinear	$y_m[k] = G_m[u[k]]$	
Plant	nonlinear	$y[k] = P[u[k]]$	

In operator representation the extended error dynamics (51) takes the form

$$D[y_t[k] - y_m[k] - y_r[k]] = D\left[G_t[z[k]] - G_m[u[k]] - G_r[\varepsilon[k]]\right] = 0.$$

Since

$$\varepsilon[k] = y[k] - y_m[k] = y[k] - G_m[u[k]]$$

we have

$$D\left[G_t[z[k]] - G_m[u[k]] - G_r\left[y[k] - G_m[u[k]]\right]\right] = 0.$$

This equation can be solved for $u[k]$ taking into account that G_r and D are linear operators. One obtains

$$u[k] = G_m^{-1} \circ (I - G_r)^{-1}\left[G_t[z[k]] - G_r[y[k]]\right] \tag{53}$$

where "$\circ$" denotes the serial connection of operators. From (53) we can deduce that the control system has the structure shown in Fig. 8. It can be seen from Fig. 8 that the nominal model is used to generate a global inverse G_m^{-1} of the plant. For the nominal case the plant is cancelled by the inverse of the model

$$P \circ G_m^{-1} = I$$

and y will be exactly equal to y_t. In the case when

- the plant differs from the nominal model, or
- external disturbance signals $\zeta[k]$ or $n[k]$ act on the control system,

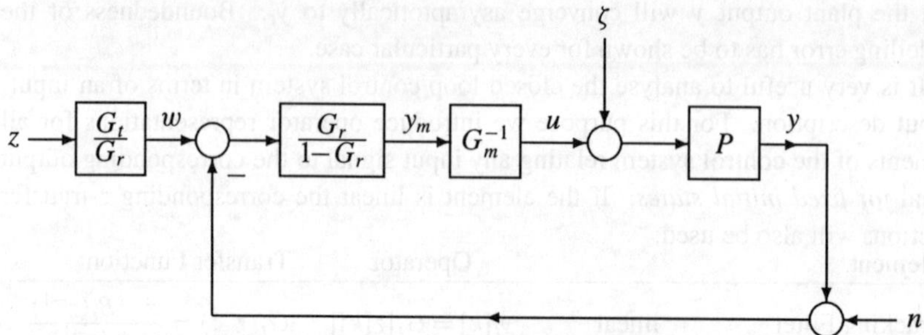

Figure 8: Feedback structure of the two-degree of freedom nonlinear controller

we have to design G_r in such a way that the closed loop system is robust against an imperfect cancellation. Thus, the design problem consists in

- selecting the coefficients of $p(z)$ in order to drive the extended system output to zero,
- selecting a tracking filter $G_t(z^{-1})$ to meet the closed loop performance requirements in compromise with the stability and robustness requirements,
- selecting the regulating filter $G_r(z^{-1})$ in such a way that
  - on the one hand its bandwidth is high enough to give a reliable estimate of the modelling error dynamics (stability and robustness requirement),
  - on the other hand its gain is rolled off to reject high frequency measurement noise $n[k]$ and model uncertainties that are beyond the bandwidth of the tracking filter.

Expertise from linear control theory can be used to shape the frequency response of tracking and regulating filters.

By analogy with [36] an *integrator* can be added to $v[k]$ in (50)

$$\begin{aligned} v[k] &= \sum_{i=1}^{d} a_{d-i}(y_t[k+d-i] - y_m[k+d-i] - y_r[k+d-i]) \\ &\quad + y_t[k+d] - y_r[k+d]) + a_{-1}I[k] \\ I[k+1] &= I[k] + T \cdot (y_t[k] - y_m[k] - y_r[k]) \end{aligned} \tag{54}$$

in which case the polynomial $p(z)$ has to be chosen such that

$$p_I(z) = (z-1)p(z) + a_{-1}T$$

is Hurwitz.

4.2.4. Relation to internal model control

It is clear that with y_E approaching zero the controller presented above renders any performance measure

$$J(u[k],\ldots,u[k+p]) = \sum_{i=0}^{p} \beta_i (y_t[k+d+i] - y_m[k+d+i] - y_r[k+d+i])^2, \quad \beta_i > 0 \tag{55}$$

minimal with respect to $u[k],\ldots,u[k+p]$ where $y_r[k+d+i]$ can be interpreted as a *prediction of the modelling error* $\varepsilon[k+d+i]$ at the time instants $k+d$ up to $k+d+p$. Controllers that are based on minimising cost indices of the form (55) are special cases of nonlinear model based predictive control (NMPC) and are usually referred to as nonlinear internal model controllers (NIMC). Such controllers have been intensively discussed in [15].

However, there are a number of important differences between the two-degrees of freedom controller presented here and the NIMC approach presented in [15]:

1. The control law of 2DFNC is entirely derived from geometric control theory.

2. The relation between various modelling error prediction schemes and the "robustness filter" used in NIMC has not been understood properly so far [15]. In 2DFNC this relation becomes obvious since the regulating filter $G_r(z^{-1})$ acts as a *modelling error predictor* with the corresponding controller ("robustness filter") given by

$$C(z^{-1}) = \frac{G_r(z^{-1})}{1 - G_r(z^{-1})}.$$

   In fact, the controller $C(z^{-1})$ will have *integral action* since $G_r(1) = 1$ is required for unity gain and, consequently, $z = 1$ will be a pole of $C(z^{-1})$.

3. Due to disturbance decoupling of y_E from the tracking input z and the plant output y the dynamics of y_E will depend only on the initial states of the extended system. It will be *invariant* with respect to the external signals acting on the extended system. This invariance property does not seem to hold for NIMC.

4. Stability/robustness issues on the one hand and performance issues on the other hand can be addressed in a natural way by designing regulating filter and tracking filter accordingly. Structural analysis of 2DFNC gives a clear idea as to how these filters should be designed.

5. The formulation of the cost index (55), which represents only an indirect specification of certain control system properties, is avoided in 2DFNC. In particular,

there is no need to specify a prediction horizon and to apply optimisation techniques.

The proposed control structure can be modified for the control of non-minimum phase plants if multirate sampling strategies are applied as suggested in [26].

4.3. *A two degrees of freedom controller for unstable systems*

Structurally, the nominal model (45) acts as an open loop observer within the two-degrees of freedom control system [25]. This is also the case for the NIMC approach [15]. It is well known that a condition for internal stability of control systems using open loop observers is that the corresponding nominal model be asymptotically stable. In particular Assumption **A2** (boundedness of modelling error) cannot be guaranteed if system and nominal model are unstable. For unstable systems it is therefore recommended [25] to modify the control law applied to the model by using separate control inputs u and u_m for model and plant respectively where the plant input is given by (49) and (50) and the model input is given by:

$$u_m[k] = \gamma(y_m[k+d-1], \ldots, y_m[k-n+d], u_m[k-1], \ldots, u_m[k-n+d], v_m[k]) \tag{56}$$

with

$$v_m[k] = \sum_{i=1}^{d} b_{d-i}(y_t[k+d-i] - y_m[k+d-i]) + y_t[k+d].$$

and the coefficients b_i forming the Hurwitz polynomial by

$$p_m(z) = z^d + b_{d-1}z^d + \ldots + b_1 z + b_0. \tag{57}$$

Since there are no uncertainties involved, the control law for the nominal model will render the model output exactly equal to the tracking output (strong model matching):

$$y_m[k] = y_t[k].$$

Consequently, the control law for the plant turns out to be

$$u[k] = \gamma(y_t[k+d-1], \ldots, y_t[k-n+d], u[k-1], \ldots, u[k-n+d], v_1[k] + y_t[k+d]) \tag{58}$$

with $v_1[k]$ being

$$v_1[k] = -\sum_{i=1}^{d} a_{d-i} y_r[k+d-i] - y_r[k+d] \tag{59}$$

Again, an analysis in terms of input-output representations is necessary to obtain more insight into the control system structure. In contrast to the continuous time case analysed in [25] a slightly different result is obtained in discrete-time due to the fact that the nominal model is in general nonlinear in u. In operator representation (58) is a nonlinear dynamic precompensator *scheduled on the states*

$$x_t[k] = \begin{bmatrix} y_t[k-n+d] \\ \vdots \\ y_t[k+d] \end{bmatrix}$$

of the tracking filter and therefore, we obtain

$$u[k] = \Gamma(x_t[k], v_1[k]).$$

On the other hand, $v_1[k]$ can be expressed as

$$v_1[k] = -D[G_r[y[k] - y_m[k]] = D \circ G_r[y_t[k] - y[k]]$$

which can be interpreted as the output of a linear controller driven by the tracking error. A combined feedforward-feedback control structure (Fig. 9) is obtained that differs considerably from the NIMC structure. The inverse design model is not part

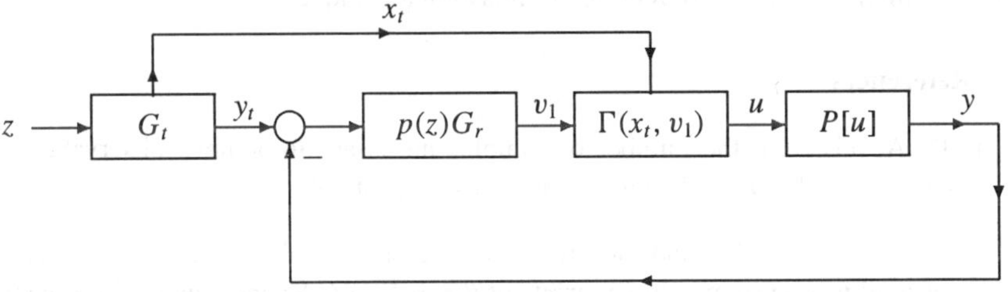

Figure 9: Structure of the controller with independent process and model inputs

of the closed loop anymore but is used to improve tracking performance by dynamic feedforward. The operator Γ still represents the nominal plant inverse but this time the inverse is taken with respect to the tracking output and internal instability of the control system cannot occur anymore. The properties of the control system can be outlined in the following way:

- If there is no uncertainty in the plant then the tracking error will be zero resulting in $v_1[k] = 0$ and there will be no feedback at all. In this case Γ will exactly represent the plant inverse driven by the tracking filter. Hence, a *feedforward tracking scheme* will be obtained.

- On the other hand, for any *constant* value y_t^0 of the tracking signal y_t (implying $y[k-n+d] = \ldots = y[k+d] = y_t^0$) a *linear regulator* will be obtained with the gain depending on the parameter y_t^0.

- The linear controller

$$C(z^{-1}) = p(z)G_r(z^{-1})$$

is used to introduce the additional feedback necessary to attenuate disturbances due to uncertainties. The parameters of C can be chosen by parameterising the error polynomial and regulating filter. Note, that C will only have integral action if $v[k]$ is calculated according to (54).

5. Conclusions

For some classes of nonlinear systems sampling theory permits the inclusion of structural information in discrete-time neural plant models. This way, more parsimonious models that better reflect the original plant dynamics can be obtained. In addition, the choice of basis function can considerably simplify the control design task. A nonlinear two-degrees of freedom discrete-time internal model controller has also been presented. This controller is based on the calculation of the right inverse of the plant model and does not require optimisation techniques.

6. References

1. D. Aeyels. On the number of samples necessary to achieve observability. *Systems and Control Letters*, 1(2):92–94, August 1981.

2. J. Alvarez. Stability and robustness of an internal-model controller for a class of nonlinear systems. In *Preprints of the 3rd IFAC Symposium on Nonlinear Control Systems Design, NOLCOS*, pages 765–770, Tahoe City, California, 1995. IFAC.

3. J. Alvarez and J. Alvarez. A tracking and regulation control scheme for discrete time nonlinear systems. In *Proceedings of the 28th Conference on Decision and Control*, pages 535–536, Tampa, Florida, December 1989. IEEE.

4. Jq. Alvarez and Jm. Alvarez. Tracking and regulation reference model control of nonlinear systems. In *Preprints of the IFAC Symposium on Nonlinear Control Systems Design, NOLCOS*, pages 466–471, Capri, 1989. IFAC.

5. E. Aranda-Bricaire, Ü. Kotta, and C. H. Moog. Accessibility and feedback linearization of discrete-time systems. In *Proceedings of the 33rd Conference*

on Decision and Control, pages 1628–1632, Lake Buena Vista, FL, December 1994. IEEE.

6. K. J. Åström, P Hagander, and J. Sternby. Zeros of sampled systems. *Automatica*, 20(1):31–38, 1984.

7. C. S. Berger. Linear splines with adaptive mesh sizes for modelling non-linear dynamic systems. *IEE Proceedings on Control Theory Applications*, 141(5):277–284, 1994.

8. M. Brown and C. Harris. *Neurofuzzy adaptive modelling and control*. Prentice Hall, New York, 1994.

9. F.-C. Chen and H. K. Khalil. Adaptive control of a class of nonlinear discrete-time systems using neural networks. *IEEE Transactions on Automatic Control*, 40(5):791–801, May 1995.

10. S. Chen and S.A. Billings. Representations of non-linear systems: the NARMAX model. *International Journal of Control*, 49(3):1013–1032, 1989.

11. D.W. Clarke. Advances in model-based predictive control, August 1993.

12. S. T. Glad. Output dead-beat control for nonlinear systems with one zero at infinity. *Systems & Control Letters*, 9:249–255, 1987.

13. W. Gröbner. *Die Lie-Reihen und ihre Anwendungen*. VEB Deutscher Verlag der Wissenschaften, Berlin, 1967.

14. Ch. Großmann and H.-G. Roos. *Numerik partieller Differentialgleichungen*. B. G. Teubner, Stuttgart, 1992.

15. M. A. Henson and D. E. Seborg. Theoretical analysis of unconstrained nonlinear model predictive control. *International Journal of Control*, 58(5):1053–1080, 1993.

16. K J Hunt, D Sbarbaro, R Żbikowski, and P J Gawthrop. Neural networks for control systems: a survey. *Automatica*, 28(6):1083–1112, November 1992.

17. A. Isidori. *Nonlinear Control Systems*. Communication and Control Engineering Series. Springer-Verlag, Berlin, second edition, 1989.

18. J. Chr. Kalkkuhl and E. Liceaga-Castro. Disturbance attenuation by discrete-time output feedback for uncertain nonlinear SISO systems. submitted for publication, 1995.

19. Ü. Kotta. Model matching of nonlinear single-input single-output discrete-time systems: formal and local solutions. *Proc. Estonian Acad. Sci. Phys. Math.*, 40:89–97, 1991.

20. Ü. Kotta. Discrete-time models of a nonlinear continuous-time system. *Proc. Estonian Acad. Sci. Phys. Math.*, 43:64–78, 1994.

21. Ü. Kotta. *Inversion Method in the Discrete-time Nonlinear Control Systems Synthesis Problems*. Springer–Verlag, London, 1995.

22. H. G. Lee, A. Arapostathis, and S. I. Marcus. Linearization of dicrete-time systems. *International Journal of Control*, 45(5):1803–1822, 1987.

23. I.J. Leontaritis and S.A. Billings. Input-output parametric models for non-linear systems part I: deterministic non-linear systems. *International Journal of Control*, 41(2):303–328, 1985.

24. A. U. Levin and K. S. Narendra. Identification using feedforward networks. *Neural Computation*, 7(2):349–357, March 1995.

25. E. Liceaga-Castro, J. Chr. Kalkkuhl, and R. Castro Linares. Disturbance attenuation by output feedback for uncertain nonlinear SISO systems. submitted for publication, 1994.

26. S. Monaco and D. Normand-Cyrot. Multirate sampling and zero dynamics: from linear to nonlinear. Dipartimento di Ingegneria Elettrica, Università di L'Aquila, 67040 L'Aquila, ITALY.

27. S. Monaco and D. Normand-Cyrot. On the sampling of a linear analytic control system. In *Proceedings of the 24th Conference on Decision and Control*, pages 1457–1463, Fort Lauderdale, FL, December 1985. IEEE.

28. S. Monaco and D. Normand-Cyrot. Minimum-phase nonlinear discrete-time systems and feedback stabilization. In *Proceedings of the 26th Conference on Decision and Control*, pages 979–986, Los Angeles, December 1987. IEEE.

29. S. Monaco and D. Normand-Cyrot. Sur la commande digitale d'un systeme non lineaire a dephasage minimal. In A. Bensoussan and J.L. Lions, editors, *Analysis and Optimization of Systems*, pages 193–204. Springer-Verlag, Berlin, 1988.

30. S. Monaco and D. Normand-Cyrot. Zero dynamics of sampled nonlinear systems. *Systems & Control Letters*, 11:229–234, 1988.

31. S. Monaco, D. Normand-Cyrot, and S. Stornelli. On the linearizing feedback in nonlinear sampled data control schemes. In *Proceedings of the 25th Conference on Decision and Control*, pages 2056–2060, Athens, Greece, December 1986. IEEE.

32. P. E. Moraal and J. W. Grizzle. Observer design for nonlinear systems with discrete-time measurement. *IEEE Transactions on Automatic Control*, 40(3):395–404, March 1995.

33. K. Narendra and K. Parthasarathy. Identification and control of dynamical systems using neural networks. *IEEE Transactions on Neural Networks*, 1(1):4–27, March 1990.

34. K.S. Narendra and J. B.D. Cabrera. Input-output representation of discrete-time dynamical systems – nonlinear ARMA models. In *Proceedings of the 33rd Conference on Decision and Control*, pages 1118–1119, Lake Buena Vista, FL, December 1994. IEEE.

35. H. Nijmeijer and A. J. van der Schaft. *Nonlinear dynamical control systems*. Springer Verlag, New York, 1990.

36. J.-J. Slotine and W. Li. *Applied nonlinear control*. Academic Press, New York, 1989.

37. R. Żbikowski, K. J. Hunt, A. Dzieliński, R. Murray-Smith, and P. J. Gawthrop. A review of advances in neural adaptive control systems. ESPRIT NACT Project: Technical Report TP-1, Glasgow University and Daimler-Benz Research, 1994. Submitted for publication.

CONTINUOUS-TIME LOCAL MODEL NETWORKS

PETER J GAWTHROP
Centre for Systems and Control, University of Glasgow,
University of Glasgow, Glasgow G12 8QQ, Scotland
E-mail: P.Gawthrop@eng.gla.ac.uk

ABSTRACT

The *continuous-time version* of the Local Model Network (LMN) approach to modelling non-linear systems of Johansen and Foss is developed and shown to provide a useful alternative to the ubiquitous NARMAX based neural network approach. It is shown how local-state LMN representations of non-linear systems can be used to design LMN-based controllers via state observer/state feedback methods. An input-output emulator-based interpretation of these LMNs is also given; it is suggested that this approach provides an alternative foundation for non-linear self-tuning control.

1. Introduction

Control system design is based on a model of the controlled system. Such design is simplified if the model has a number of desirable attributes; it should be:

1. an *accurate* model of the system capturing the essential features of the actual system;
2. a *robust* model in the sense that small perturbations in model parameters lead to small errors between system and model states and outputs;
3. of a form lending itself to the design method being used;
4. of a form lending itself to *adaptation* and;
5. its structure and parameters should have *physical significance*.

Models with such desirable characteristics are well developed in the context of *linear* systems; but are not generally available in the context of *non-linear* systems.

Physical system modelling leads, in general, to a set of non-linear differential algebraic equations (DAE). In control systems work, it is usually assumed that this set of equations is, or can be reduced to, a non-linear ordinary differential equation of the form

$$\begin{aligned} \dot{x} &= f(x,u) \\ x(0) &= x_0 \\ y &= g(x) \end{aligned} \tag{1}$$

where x is a $n_x \times 1$ vector of system states, u is a $n_u \times 1$ vector of exogenous system inputs (including control signals, disturbances and external variables affecting system dynamics), and y is a $n_y \times 1$ vector of system outputs which are assumed to be measurable. For simplicity, it will be assumed that $n_y = 1$.

For the purposes of this paper, Equation 1 is taken to represent a physical system and is therefore the starting point of the paper.

Although Equation 1 has the advantage of representing a wide range of physical systems, it has the disadvantage that there is no general control design procedure for such systems. For this reason, other system representations have been used as a basis for non-linear control. In particular the Non-linear Autoregressive Moving-Average model of Leonitaris and Billings [1] has proved popular both as a generalisation of (linear) self-tuning control methods and as a basis of much activity in the context of Artificial Neural Networks (ANN)[2,3,4,5,6]. In its deterministic form, it is a discrete-time model given by Equation 2

$$y(t) = F(y(t-1), \ldots, y(t-n_y), u(t-k), \ldots, u(t-k-n_u)) \tag{2}$$

Equation 2 has the advantage of having a number of obvious representations in terms of both polynomials and neural nets. However there are two significant problems

1. Equation 2 is a *locally* (not globally) valid representation of Equation 1 [1], see also Chapter 1 by Kalkkuhl & Hunt;

2. all physical significance is lost.

For these reasons, the formulation of Equation 1 is prefered to that of Equation 2 for the purposes of this paper.

Local model networks (LMN) were introduced by Johansen and Foss [7,8] as a means of decomposing NARMAX models like Equation 2 into an insightful structure for system identification and control. Further developments have been reported by Murray-Smith [9,10].

More recently, Johansen and Foss have investigated a local model network representation of the ODE formulation of Equation 1. Because of the clear physical relevance of Equation 1, this paper paper builds on the ODE formulation of LMN.

The aim of this research is to provide a natural generalisation of (linear) self-tuning control, in a continuous-time setting, to non-linear problems. The purpose of this paper is to provide a foundation for this work by examining networks of continuous-time observers each with its own local state. Although most of the classical work on Self-tuning Control [11,12,13,14] based on the linear ARMA model and a discrete-time representation, a continuous time formulation has been developed [15,16,17,18] which has been rationalised using the concept of an *emulator*. The purpose of this paper is to combine the LMN concept with the emulator concept to give a system representation which, whilst being founded on the physical system of Equation 1 and thus retaining physical insight, is of convenient form for emulator-based controller design and allows linear-in-the-parameters tuning as a basis for non-linear control. It thus provides a natural extension of continuous-time Self-tuning Control to non-linear systems.

There has been considerable activity over the past few years in combining linear control and ANN to give non-linear control. The standard approach [2,3,4,5,6,19] to combining linear control and ANN to give non-linear control is to embed the ANN within an otherwise linear control structure. In contrast, *the approach taken here is to embed a number of linear controllers within a network.*

The outline of the chapter is as follows.

- Section 2. gives a detailed exposition of the construction of LMNs; a continuous-time state-space setting is used to clarify the issues involved. It is emphasised that linearisation is associated with constants associated with the corresponding equilibria and that this naturally leads to local models with an explicit integral term.

- The state-space formulation of the (linear) local models is reinterpreted using transfer functions in Section 3..

- Section 4. shows how each local model can be associated with an *emulator* - a transfer-function form of an observer [15,16] which forms the basis of a number including standard algorithms such as generalised minimum variance [20,21,22,23], generalised predictive [13,14,24,17,18], pole-placement and their self-tuning versions; again in a continuous-time setting. In Section 4.1., it is shown how multiple local states can be combined.

- Section 5. explicitly gives a Local Controller network based on the generalised minimum variance [20,21,22,23] controller.

- Section 6. briefly compares the LMN approach to related approaches, in particular:

  - gain-scheduling,
  - sliding-mode control,
  - modular neural networks and
  - logic-based switching control.

- Section 7. provides illustrative examples and Section 8. concludes the paper.

2. State-space Local Model Networks

The non-linear state-space model of Equation 1 is the starting point of derivation of the LMN models of this paper; this Section considers the state-space foundations of the continuous-time ARMA models. In so doing, some state-space LMN control designs are developed along the way.

The two important issues of *equilibria* and *linearisation* of non-linear systems are considered in Sections 2.1. and 2.2. respectively. These sections can, however, be omitted on a first reading.

2.1. *Equilibria*

The *equilibria* of the system of Equation 1 are given by the solutions for x^e, y^e and u^e of the nonlinear algebraic Equations obtained by setting $\dot{x}$ to zero in Equation 1 to give:

$$\begin{aligned} 0 &= f(x^e, u^e) \\ y^e &= g(x^e) \end{aligned} \tag{3}$$

It is of particular interest to *parameterise* the solutions in some way. A complete solution to this problem is beyond the scope of this chapter; but rather an indication of the issues involved are given.

If Equation 1 were linear, Equation 3 would become:

$$\begin{aligned} 0 &= Ax^e + Bu^e \\ y^e &= Cx^e \end{aligned} \tag{4}$$

There are two cases to consider.

1. If $\det A \neq 0$, ie the system has no poles at $s = 0$, then

   $$\begin{aligned} y^e &= G_{ss}u^e \\ x^e &= A^{-1}Bu^e \end{aligned} \tag{5}$$

   where the system *steady-state gain* matrix G is given by:

   $$G_{ss} = CA^{-1}B \tag{6}$$

   In this case, the equilibria of the system may be parameterised by the n_u dimensional vector u^e. As, usually, $n_x > n_u$ the equilibrium states of the system lie in a lower dimensional subspace of the state-space of the system. Other parameterisations are possible, for example incorporating elements of y^e or x^e; nevertheless an n_u dimensional vector suffices.

2. If, on the other hand, det $A = 0$, ie the system has at least one pole at $s = 0$ then the choice of u^e then the steady-state gain does not exist. One possible set of equilibria is defined by:

$$\begin{aligned} u^e &= 0 \\ Ax^e &= 0 \end{aligned} \tag{7}$$

That is x^e is an eigenvector of A corresponding to a zero eigenvalue. In this case u^e cannot be used as a parameter, and the subspace of x containing x^e may be of even lower dimension. Other solutions may be possible if the system is uncontrollable.

Two particular cases of nonlinear system are discussed in Section 7.. In these cases, the equilibria are parameterised by vectors of dimension n_u; but this is by no means the case in general.

2.2. *Linearisation*

Given a particular equilibrium, local linearisation of non-linear dynamic systems of the form of Equation 1 about such an equilibrium, is a standard procedure [25]. The first step is to consider the (non-dynamic) linearisation of $f(x, u)$ with respect to its arguments by performing a first-order Taylor expansion about $x = x_i^e$ and $u = u_i^e$ (see Equation 3):

$$\begin{aligned} f(x,u) &\approx f(x_i^e, u_i^e) + A_i^d x_i^d + B_i^d u_i^d \\ g(x,u) &\approx g(x_i^e, u_i^e) + C_i^d x_i^d \end{aligned} \tag{8}$$

where the state deviation x_i^d and input deviation u_i^d are

$$\begin{aligned} x_i^d &= x - x_i^e \\ u_i^d &= u - u_i^e \end{aligned} \tag{9}$$

and the jk elements of the matrices A_i^d, B_i^d and C_i^d are:

$$\begin{aligned} A_{ijk}^d &= \frac{\partial f_j}{\partial x_k} \\ B_{ijk}^d &= \frac{\partial f_j}{\partial u_k} \\ C_{ijk}^d &= \frac{\partial g_j}{\partial x_k} \end{aligned} \tag{10}$$

evaluated at the point $x = x_i^e$ and $u = u_i^e$.

This (non-dynamic) linearisation is with respect to a *point* $x = x_i^e$ and $u = u_i^e$; in contrast *dynamic* linearisation of the ODE of Equation 1 is with respect to a *trajectory* $x_i^e(t)$ forming the solution of Equation 1 for a given input function $u_i^e(t)$. In general, dynamic linearisation leads to a time varying linear system [25].

The resultant linearised system is thus time-invariant. Substituting Equations 8 and 9 into Equations 1 the linearised system is:

$$\begin{aligned} \dot{x} &\approx f(x_i^e, u_i^e) + A_i^d x_i^d + B_i^d u_i^d \\ y &\approx g(x_i^e) + C_i^d x_i^d \end{aligned} \tag{11}$$

Defining the output deviation y_i^d

$$y_i^d = y - y_i^e \tag{12}$$

and using Equations 3, it follows that:

$$\begin{aligned} \dot{x}_i^d &= A_i^d x_i^d + B_i^d u_i^d + \epsilon_i^x \\ x_i^d(0) &= x_0 - x_i^e \\ y_i^d &= C_i^d x_i^d + \epsilon_i^y \end{aligned} \tag{13}$$

where ϵ_i^x and ϵ_i^y are the errors implied by the approximations in Expression 11.

2.3. *Implicit Equilibrium Offsets*

Equation 13 contains the equilibrium constants y_i^e and u_i^e explicitly via Equations 9 and 12. As far as each linear system (Equation 13) is concerned, these equilibrium constants appear as constant disturbances or offsets. These constants are, in practice, often poorly known and so lead to sensitivity problems if they explicitly appear in a controller. For this reason, it is prudent to reformulate the equations so that these constants are transformed to states and can thus be estimated via a state observer. As will be seen in Section 4., this leads to controllers with integral action [26,27].

Firstly, the equilibrium input u_i^e is considered. If the system matrix A_i^d has at least one eigenvalue at 0 then u_i^e must be zero and does not need to be considered further. If this is not so, then the state equation of Equations 13 may be rewritten as:

$$\dot{x}_i^d = A_i^d (x_i^d - (A_i^d)^{-1} B_i^d u_i^e) + B_i^d u + \epsilon_i^x \tag{14}$$

Defining the new state vector $\tilde{x}_i$

$$\tilde{x}_i = x_i^d - (A_i^d)^{-1} B_i^d u_i^e \tag{15}$$

Equations 13 may be rewritten as:

$$\dot{\tilde{x}}_i = A_i^d \tilde{x}_i + B_i^d u + \epsilon_i^x$$

$$\begin{aligned}\tilde{x}_i(0) &= x_0 - x_i^e - (A_i^d)^{-1} B_i^d u_i^e \\ y &= C_i \tilde{x}_i + y_i^e + C_i^d (A_i^d)^{-1} B_i^d u_i^e + \epsilon_i^y \end{aligned} \tag{16}$$

Equations 16 now contain u in place of u_i^d, but have a modified initial condition and an additional output offset $C_i^d A_i^{d-1} B_i^d u_i^e$.

Secondly, the equilibrium output y_i^e, together with the additional output term $C_i^d (A_i^d)^{-1} B_i^d u_i^e$, is considered. The state vector $\tilde{x}_i$ is augmented with an additional state corresponding to the (constant) total output offset to give a new state x_i^a where

$$x_i^a = \begin{bmatrix} \tilde{x}_i \\ y_i^e + C_i^d A_i^{d-1} B_i^d u_i^e + \epsilon_i^y \end{bmatrix} \tag{17}$$

(if $u_i^e = 0$, the term containing u_i^e is omitted). Defining a new set of system matrices by:

$$A_i = \begin{bmatrix} A_i^d & 0 \\ 0 & 0 \end{bmatrix}; \; B_i = \begin{bmatrix} B_i^d \\ 0 \end{bmatrix}; \; C_i = [\, C_i^d \quad 1 \,] \tag{18}$$

Equation 16 can be rewritten as

$$\begin{aligned}\dot{x}_i^a &= A_i x_i^a + B_i u + \epsilon_i^x \\ x_i^a(0) &= \begin{bmatrix} \tilde{x}_i(0) \\ y_i^e + C_i^d (A_i^d)^{-1} B_i^d u_i^e \end{bmatrix} \\ y &= C_i x_i^a + \epsilon_i^y \end{aligned} \tag{19}$$

Equation 16 gives the system output y in terms of the system input u; the equilibrium offset variables x_i^e, y_i^e and u_i^e are now implicitly banished to the initial (augmented) state $x_i^a(0)$.

2.4. *Local-state LMN*

For each equilibrium indexed by i, Equations 19 each exactly describe the non-linear system of Equation 1. With the error terms ϵ_i^x and ϵ_i^y; each becomes an approximation to the non-linear system of Equation 1. With this in mind, the ith *local state local model*, with state x_i^l, is defined by the following *approximation* to the system of Equations 19 with ϵ_i^x and ϵ_i^y deleted

$$\begin{aligned}\dot{x}_i^l &= A_i x_i^l + B_i u_i^d \\ y_i^l &= C_i x_i^l \end{aligned} \tag{20}$$

The initial state is left unspecified; its choice leads to further error.

Following a similar idea to that of Johansen and Foss [7,8], the non-linear system of Equation 1 is approximated by the network of local models of the form of Equation

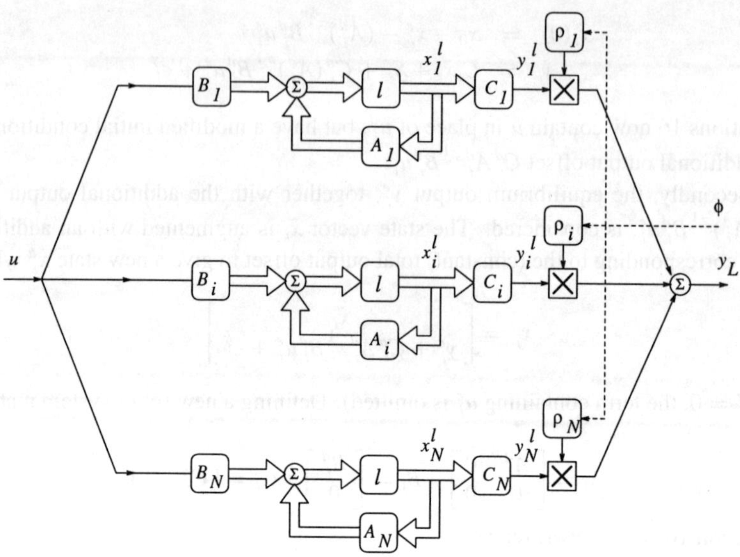

Figure 1: Local-state LMN.

20 together with a set of *validity functions* $\rho_i(\phi)$ depending on the *operating point vector* ϕ. These validity function are a partition of unity in the sense that:

$$\sum_{i=1}^{N} \rho_i(\phi) = 1 \tag{21}$$

That is, the system output y is approximated by the local model output y_L as

$$y_L = \sum_{i=1}^{N} y_i^l \rho_i(\phi) \tag{22}$$

Equations 1 and 22 are shown in diagrammatic form in Figure 1.

The relationship between this LMN and that of Johansen and Foss (which uses a single global state) is discussed elswhere [28]. In summary:

1. The local-state model has the advantage that each local state can be used in a local controller and can be observed by a local observer each based on a linear design. On the other hand, observer design for global-state local model network is a non-linear problem.

2. The global-state model has the advantage of performing static—as opposed to dynamic—linearisation and thus avoids issues of dynamic linearisation.

3. The global-state model has the advantage of having less states to store and update.

4. Except for the currently active local model, the local-state model states may not be close to the true system state; and this is a disadvantage compared to the global state model. But this is counteracted by item 1., as the (easily designed) observer feedback drives the local state towards the true system state. This is the motivation for Section 2.5. below.

2.5. *Observer-based LMN*

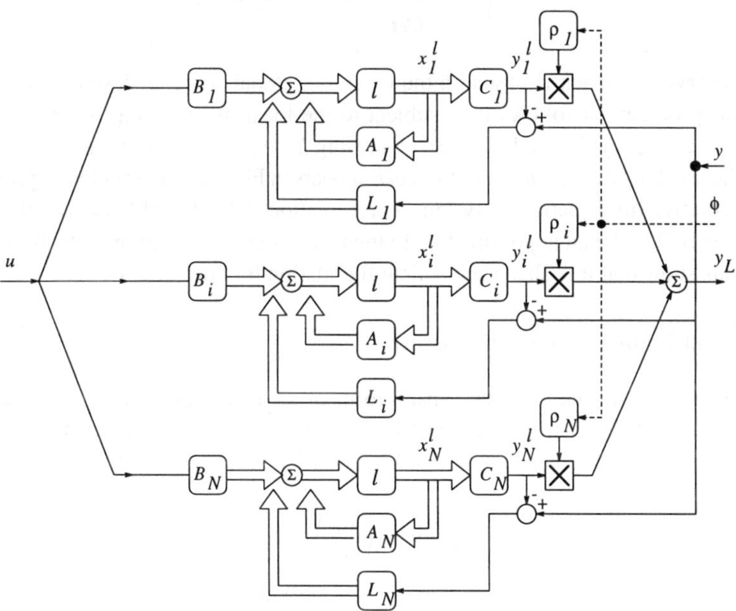

Figure 2: Local-state Observer Network.

The approximation errors ϵ_i^x and ϵ_i^y appearing in Equation 20 may be large for the models operating away from their respective equilibria; thus as *open-loop* equations the states will not in general provide a good approximation to the system of Equation 1; this is particularly true if the local systems are unstable. It is therefore natural to incorporate (local) feedback via classical observer theory to each of the local models.

Choosing an observer gain vector L_i, each of the linear systems of Equation 20 can be associated with a state observer:

$$\begin{aligned} \dot{\hat{x}}_i^l &= A_i\hat{x}_i^l + B_iu_i^l + L_ie_i \\ \hat{y}_i^l &= C_i\hat{x}_i^l \end{aligned} \tag{23}$$

where the *output error* e_i is given by:

$$e_i = y - \hat{y}_i^l \tag{24}$$

Equations 23 can also be written as:

$$\begin{aligned} \dot{\hat{x}}_i^l &= \Gamma_i\hat{x}_i^l + F_iy_i^l + G_iu_i^l \\ \hat{y}_i^l &= C_i\hat{x}_i^l \end{aligned} \tag{25}$$

The observer feedback is based on the local, not global state; its design is therefore a linear problem. It follows that, subject to the local model being observable, the observer can always be made stable by choosing L_i to give Γ_i with stable eigenvalues.

The resultant *Local Observer Network* appears in Figure 2. This network provides an alternative (to Equation 1) system representation. This Local Observer Network has been built to be an approximation to the true system of Equation 1; the following Sections show that it is also a convenient foundation for controller design.

3. Local Transfer Function Networks

Section 2.4. provides a state-space form of an LMN; in contrast, this Section provides a transfer-function interpretation. To avoid cumbersome notation, the single-output, multi-input dynamic system of Equation 1 is specialised to the two-input case ($n_u = 2$) and:

$$u = \begin{pmatrix} u_c \\ v \end{pmatrix} \tag{26}$$

where u_c is a (scalar) control input and v a (scalar) disturbance input.

The ith local model (in state-space form) (Equation 20) is linear and therefore has the transfer function representation

$$Y_i^l(s) = \frac{b_i(s)}{a_i(s)}U_c(s) + \frac{d_i(s)}{a_i(s)}V(s) \tag{27}$$

where $Y_i^l(s)$, $U_c(s)$ and $V(s)U_c$ are the Laplace transformed ith local model output and system input respectively and

$$\frac{b_i(s)}{a_i(s)} = C_i\left[sI - A_i\right]^{-1} B_{i1}$$

$$\frac{d_i(s)}{a_i(s)} = C_i\,[sI - A_i]^{-1}\,B_{i2} \tag{28}$$

where B_{i1} and B_{i2} are the first and second columns of the $n_x \times 2$ B_i matrix.

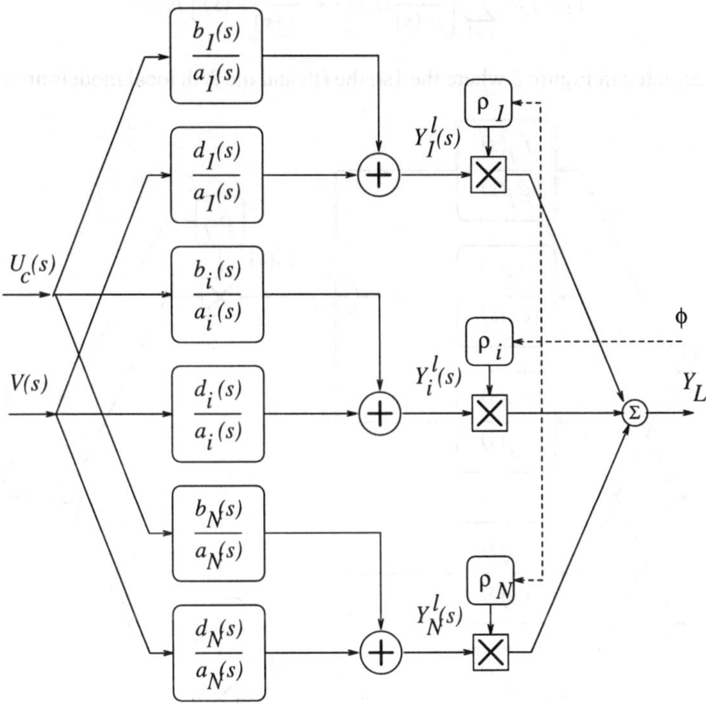

Figure 3: Transfer Function LMN.

By construction A_i has at least one zero eigenvalue so

$$a_i(0) = \det(-A_i) = 0 \tag{29}$$

However, the additional state (corresponding to the zero eigenvalue) is uncontrollable and so there is a corresponding pole/zero cancellation implying that both $b_i(0) = 0$ and $d_i(0) = 0$. In other words

$$\begin{aligned} a_i(s) &= s\tilde{a}_i(s) \\ b_i(s) &= s\tilde{b}_i(s) \\ d_i(s) &= s\tilde{d}_i(s) \end{aligned} \tag{30}$$

where $\tilde{a}_i(s)$, $\tilde{b}_i(s)$ and $\tilde{d}_i(s)$ are polynomials in s. As discussed in [27] and [15] this naturally leads to controllers with integral action.

A third approximate representation of the non-linear system of Equation 1 is thus given (in Laplace transform terms) by:

$$Y_L(s) = \sum_{i=1}^{N} \left[\frac{b_i(s)}{a_i(s)} U_c(s) + \frac{d_i(s)}{a_i(s)} V(s) \right] \rho_i(\phi) \tag{31}$$

This is depicted in Figure 3 where the 1st, the ith and the Nth local models are shown.

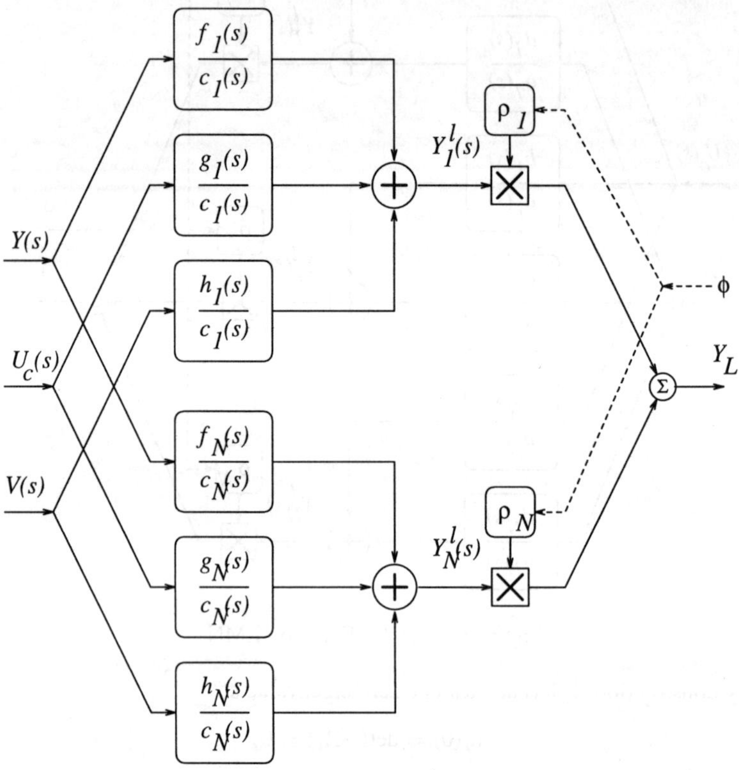

Figure 4: Emulator LMN.

4. Local Emulator Networks

Emulators [15,16,17,18] provide the basis for various flavours of self-tuning control of linear (possibly linear in the parameter) systems; *local emulator networks* thus provide the basis for self-tuning control of non-linear systems. This Section provides a transfer function based account of the emulator-based approach in the context of local-model networks.

Section 2.5. provides a state-space form of an LMN; in contrast, this Section provides a transfer-function interpretation. As discussed in Section 2.5., the approximation errors inherent in each local model may be large for the models operating away from their respective equilibria. Thus as *open-loop* equations the states will not in general provide a good approximation to the system of Equation 1; this is particularly true if the local systems are unstable. The input-output equivalent of an observer is an emulator [15]. A particular version of such an emulator provides an alternative approximation y_i^l *making use of the system output measurement.*

The basis of an emulator is a polynomial $c_i(s)$ chosen as follows:

1. $c_i(s)$ is stable ($c_i(s) \neq 0 \; \forall \; \Re s > 0$),
2. $c_i(s)$ has same degree as $a_i(s)$.

The following polynomial identity can then be solved for $f_i(s)$ and $e_i(s)$

$$c_i(s) = e_i(s)a_i(s) + f_i(s) \tag{32}$$

where $f_i(s)/a_i(s)$ is strictly proper—this is essentially polynomial division. The polynomials $g_i(s)$ and $h_i(s)$ are defined as:

$$\begin{aligned} g_i(s) &= e_i(s)b_i(s) \\ h_i(s) &= e_i(s)d_i(s) \end{aligned} \tag{33}$$

The local transfer-function model of Equation 27 can then be rewritten as:

$$Y_i^l(s) = \frac{f_i(s)}{c_i(s)}Y(s) + \frac{g_i(s)}{c_i(s)}U_c(s) + \frac{h_i(s)}{c_i(s)}V(s) \tag{34}$$

This model is, by construction, stable and the local outputs can be combined using Equation 31.

This is depicted in Figure 4 where only the 1st and the Nth local models are shown.

From Equations 30, 32 and 33, it follows that:

$$\begin{aligned} f_i(0) &= 1 \\ g_i(0) &= 0 \\ h_i(0) &= 0 \end{aligned} \tag{35}$$

and so the three polynomials are of the form:

$$f_i(s) = 1 + s\tilde{f}_i(s)$$

$$\begin{aligned} g_i(s) &= s\tilde{g}_i(s) \\ h_i(s) &= s\tilde{h}_i(s) \end{aligned} \tag{36}$$

Equation 34 may then be rewritten as:

$$\tilde{Y}_i^l(s) = \frac{s\tilde{f}_i(s)}{c_i(s)}Y(s) + \frac{s\tilde{g}_i(s)}{c_i(s)}U_c(s) + \frac{s\tilde{h}_i(s)}{c_i(s)}V(s) \tag{37}$$

and

$$Y_i^l(s) = \tilde{Y}_i^l(s) + \frac{1}{c_i(s)}Y(s) \tag{38}$$

This form is significant in that each local model creating $\tilde{Y}_i^l(s)$ *blocks all constant values* relating to y, u and v.

4.1. Combining the Local States

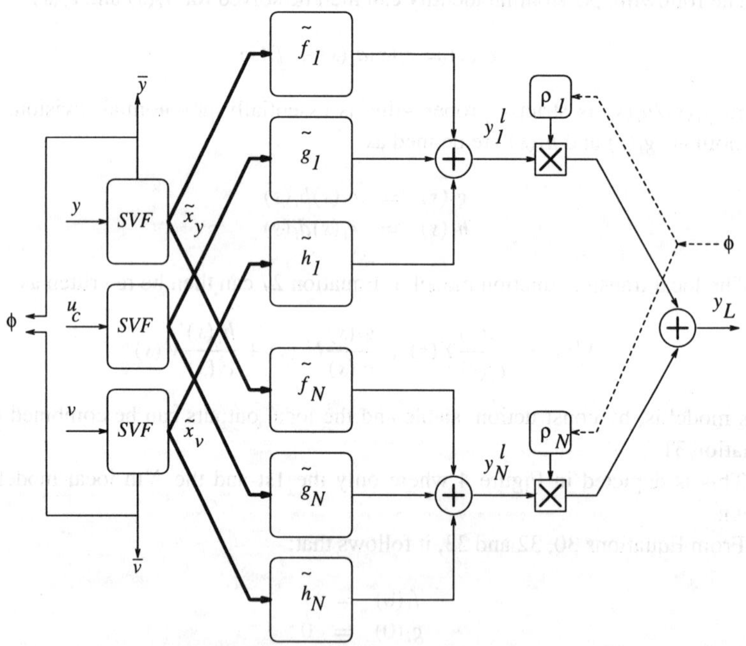

Figure 5: Emulator LMN with common state.

An interesting special case arises if all of the polynomials $c_i(s)$ are identical: $c_i(s) = c(s)$. In this case, we can define three *state-variable filters*, each with characteristic polynomial $c(s)$, generating the three states which, in Laplace transform terms are given by:

$$X_y(s) = \frac{1}{c(s)} \begin{pmatrix} 1 \\ s \\ s^2 \\ \vdots \\ s^{n-1} \end{pmatrix} Y(s); \qquad X_u(s) = \frac{1}{c(s)} \begin{pmatrix} 1 \\ s \\ s^2 \\ \vdots \\ s^{n-1} \end{pmatrix} U_c(s);$$

$$X_v(s) = \frac{1}{c(s)} \begin{pmatrix} 1 \\ s \\ s^2 \\ \vdots \\ s^{n-1} \end{pmatrix} V(s).$$

Each state is further decomposed in to a mean-value term and derivative terms as:

$$x_y = \begin{pmatrix} \bar{y} \\ \tilde{x}_y \end{pmatrix}; x_u = \begin{pmatrix} \bar{u} \\ \tilde{x}_u \end{pmatrix}; x_v = \begin{pmatrix} \bar{v} \\ \tilde{x}_v \end{pmatrix}; \tag{39}$$

Each local emulator can be written in (nonminimal) state-space form as:

$$\tilde{y}_i^l = \tilde{f}_i \tilde{x}_y + \tilde{g}_i \tilde{x}_u + \tilde{h}_i \tilde{x}_v \tag{40}$$

$$y_i^l = \tilde{y}_i^l + \bar{y} \tag{41}$$

where the vectors $\tilde{f}_i$, $\tilde{g}_i$ and $\tilde{h}_i$ contain the appropriate coefficients of the polynomials $\tilde{f}_i(s)$, $\tilde{g}_i(s)$ and $\tilde{h}_i(s)$ respectively.

The crucial fact about this representation is that x_y, x_u and x_v are *independent of* i: each local model shares a common state.

This is depicted in Figure 5 where only the 1st and the Nth local models are shown and vector signals are shown by wide lines.

5. A Local Controller Network (LCN)

As discussed in the introduction, the purpose of developing the Local Model Network (LMN) is to provide a basis for the control of non-linear systems. There has been considerable activity over the past few years in combining linear control and ANN to give non-linear control. The standard approach [2,3,4,5,6,19] to combining linear control and ANN to give non-linear control is to embed the ANN within an otherwise linear control structure. In contrast, *the approach taken here is to embed a number of linear controllers within an network.*

One approach to linear control design uses the concept of an emulator as a unifying theme [15,16,17,18]. This concept has been shown [29] to correspond to certain model predictive control schemes [30] including that of [31]. In this Section, one special, and simple case is considered for the purposes of illustration: pole placement control (see [15] for more detail). In the context of linear systems, the aim to design place the closed-loop system poles to give a closed loop response:

$$Y(s) = \frac{1}{p(s)} R(s) \tag{42}$$

where $R(s)$ is the (Laplace transformed) reference signal and $p(s)$ a prespecified polynomial with degree corresponding to the relative degree of the system.

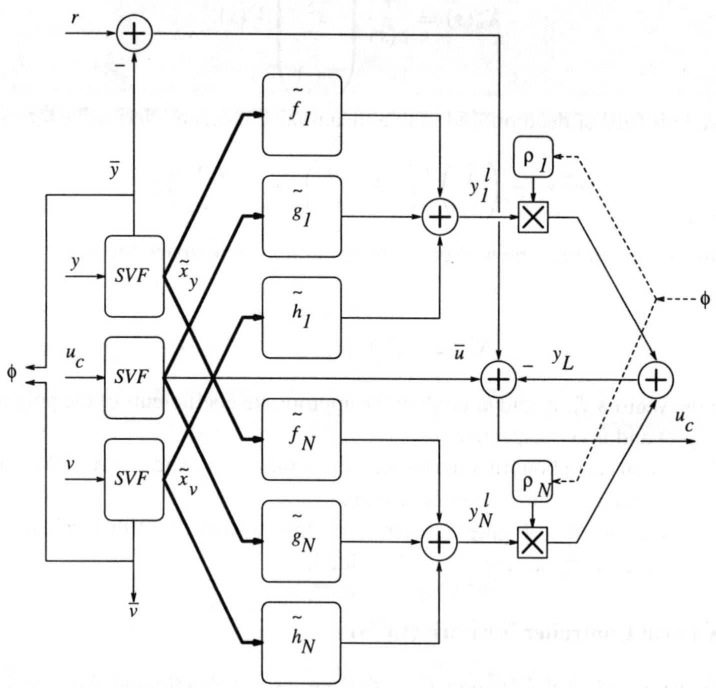

Figure 6: A local controller network.

The design procedure is based on the emulator design of Equations 32, 33 and 34. In particular, Equation 32 is replaced by Equation 43

$$p(s)c(s) = e_i(s)a_i(s) + f_i(s) \tag{43}$$

where a common $C_i(s) = c(s)$ has been chosen for the reasons given in Section 4.1.

The emulator (with output $Y^p(s)$) then becomes:

$$Y_i^p(s) = \frac{s\tilde{f}_i(s)}{c(s)}Y(s) + \frac{s\tilde{g}_i(s)}{c(s)}U_c(s) + \frac{s\tilde{h}_i(s)}{c(s)}V(s) + \bar{Y}(s) \tag{44}$$

where the polynomials $f_i(s)$, $g_i(s)$ and $h_i(s)$ are now based on Equation 43 rather than Equation 32. The main difference between the polynomials is that the transfer function $\tilde{g}_i(s)/c(s)$ is no longer strictly proper; in fact, the degree of $\tilde{g}_i(s)$ is the same as that of $c(s)$. It is convenient to rewrite this transfer function as:

$$\frac{s\tilde{g}_i(s)}{c(s)} = \gamma_i\left(1 - \frac{1}{c(s)}\right) + \frac{s\tilde{g}_i'(s)}{c(s)} \tag{45}$$

Using the common-state emulator formulation of Equation 40, together with values of f_i, g_i and h_i arising from Equation 43, and using Equations 45 and 21, Equation 44 becomes:

$$y_p = \bar{y} + \gamma(u - \bar{u}) + \sum_{i=1}^{N}\left[\tilde{f}_i\tilde{x}_y + \tilde{g}_i'\tilde{x}_u + \tilde{h}_i\tilde{x}_v\right]\rho_i(\phi) \tag{46}$$

where

$$\gamma = \sum_{i=1}^{N}\gamma_i\rho_i(\phi) \tag{47}$$

Following the methods of [15], the emulator-based controller which leads to the desired response of Equation 42 is obtained by setting y_p to the desired reference r

$$y_p = r \tag{48}$$

Combining Equations 46 and 48 finally gives the controller in the form:

$$u = \bar{u} + \frac{1}{\gamma}(r - \bar{y}) + u_h \tag{49}$$

where u_h is given by

$$u_h = -\frac{1}{\gamma}\sum_{i=1}^{N}\left[\tilde{f}_i\tilde{x}_y + \tilde{g}_i'\tilde{x}_u + \tilde{h}_i\tilde{x}_v\right]\rho_i(\phi) \tag{50}$$

This controller depicted in Figure 6 where only the 1st and the Nth local models are shown and vector signals are shown by wide lines.

Remarks

1. The state-variable filters not only provide the inputs to each local controller but also generates the operating point vector ϕ. As discussed in Section 6.3., this provides a natural frequency-domain decomposition of the measured variables. There are two fundamentally different elements appearing in ϕ:
   (a) A (smoothed) *external* signal $\bar{v}$ and
   (b) A (smoothed) *internal* signal $\bar{y}$.
2. The controller of Equation 49 has integral action. This can be seen as follows. If all signals are constant then $u_h = 0$, $\bar{y} = y$ and $\bar{u} = u$. Equation 49 then implies that $\bar{y} = y = r$.
3. The control output u is applied to the actual system as well as to the input of the LCN. These signals are not shown connected on the diagram as, in practice, the actual control may be limited by, for example a saturation element. In such circumstances, the LCN input would be the control applied to the plant, not the controller output. This is a standard way of avoiding integral saturation
4. Equation 50 only generates a control signal when signals are changing with time; in this sense, it represents a dynamic *high-frequency* compensator working together with the *low-frequency* compensator implied by Equation 49.
5. Each local model appearing in Equation 46 is *linear in the parameters.* This potentially leads to simple adaptive algorithms.

6. Related approaches

No technique exists in isolation; indeed, the convergence of apparently different approaches towards a common algorithm is evidence for the importance of the technique.

This section briefly surveys some approaches which appear to be related to local-model networks. The purpose is not to give a detailed exposition (this would take up too much space) but rather to give pointers to further reading. These approaches are:

- gain-scheduling,
- sliding-mode control,
- modular neural networks and
- logic-based switching control.

6.1. Gain-scheduling

The technique by which controller parameters or *gains* are adjusted according to an *operating point* is know as *gain-scheduling*. This is a well established technique in practice, but has only relatively recently received theoretical attention. A modern treatment is given by [32].

The techniques presented in this chapter can indeed be viewed as a form of gain-scheduling, but from a rather different perspective than usual; further discussion of this point is given by [10].

6.2. Sliding-mode control

Sliding-mode control [33] (sometimes called variable-structure, or switched or chattering or relay control) is a well-established non-linear control technique. In its simplest form, it is assumed that the system state is directly measured. The state space is divided into regions by "switching surfaces", and these regions determine the value of the system control signal. If the state is not available, a self-tuning emulator-based approach can be used [34].

Approach is related to that of this Chapter in that the state-space is divided into regions; in the case of Sliding-mode control by appropriate switching surfaces and in the case of LCN by the validity functions.

The approach is different in that the Sliding-mode control regions determines the *value of the control signal* whereas the LCN validity functions determines the *feedback control law*.

6.3. Modular neural networks

The networks displayed in Figures 5 and 6 have a clear structure; moreover, each element of the structure has a clear function.

The state-variable filters partition signals by frequency content: e.g. $\bar{y}$ contains low frequency information and x_y contains high frequency information.

The local models emulate the high frequency system dynamics *at a given operating point* using the high-frequency information.

The operating point vector encapsulates the low-frequency information contained in $\bar{y}$ and $\bar{v}$ as

$$\phi = \begin{pmatrix} \bar{y} \\ \bar{v} \end{pmatrix} \tag{51}$$

The validity functions ρ_i coordinate the outputs of the local models.

The controller uses the low frequency information to give correct steady-state performance (integral action) and the high-frequency information to give good dynamic performance.

This decomposition by function is a characteristic feature of Modular Neural Networks [35]. MNN have been used in the context of control [36] and display some similarities with the LMN approach used here. In particular, their 'gating function' has a close resemblance to the validity functions and the 'expert networks' bear a close resemblance to the local models.

It is a matter of future research to investigate the learning algorithms of [36] in this context.

6.4. *Logic-based switching control*

In a recent paper, [37] gives a survey of a class of controllers, logic-based switching controllers. As discussed in that paper, such controllers can be viewed as *hybrid controllers* involving a parameterised set of controllers together with switching logic. The controllers are parameterised by a switching signal σ which is generated by the switching logic.

In particular, the *state-shared estimator-based supervisor* approach appears to have much in common with with LCN.

7. Examples

7.1. *A Heated Tank*

The following non-linear dynamic system has two inputs, u and v and output $y = x$.

$$\begin{aligned} \dot{x} + vx &= u \\ y &= x \end{aligned} \tag{52}$$

As discussed previously [10], this has a physical interpretation as a heated tank of liquid with variable liquid flow rate. The state x may be interpreted as temperature; v as the flowrate though the tank, and u as a source of heat. The product of temperature and flow xv can be interpreted as the (convected) heat flow from the tank.

The equilibrium corresponding to $\phi_i = [y_i^e, v_i^e]^T$ is:

$$\begin{aligned} x_i^e &= y_i^e \\ u_i^e &= v_i^e y_i^e \end{aligned} \tag{53}$$

The corresponding linearised system is described by:

$$\begin{aligned} A_i^d &= -v_i \\ B_i^d &= [1 - y_i^e]^T \\ C_i^d &= 1 \end{aligned} \tag{54}$$

This system is first-order and of relative degree 1; after augmentation and converting to transfer function form it can be rewritten in emulator form (Section 5.) as:

$$\phi^* = \frac{1 + f_i s}{1 + c_1 s + c_2 s^2} y + \frac{1 + g_{ui} s}{1 + c_1 s + c_2 s^2} + \gamma_{ui} u + \frac{1 + g_{vi} s}{1 + c_1 s + c_2 s^2} \tag{55}$$

The control is designed to give a desired response:

$$y_r = \frac{1}{1 + ps} r \tag{56}$$

with $p = 0.5$. The emulator denominator was chosen as $c(s) = (1 + 0.25s)^2$.

One hundred operating points ϕ_i were specified by

$$\begin{aligned} y_i^e &= 0.5, 1.0, 1.5, ..., 5.0 \\ v_i &= 0.5, 1.0, 1.5, ..., 5.0 \end{aligned} \tag{57}$$

The validity function ρ was a 2-D Gaussian function with $\sigma = 0.5$.

The non-linear system, Equation 52, was simulated using Simulink together with the CLMN emulator-based controller. For the purposes of comparison, a single linear controller corresponding to the single operating point $y^e = v^e = 1$ was also simulated. Figure 7 has two sets of graphs. The first shows three signals plotted against time: the system output y, the output corresponding to the linear controller y_{lin}, and the reference model output y_r. The second shows the corresponding control signals.

The performance of the CLMN-based controller is superior to that of the linear controller.

Figure 8 shows the validity functions evolving with time. This gives an idea of how and when switching occurs between the various controllers within the network.

7.2. *An Unstable Non-linear System*

The following non-linear dynamic system has a control input u, and output $y = x_1$.

$$\begin{aligned} \dot{x}_1 &= x_2 \\ \dot{x}_2 &= x_1^2 + x_2^2 + u \end{aligned} \tag{58}$$

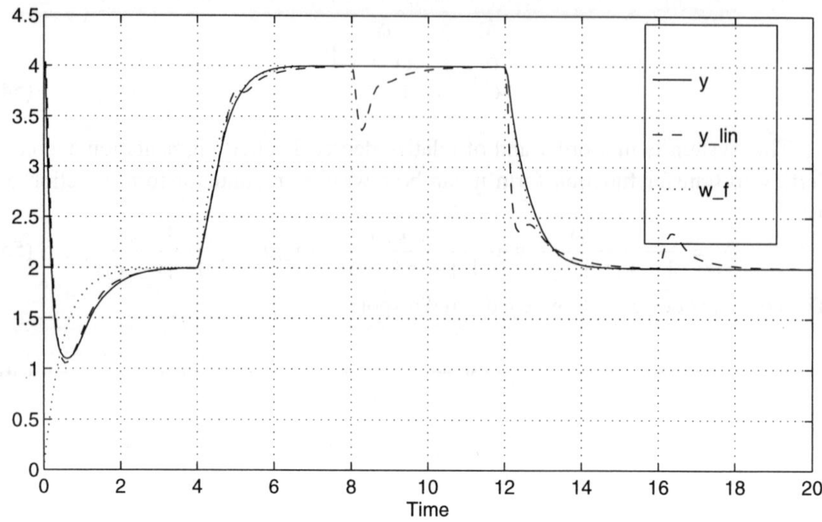

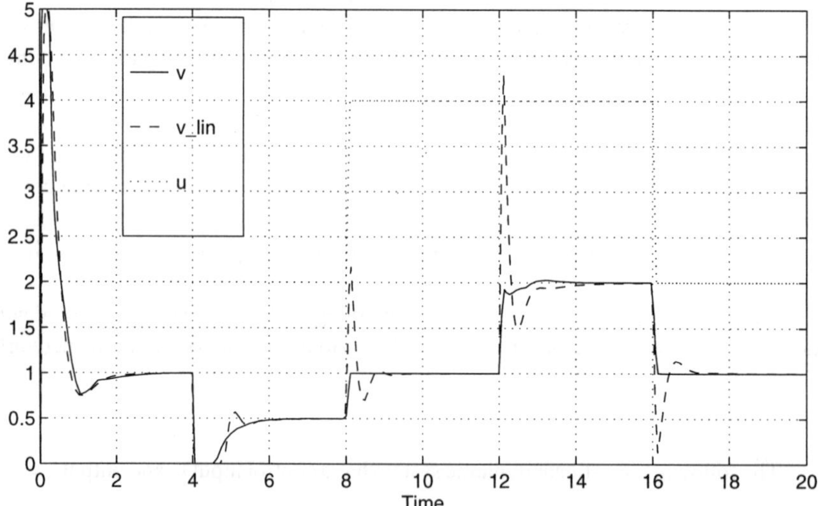

Figure 7: CLMN emulator-based control: heated tank.

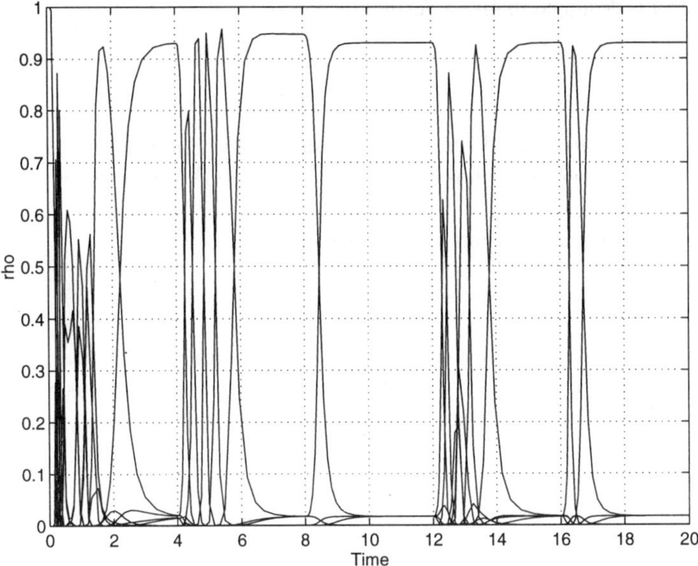

Figure 8: Validity functions: heated tank.

This system has been used in the literature as an example of a difficult non-linear system. The equilibrium corresponding to $\phi_i = [y_i^e]^T$ is:

$$\begin{aligned} x_i^e &= [y_i^e, 0]^T \\ u_i^e &= -(y_i^e)^2 \end{aligned} \tag{59}$$

The corresponding linearised system is described by:

$$\begin{aligned} A_i^d &= \begin{bmatrix} 0 & 1 \\ 2y_i^e & 0 \end{bmatrix} \\ B_i^d &= \begin{bmatrix} 0 \\ 1 \end{bmatrix} \\ C_i^d &= [1 \quad 0] \end{aligned} \tag{60}$$

This system is second-order and of relative degree 2; after augmentation and converting to transfer function form it can be rewritten in emulator form (Section 5.) as:

$$\phi^* = \frac{1 + f_{1i}s + f_{2i}s^2}{1 + c_1 s + c_2 s^2 + c_3 s^3} y + \frac{1 + g_{1i}s + g_{2i}s^2}{1 + c_1 s + c_2 s^2 + c_3 s^3} + \gamma_i u \tag{61}$$

The control is designed to give a desired response:

$$y_r = \frac{1}{1 + p_1 s + p_2 s^2} r \tag{62}$$

with $p_1 = 1$ and $p_2 = 0.25$ (this gives two desired system poles at $s = -2$). The emulator denominator was chosen as $c(s) = (1 + 0.2s)^3$ (this gives three emulator poles at $s = -5$).

Twenty-six operating points ϕ_i were specified by

$$y_i^e = 0.0, 0.2, 0.4..., 5.0 \tag{63}$$

The validity function ρ was a 1-D Gaussian function with $\sigma = 0.5$.

The unstable non-linear system, Equation 58, was simulated using Simulink together with the CLMN emulator-based controller. For the purposes of comparison, a single linear controller corresponding to the single operating point $y^e = 2.5$ was also simulated. Figure 9 has two sets of graphs. The first shows three signals plotted against time: the system output y, the output corresponding to the linear controller y_{lin}, and the reference model output y_r. The second shows the corresponding control signals.

The performance of the CLMN-based controller is superior to that of the linear controller.

Figure 10 shows the validity functions evolving with time. This gives an idea of how and when switching occurs between the various controllers within the network.

8. Conclusions

In the introduction, the following desirable features of a model were noted:

1. an *accurate* model of the system capturing the essential features of the actual system;
2. a *robust* model in the sense that small perturbations in model parameters lead to small errors between system and model states and outputs;
3. of a form lending itself to the design method being used;
4. of a form lending itself to *adaptation* and
5. its structure and parameters should have *physical significance*.

Although detailed analysis remains to me done, the continuous-time LMN approach appears to satisfy these requirements, to some extent as follows:

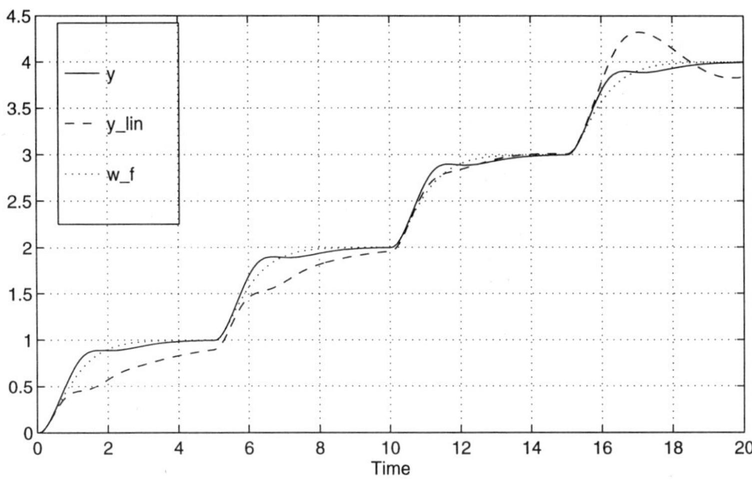

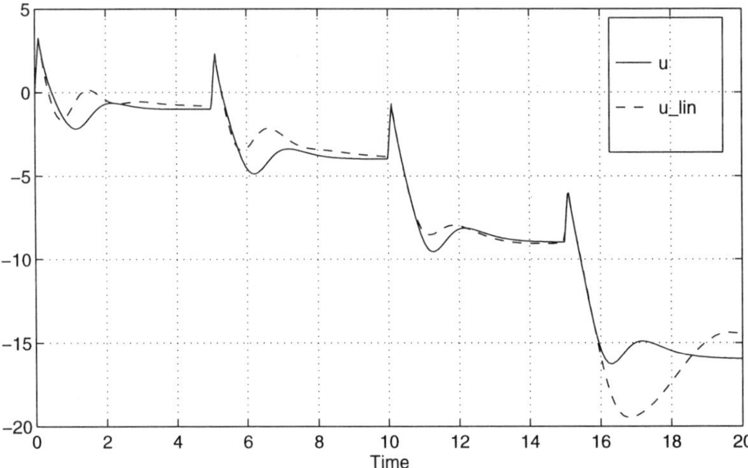

Figure 9: CLMN emulator-based control: unstable non-linear system.

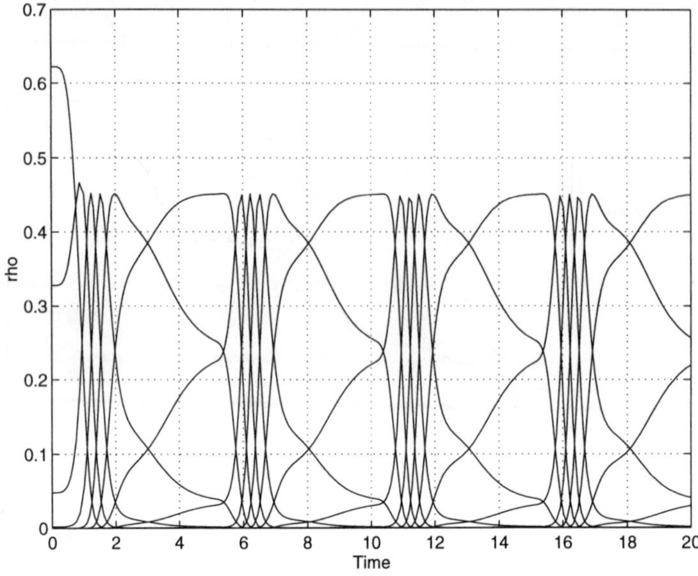

Figure 10: Validity functions: unstable non-linear system.

1. with sufficient local models, the accuracy is good near to equilibria; the performance away from equilibria needs further investigation.
2. a continuous-time approach is inherently more robust to parameter variation than a discrete-time approach;
3. it naturally lends itself to emulator-based design;
4. the local models are linear in the parameters; this makes for simple adaptive algorithms,
5. as noted in Section 6.3., the LMN structure has a clear decomposition by function.

This approach to the control of non-linear systems can be summarised by the following contrast:

- The standard approach is to embed a non-linear network within an otherwise linear control structure, whereas
- the Local-State Local-Model approach is to embed a number of linear controllers within a non-linear network.

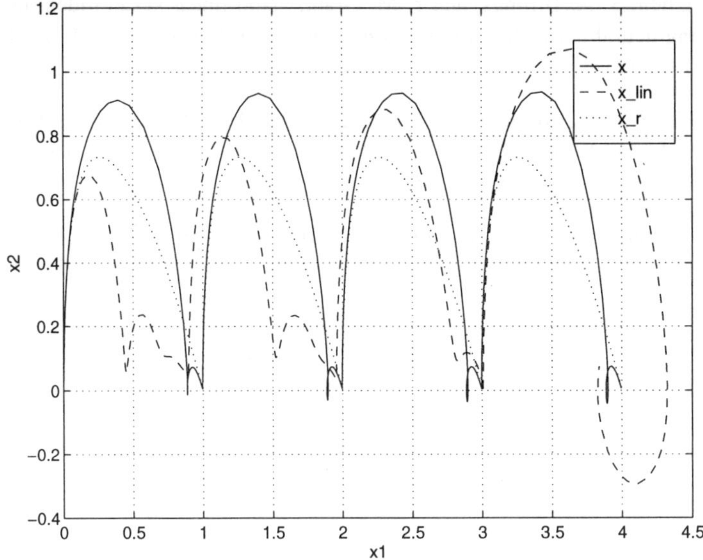

Figure 11: Phase-plane: unstable non-linear system.

Key issues for future work in this area include:

- stability of the non-linear feedback loop formed by the controlled system and the CLMN;
- the (automated) decomposition of the space of ϕ into operating regions and the role of equilibria in this context;
- development and analysis of networks of self-tuning controllers;
- investigation of the relationship with Gain Scheduling and
- further investigation of the relationship with Modular Neural Networks.

9. References

1. I. J. Leontaritis and S. A. Billings, "Input-output parametric models for non-linear systems. Part I: Deterministic non-linear systems," *International Journal of Control*, vol. 41, pp. 303–328, 1985.

2. S. Chen, S. A. Billings, and P. M. Grant, "Non-linear system identification using neural networks," *International Journal of Control*, vol. 51, pp. 1191–1214, 1990.

3. S. Chen, S. A. Billings, C. F. Cowan, and P. M. Grant, "Practical identification of NARMAX models using radial basis functions," *International Journal of Control*, vol. 52, pp. 1327–1350, 1990.

4. K. S. Narendra, *Neural Networks for Control*, ch. 5, pp. 115–142. MIT Press, 1990.

5. K. S. Narendra and K. Parthasarathy, "Identification and control of dynamic systems using neural networks," *IEEE Transactions on Neural Networks*, vol. 1, pp. 4–27, 1990.

6. K. Narendra and K. Parthasarathy, "Gradient Methods for the Optimization of Dynamical Systems Containing Neural Networks," *IEEE Trans. on Neural Networks*, vol. 2, pp. 252–262, 1991.

7. T. A. Johansen and B. A. Foss, "Constructing NARMAX models using ARMAX models," *IJC*, vol. 58, pp. 1125–1153, 1993.

8. T. A. Johansen and B. A. Foss, "A NARMAX model representation for adaptive control based on local models," *Modeling, Identification, and Control*, vol. 13, no. 1, pp. 25–39, 1992.

9. R. Murray-Smith, "Local Model Networks and Local Learning," in *Fuzzy Duisburg, '94*, pp. 404–409, 1994.

10. R. Żbikowski, K. J. Hunt, A. Dzieliński, R. Murray-Smith, and P. J. Gawthrop, "A review of advances in neural adaptive control systems," Technical Report of the ESPRIT NACT Project TP-1, Glasgow University and Daimler-Benz Research, 1994. (Available from FTP server `ftp.mech.gla.ac.uk` as PostScript file `/nact/nact_tp1.ps`).

11. K. J. Åström and B. Wittenmark, "On self-tuning regulators," *Automatica*, vol. 9, pp. 185–199, 1973.

12. K. J. Åström and B. Wittenmark, "Self-tuning controllers based on pole-zero placement," *Proceedings IEE Part D*, vol. 127, pp. 120–130, 1980.

13. D. W. Clarke, C. Mohtadi, and P. S. Tuffs, "Generalized Predictive Control—Part I. The basic algorithm," *Automatica*, vol. 23, no. 2, pp. 137–148, 1987.

14. D. W. Clarke, C. Mohtadi, and P. S. Tuffs, "Generalized Predictive Control—Part II. Extensions and interpretations," *Automatica*, vol. 23, no. 2, pp. 149–160, 1987.

15. P. J. Gawthrop, *Continuous-time Self-tuning Control. Vol 1: Design.* Lechworth, England: Research Studies Press, Engineering Control Series, 1987.

16. P. J. Gawthrop, *Continuous-time Self-tuning Control. Vol 2: Implementation.* Taunton, England: Research Studies Press, Engineering Control Series, 1990.

17. H. Demircioglu and P. J. Gawthrop, "Continuous-time generalised predictive control," *Automatica*, vol. 27, pp. 55–74, January 1991.

18. H. Demircioglu and P. J. Gawthrop, "Multivariable continuous-time generalised predictive control," *Automatica*, vol. 28, 1992.

19. K. J. Hunt, D. Sbarbaro, R. Żbikowski, and P. J. Gawthrop, "Neural networks for control systems—A survey," *Automatica*, vol. 28, no. 6, pp. 1083–1112, 1992.

20. D. W. Clarke and P. J. Gawthrop, "Self-tuning controller," *IEE Proceedings Part D*, vol. 122, no. 9, pp. 929–934, 1975.

21. P. J. Gawthrop, "Some interpretations of the self-tuning controller," *IEE Proceedings Part D*, vol. 124, no. 10, pp. 889–894, 1977.

22. D. W. Clarke and P. J. Gawthrop, "Self-tuning Control," *IEE Proceedings Part D*, vol. 126, no. 6, pp. 633–640, 1979.

23. D. W. Clarke and P. J. Gawthrop, "Implementation and application of microprocessor-based self-tuners," *Automatica*, vol. 17, no. 1, pp. 233–244, 1981.

24. P. J. Gawthrop and H. Demircioglu, "Continuous-time generalised predictive control," in *Preprints of the 3rd IFAC Workshop on Adaptive Systems in Control and Signal Processing, Glasgow, U.K.*, 1989.

25. T. Kailath, *Linear Systems.* Prentice-Hall, 1980.

26. P. J. Gawthrop, "Self-tuning PI and PID controllers," in *Proceedings of the IEEE Conference on Applications of Adaptive and Multivariable Control*, (Hull), 1982.

27. P. J. Gawthrop, "Self-tuning PID controllers: Algorithms and implementation," *IEEE Transactions on Automatic Control*, vol. AC-31, no. 3, pp. 201–209, 1986.

28. P. J. Gawthrop, "Continuous-time local state local model networks," in *Proceedings of 1995 IEEE conference on Systems Man and Cybernetics*, (Vancouver, British Columbia), 1995.

29. P. J. Gawthrop, R. W. Jones, and D. G. Sbarbaro, "Emulator-based control and internal model control: Complementary approaches to robust control design," *Aut*, 1995. (Submitted).

30. D. W. Clarke, ed., *Advances in Model-based Predictive Control.* Oxford University Press, 1994.

31. M. Morari and E. Zafiriou, *Robust Process Control.* Englewood Cliffs: Prentice-Hall, 1989.

32. D. A. Lawrence and W. J. Rugh, "Gain scheduling dynamic linear controllers for a nonlinear plant," *Automatica*, vol. 31, no. 3, pp. 381–390, 1995.

33. V. Utkin, *Sliding Modes and Their Application in Variable Structure Systems.* Moscow: Mir Publishers, 1978.

34. H. Demircioglu and P. J. Gawthrop, "Continuous-time relay self-tuning control," *International Journal of Control*, vol. 47, no. 4, pp. 1061–1080, 1988.

35. F. Fogelman-Soulie, "Multi-modular neural network-hybrid architectures: a review," in *Proceedings of 1993 International Joint Conference on Neural Networks*, 1993.

36. R. A. Jacobs and M. I. Jordan, "Learning piecewise control strategies in a modular neural network architecture," *IEEE Transaction on Systems, Man, and Cybernetics*, vol. 23, no. 2, pp. 337–345, 1993.

37. A. S. Morse, "Control using logic-based switching," in *Trends in Control—A European Perspective* (A. Isidori, ed.), pp. 69–114, Springer, 1995.

NONUNIFORM SAMPLING APPROACH TO CONTROL SYSTEMS MODELLING WITH FEEDFORWARD NEURAL NETWORKS

R. Żbikowski and A. Dzieliński
Department of Mechanical Engineering, University of Glasgow
Glasgow G12 8QQ, Scotland, UK
E-mail: rafal@mech.gla.ac.uk

ABSTRACT

This Chapter describes a novel approach to modelling of non-linear control systems (given by NARMA models) with feedforward neural networks. The method is based on Fourier analysis and nonuniform multi-dimensional (N-D) sampling. In order to obtain an approximation of system's dynamics with a neural network the right-hand side (RHS) of the NARMA model is reconstructed from irregularly spaced N-D samples. The algorithm uses an approximation to the Fourier transform and the distortions involved are discussed in detail. Real-world constraints on the system's inputs and outputs make the function defining the RHS have a bounded support and thus its Fourier transform (which is a Paley-Wiener function) can be approximately reconstructed and inverted. Then an RBF neural network may be applied as an implementation of a multi-dimensional interpolating filter. The neural model obtained this way is smooth and suitable for the purpose of non-linear geometric control. A tutorial survey of the theory of Paley-Wiener functions, with emphasis on the neural modelling aspects, completes the presentation.

1. Introduction

This Chapter discusses and analyses in detail a new neural modelling strategy for deterministic, non-linear, single-input single-output (SISO) control systems given by the discrete-time, $t \in \mathbb{Z}_+$, input-output NARMA [1] model

$$y_{t+1} = f(y_t, \ldots, y_{t-n+1}, u_t, \ldots, u_{t-m+1}) \tag{1}$$

with $y \in [a, b] \subset \mathbb{R}$, $u \in [c, d] \subset \mathbb{R}$ and $f\colon D \to [a, b]$ with the domain of definition $D = [a, b]^n \times [c, d]^m$. It is physically natural that output y and input u assume only finite values on a connected set and can attain their bounds. In fact, D is a compact, connected and convex subset of $\mathbb{R}^N$, where $N = n + m$.

It should be mentioned that model (1) is usually obtained by discretisation (see Chapter 1 by Kalkkuhl & Hunt) of a deterministic non-linear (Lipschitz) continuous-time, $t \in \mathbb{R}_+$, SISO control system. This, in general, is an approximation process and may not result in any mathematically nice properties of f, even if the underlying continuous-time model has them. Hence the need of a modelling strategy assuming as little about f as possible. Finally, the input-output model (1)—obtained from the discrete-time state-space description—is valid only locally and therefore it is not the 'ultimate black box'.

As argued in [2,3], the unknown f in (1) is given by (nonuniform) multi-dimensional samples. Therefore Fourier analysis is a natural theoretical framework in this context. From engineering considerations it follows that the function f is defined only on a (multi-dimensional) rectangle $D \subset \mathbb{R}^N$. To apply Fourier analysis we have to extend f to the whole of $\mathbb{R}^N$. The extension should result in theoretically and computationally useful representation on f without imposing too restrictive requirements on smoothness of f. The space-limited extension $\tilde{f}: \mathbb{R}^N \to \mathbb{R}$, dual to the band-limited one,

$$\tilde{f}(x) = \begin{cases} f(x), & \text{if } x \in D; \\ 0, & \text{otherwise,} \end{cases} \tag{2}$$

possesses [2,3] all the desirable features. In order for $\tilde{f}$ to be direct and inverse Fourier transformable it suffices that f in (1) is absolute and square (Lebesgue) integrable, $f \in L^1(D) \cap L^2(D)$, a very mild assumption.

The Chapter is organised as follows. Section 2. explains what approximations have to be made to the Fourier transform of a multi-variable function, when only finite discrete data are available. This provides background for a new reconstruction algorithm of the space limited extension (2) of the unknown function in (1), given the past inputs and outputs. The novel method of neural modelling of NARMA systems is described in Section 3. and compared with the Sanner & Slotine [4] approach. The Fourier transform $\tilde{F}$ of the space-limited extension $\tilde{f}$ is a special kind of function of several complex variables, a Paley-Wiener function. Section 4. is a tutorial survey of the mathematical theory with emphasis on the aspects relevant to the neural modelling of $\tilde{f}$. While the apparatus does not directly enter into the algorithm of Section 3., it has considerable theoretical importance, especially for the approach to the reconstruction of $\tilde{F}$ (and thus $\tilde{f}$) involving the ill-posed problems theory [5,6]. The Chapter ends with conclusions.

2. Fourier Transform Approximation

In this Section we explain the interplay between real-world engineering and computational constraints and mathematical techniques involved in modelling of (1). More precisely we shall consider $\tilde{f}$ of (2), as it is—unlike f—defined over the *whole* of $\mathbb{R}^N$, but we drop the tilde, because all the comments apply to any multi-variable, Fourier transformable function.

We discuss fundamental problems which arise when Fourier Analysis is applied in this context and show what approximations have to be made. We briefly summarise the simplifications resulting from using a computationally acceptable version of the Fourier transform, i.e., Discrete Fourier Transform (or its efficient computer algorithm, Fast Fourier Transform) instead of the transform itself.

The Fourier transform of a function $f \in L^1(\mathbb{R}^N) \cap L^2(\mathbb{R}^N)$ of continuously varying arguments is defined as follows

$$F_c(\omega^c) = \int_{\mathbb{R}^N} f(x) e^{-j\omega^c \cdot x} dx, \tag{3}$$

where $\omega^c = (\omega_1^c, \ldots, \omega_N^c) \in \mathbb{R}^N$ and $x = (x_1, \ldots, x_N) \in \mathbb{R}^N$ and $\cdot$ denotes the Euclidean inner product $\omega^c \cdot x = \sum_{k=1}^N \omega_k^c x_k$. Finally, $j = \sqrt{-1}$.

In the classical one-dimensional case, i.e., Fourier Analysis of time signals, (3) is called the Continuous Time Fourier Transform (CTFT). This means that both the original function and its transform are of continuously varying arguments (in signal processing applications they are called time and frequency, respectively). In the multi-dimensional (N-D) case we have to deal with variables of different nature and neither the original domain has anything to do with time, nor its transformed counterpart is an analogue of frequency. This is a purely *formal* application of Fourier integral to multi-variable functions. Even though we make use of tools and techniques typical for signal processing we apply them to objects of completely different nature. The variables of the NARMA model (1), which in our approach play the role of x in (3), are the values of system inputs and outputs in consecutive time instants. Their nature may be diverse and interpretation of the variables of the transform ('frequencies') is not as clear as for the time signals. Therefore we shall refer to (3) as the Continuous Fourier Transform (CFT) to emphasise that both the original and transformed variables vary continuously.

Continuous transforms are of fundamental theoretical importance. However, in order to make their application computationally feasible we have to approximate them by notions enabling the use of the Digital Signal Processing tools. This Section is devoted to the explanation of the way we proceed from the Continuous Fourier Transform to the Discrete Fourier Transform which is the actual processing tool (or rather its efficient computer implementation FFT—Fast Fourier Transform). We follow the chain of reasoning introduced by Cadzow [7] for the one-dimensional (time signals) case, generalising it to N-D.

The consecutive steps of necessary simplifications and the resulting approximations are given in Figure 1. Starting from CFT we pass to the Discrete Continuous Fourier Transform (DCFT) which is CFT applied to the function of discrete arguments. Then, as the data are of finite length in any practical application, we introduce the Truncated DCFT, i.e., DCFT defined for a function of arguments restricted to a finite set only. Note that both DCFT and the Truncated DCFT, while defined for a discrete function, are themselves functions of continuous arguments. If the Truncated DCFT is to be of any practical value, a systematic procedure of its *numerical evaluation* must be developed. The *Discrete Fourier Transform (DFT)* provides such a mechanism, whereby sampled values of the Truncated DCFT are computed on a prescribed *finite* set of discrete transformed arguments.

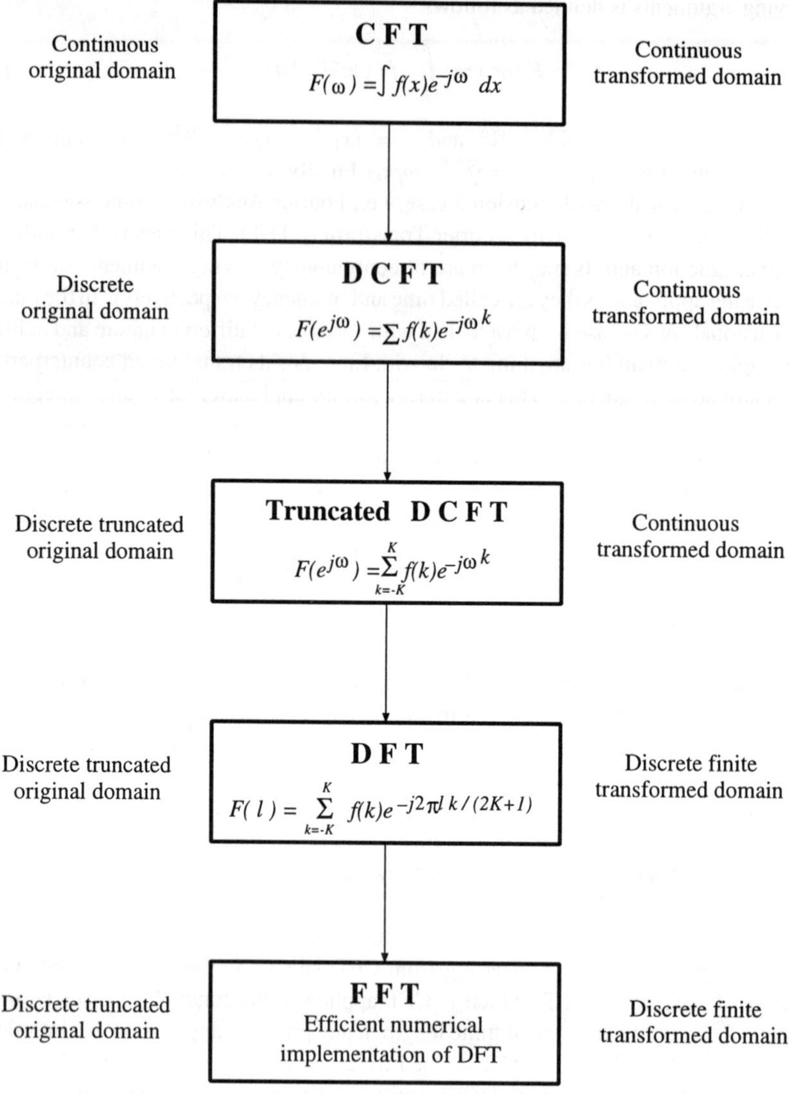

Figure 1: Consecutive steps of approximation from CFT to FFT.

The values on the set are found through a finite number of (possibly complex) additions and multiplications and is therefore amenable to digital implementation. Furthermore, the Fast Fourier Transform algorithm can be used to obtain a computationally effective means for computing the sampled values.

In the sequel we call the arguments of f in (3) the *space* variables and the arguments of F the *frequencies*. As mentioned before, even in the one-dimensional case the analogy with the terminology of time signals is false, but we nevertheless use the names in order to avoid excessive wording.

2.1. Continuous Fourier Transform Approximation: Discrete Continuous Fourier Transform (DCFT)

In order to apply the efficient tools of Digital Signal Processing we have to assume that functions under consideration are of discrete arguments rather then continuous ones. Firstly, we decompose the region of integration into the sets of contiguous intervals, each of width T_i, specified by $k_i T_i \leq x_i < (k_i + 1)T_i$ for $k_i \in \mathbb{Z}$ and $i = 1, \ldots, N$. The intervals' widths must be chosen small enough to ensure good quality of approximation. Thus, instead of integrals as in (3) we obtain

$$F_c(\omega^c) = \sum_{k \in \mathbb{Z}^N} \int_{kT}^{(k+1)T} f(x) \mathrm{e}^{-j\omega^c \cdot x} dx, \tag{4}$$

where $k = (k_1, \ldots, k_N)$ and $\sum_{k \in \mathbb{Z}^N} = \sum_{k_1 \in \mathbb{Z}} \cdots \sum_{k_N \in \mathbb{Z}}$, while $\int_{kT}^{(k+1)T}$ stands for $\int_{k_1 T_1}^{(k_1+1)T_1} \cdots \int_{k_N T_N}^{(k_N+1)T_N}$; also, $\omega^c = (\omega_1^c, \ldots, \omega_N^c) \in \mathbb{R}^N$ is introduced to mark the distinction from the variable ω in (6). Using the rectangular approximation of integration we may approximate (4) as

$$F_c(\omega^c) \approx \sum_{k \in \mathbb{Z}^N} T f(kT) \mathrm{e}^{-j\omega^c \cdot kT}, \tag{5}$$

where $T = \prod_{i=1}^N T_i$ and $kT = (k_1 T_1, \ldots, k_N T_N)$. Thus, we may formally introduce the Fourier Transform of a function of discrete arguments as follows

$$F(\mathrm{e}^{j\omega}) = \sum_{k \in \mathbb{Z}^N} f(k) \mathrm{e}^{-j\omega \cdot k} \tag{6}$$

with $f(k) \stackrel{\text{def}}{=} T f(kT)$. This formula defines the Discrete Continuous Fourier Transform (DCFT), F, which provides an insightful representation (in the multi-dimensional frequency domain) of the function $f = f(k)$ of discrete arguments and is itself a function of continuous arguments $\omega = (\omega_1, \ldots, \omega_N) = (T_1\omega_1^c, \ldots, T_N\omega_N^c)$. The transform given by (6) is a sum of periodically shifted copies of (3) and is a periodic function of ω (with period 2π).

DCFT defined by (6) requires infinite (discrete) data $\{f(k)\}_{k \in \mathbb{Z}^N}$, while in practice we have only a *finite* set of samples. This motivates introduction of the Truncated DCFT in Section 2.2. below.

2.2. *Truncated Discrete Continuous Fourier Transform*

Availability of a finite data set may be interpreted as $f(k) = 0$ for $|k| \geq K$, i.e., $|k_i| \geq K_i \in \mathbb{N}$, for $i = 1, \ldots, N$. Under this assumption we can formally define the *Truncated Discrete Continuous Fourier Transform* as

$$F_K(\mathrm{e}^{j\omega}) = \sum_{k=-K}^{K} f(k)\mathrm{e}^{-j\omega\cdot k}, \tag{7}$$

where $\sum_{k=-K}^{K}$ stands for the multiple sum $\sum_{k_1=-K_1}^{K_1} \cdots \sum_{k=-K_N}^{K_N}$. The relationship between DCFT and its truncated version is established by finding $f(k)$ from (6) and substituting it in (7), that is,

$$F_K(\mathrm{e}^{j\omega}) = \sum_{k=-K}^{K} \left(\frac{1}{(2\pi)^N} \int_{-\pi}^{\pi} F(\mathrm{e}^{j\nu})\mathrm{e}^{j\nu\cdot k} d\nu \right) \mathrm{e}^{-j\omega\cdot k} \tag{8}$$

with $\int_{-\pi}^{\pi}$ meaning the multiple integral over $(-\pi, \pi]^N$. Interchanging the order of summation and integration (recall that $f \in L^1(\mathbb{R}^N) \cap L^2(\mathbb{R}^N)$) we get

$$F_K(\mathrm{e}^{j\omega}) = \frac{1}{(2\pi)^N} \int_{-\pi}^{\pi} F(\mathrm{e}^{j\nu}) W_h(\mathrm{e}^{j(\omega-\nu)}) d\nu = \frac{1}{(2\pi)^N} F(\mathrm{e}^{j\omega}) * W_h(\mathrm{e}^{j\omega}), \tag{9}$$

where $*$ denotes the (multi-dimensional) convolution operator and W_h is referred to as the *hyperrectangular window transform*, that is,

$$W_h(\mathrm{e}^{j\omega}) = \sum_{k=-K}^{K} \mathrm{e}^{-j\omega\cdot k} = \prod_{i=1}^{N} \frac{\sin(\omega_i(K_i + 1/2))}{\sin(\omega_i/2)}. \tag{10}$$

In other words, truncated data $\{f(k)\}_{k=-K}^{K}$ are specified by the product of the infinite-extent data $\{f(k)\}_{k\in\mathbb{Z}^N}$ and hyperrectangular window function $w_h = w_h(k)$, whose DCFT, computed from (6), is W_h in (10).

2.3. *Spectral Distortions Caused by Using Truncated DCFT*

We now give an analysis of the convolution integral (9) showing the effects of using finite data to evaluate the underlying DCFT. For illustration purposes we present all results in one- and two-dimensional cases (1-D and 2-D, respectively).

The Truncated DCFT $F_K(\exp(j\omega))$ is equal to DCFT $F(\exp(j\omega))$ exactly, as is desired, only when the DCFT $W_h(\exp(j\omega))$ is equal to the Dirac delta function. The degree to which this is achieved may be estimated by observing the plot of the actual window's DCFT (see Figures 2 and 3). These DCFTs are real functions taking positive and negative values (reflected through the ω axis/plane in Figures 2 and 3).

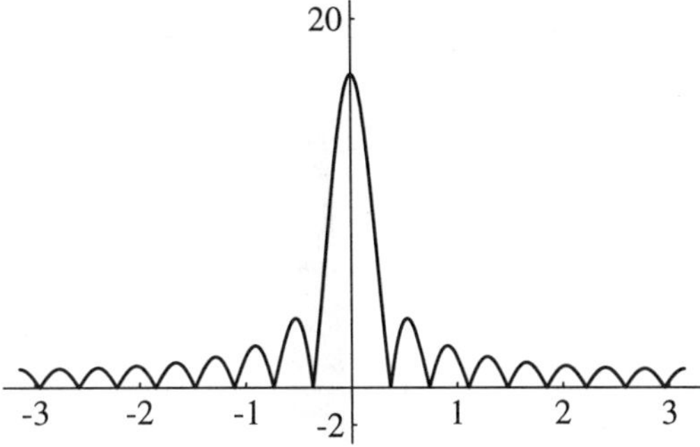

Figure 2: Modulus of one-dimensional rectangular window's DCFT.

In general, DCFT is complex-valued and its 'phase', Arg F, is a more complicated function of ω. The window's transform functions are seen to be composed of a number of pulselike waveforms, which are generally referred to as *lobes*. The largest one, centred at $\omega = 0$ (1-D) or $\omega_1 = \omega_2 = 0$ (2-D), is called the *main lobe*, while the others *side lobes*.

The window's transforms tend to the Dirac delta functions as K (or K_1 and K_2) approach infinity. However, in most real-world applications these values are relatively small, making the Truncated DCFT a distorted version of DCFT. We now discuss in some detail two most important consequences of this distortion.

2.3.1. Leakage

The first result of using finite data for DCFT evaluation is the spectral *leakage* or *smearing*, i.e., the spectrum of a sinusoid appears as a continuum of spectral components in a region around the sinusoid's two frequencies. To analyse this effect let us consider the 1-D and 2-D sinusoidal functions of the form

$$x(k) = A\cos(vk) \quad \text{and} \quad x(k_1, k_2) = A\cos(v_1 k_1 + v_2 k_2). \tag{11}$$

The sinusoids radian frequencies (see page 19 in [8]) are chosen to be $v = v_1 = 2$ and $v_2 = 1$. The corresponding DCFTs are

$$X(e^{j\omega}) = \pi A(\delta(\omega - v) + \delta(\omega + v)) \quad \text{for} \quad -\pi < \omega \leq \pi \tag{12}$$

and

$$X(e^{j\omega_1}, e^{j\omega_2}) \;=\; 2\pi^2 A(\delta(\omega_1 - v_1)\delta(\omega_2 - \frac{v_2}{v_1}\omega_1) + \delta(\omega_1 + v_1)\delta(\omega_2 - \frac{v_2}{v_1}\omega_1))$$

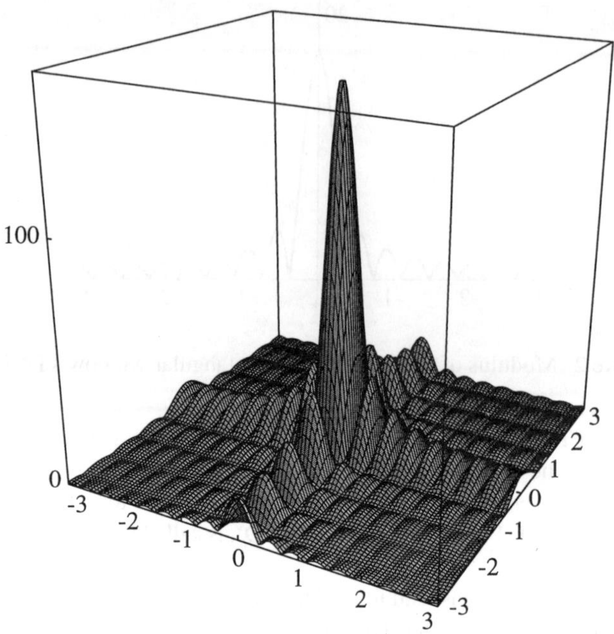

Figure 3: Modulus of two-dimensional rectangular window's DCFT.

$$\text{for} \quad -\pi < \omega_1 \leq \pi \quad \text{and} \quad -\pi < \omega_2 \leq \pi. \tag{13}$$

The associated Truncated DCFTs are obtained by convolving the DCFTs with the window's DCFT W_h:

$$X_K(\mathrm{e}^{j\omega}) = A\mathrm{e}^{-j(\omega-v)(K-1)/2}\frac{\sin[(\omega-v)K/2]}{\sin[(\omega-v)/2]} + A\mathrm{e}^{-j(\omega+v)(K-1)/2}\frac{\sin[(\omega+v)K/2]}{\sin[(\omega+v)/2]} \tag{14}$$

for the 1-D case and

$$X_{K_1K_2}(\mathrm{e}^{j\omega_1}, \mathrm{e}^{j\omega_2}) = \frac{A}{2}\mathrm{e}^{-j(\omega_1-v_1)(K_1-1)/2}\frac{\sin((\omega_1-v_1)K_1/2)}{\sin((\omega_1-v_1)/2)}\times \mathrm{e}^{-j(\omega_2-\frac{v_2}{v_1}\omega_1)(K_2-1)/2}\frac{\sin((\omega_2-\frac{v_2}{v_1}\omega_1)K_1/2)}{\sin(\omega_2-\frac{v_2}{v_1}\omega_1)}$$

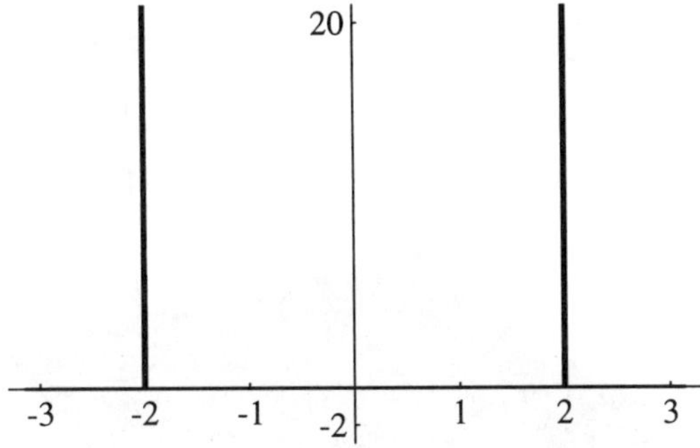

Figure 4: Modulus of the cosine function's DCFT: 1-D case.

$$+\frac{A}{2}e^{-j(\omega_1+v_1)(K_1-1)/2}\frac{\sin((\omega_1+v_1)K_1/2)}{\sin((\omega_1+v_1)/2)}\times$$

$$e^{-j(\omega_2+\frac{v_2}{v_1}\omega_1)(K_2-1)/2}\frac{\sin((\omega_2+\frac{v_2}{v_1}\omega_1)K_1/2)}{\sin(\omega_2+\frac{v_2}{v_1}\omega_1)} \tag{15}$$

for the 2-D case. These functions, compared with the Dirac deltas, are given in Figures 4 and 5 for the 1-D case and Figures 6 and 7 for 2-D. Both in the 1-D and 2-D case usage of finite data results in a version of the original impulse-like spectrum, smeared around the radian frequencies of the sinusoids.

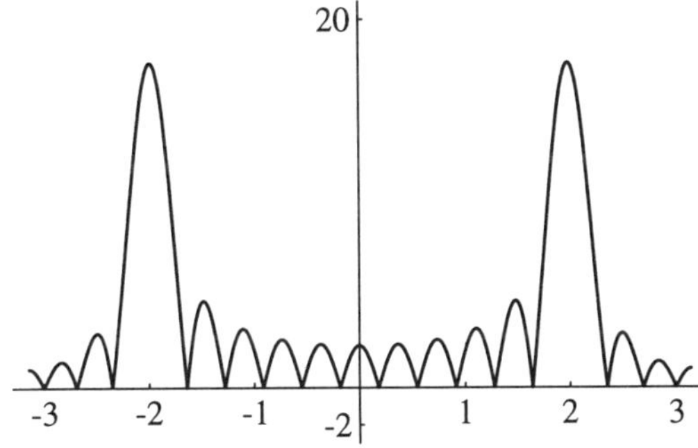

Figure 5: Modulus of the cosine function's Truncated DCFT (leakage effect): 1-D case.

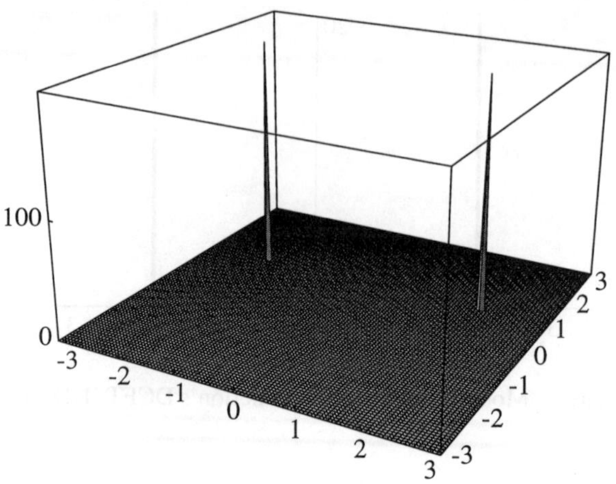

Figure 6: Modulus of the cosine function's DCFT: 2-D case.

2.3.2. Spectral Smoothing and the Ripple Effect

Use of finite data has also a distorting effect on functions whose DCFT has sharp discontinuities. In such cases, the Truncated DCFT exhibits smoothing of the discontinuities and a relatively low amplitude ripple throughout the entire frequency range. We illustrate these effects with the following examples:

$$x(k) = \frac{\sin(vk)}{\pi k} \tag{16}$$

for 1-D and

$$x(k_1, k_2) = \frac{\sin(v_1 k_1)}{\pi k_1} \frac{\sin(v_2 k_2)}{\pi k_2}, \tag{17}$$

for 2-D. The corresponding DCFTs are:

$$F(e^{j\omega}) = \begin{cases} 1, & \text{for } |\omega| \leq v; \\ 0, & \text{for } v < |\omega| \leq \pi, \end{cases} \tag{18}$$

and

$$F(e^{j\omega_1}, e^{j\omega_2}) = \begin{cases} 1, & \text{for } |\omega_1| \leq v_1 \text{ and } |\omega_2| \leq v_2; \\ 0, & \text{for } v_1 < |\omega_1| \leq \pi \text{ and } v_2 < |\omega_2| \leq \pi. \end{cases} \tag{19}$$

Plots of (18) and its distortions are shown in Figures 8 and 9. The 2-D case of (19) is depicted in Figures 10 and 11.

Truncated DCFTs associated with the functions (16) and (17) are computed and presented in Figure 8 and Figure 10, respectively, and they are distorted versions of exact DCFTs.

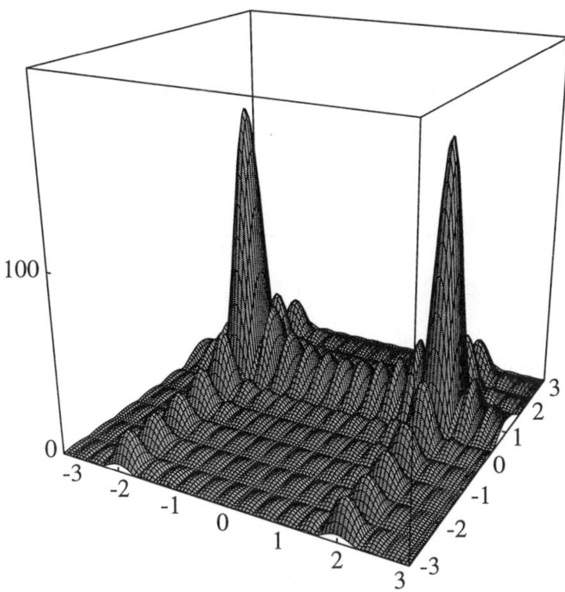

Figure 7: Modulus of the cosine function's Truncated DCFT (leakage effect): 2-D case.

The smoothing effect is related to the main lobe of the hyperrectangular window's DCFT, while ripple is due to the side lobes of this transform. The decomposition of the window's transform into main lobe and side lobes and their separate effects on DCFTs of (16) and (17) is also shown. Figures 9a, 9b, 11a, 11b present the influence of main lobe on the given hyperrectangular 1-D and 2-D functions. Note that the main lobe is the only cause of the smoothing effect, where sharp discontinuities became smoothly rising (declining) slopes. The influence of side lobes is given in Figures 9c, 9d and 11c, 11d and it is clear that they cause the ripple effect. It is important to notice that the ripple's values are in fact negative, while the figures show its modulus. Remembering that, we can superimpose the plots of smoothed and rippled DCFTs to obtain the distorted DCFTs of the original functions.

The effects may be justified by the following analysis. The Truncated DCFT is obtained from the convolution of DCFT and hyperrectangular window's DCFT. Decomposing the window transform as

$$W_h(e^{j\omega}) = W_m(e^{j\omega}) + W_s(e^{j\omega}), \tag{20}$$

where $W_m(e^{j\omega})$ and $W_s(e^{j\omega})$ denote the main-lobe and side-lobes components, re-

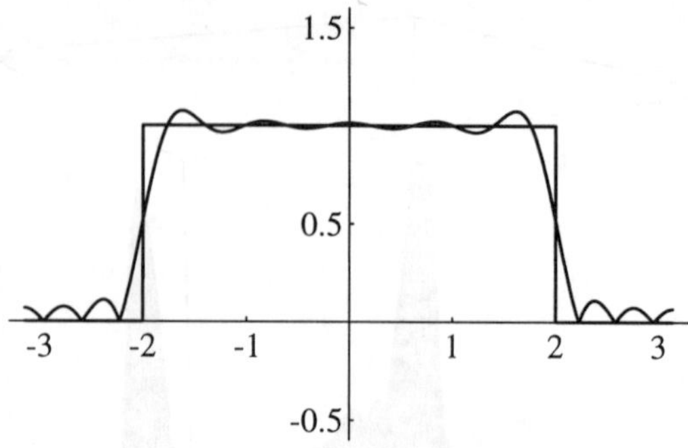

Figure 8: Modulus of Truncated DCFT of (16).

spectively, the Truncated DCFT convolution relation is given by

$$F_K(\mathrm{e}^{j\omega}) = \frac{1}{(2\pi)^N}\Big(F(\mathrm{e}^{j\omega}) * W_m(\mathrm{e}^{j\omega}) + F(\mathrm{e}^{j\omega}) * W_s(\mathrm{e}^{j\omega})\Big). \tag{21}$$

The first term on the right-hand side of (21) gives rise to the smoothing distortion, while the second term produces the rippling distortion.

2.3.3. Other Data Windows

Approximation of the DCFT of a function f from a finite set of samples may be considered in a general setting. Finiteness of data can be viewed as a result of application of *any* window function w, which is zero except for a bounded region, giving the product $f(k)w(k)$ for all $k \in \mathbb{Z}^N$. As before, the thus obtained truncated (windowed) DCFT is equal to the convolution of the underlying DCFT and the window transform. The smoothing and ripple effects are also present in the general case, but different windows have them in different proportions. The hyperrectangular window provides the least smoothing effect, while some other (e.g., Hamming or Blackman window, see [7]) exhibit significantly less ripple effect. Due to their relative implementational complexity we shall not consider them in our applications.

2.4. *Discrete Fourier Transform (DFT)*

Finiteness of real-world data forced the introduction of the Truncated DCFT, a (complex-valued) function of continuously varying variables. To make it practically useful, a systematic procedure for its numerical evaluation must be given. The Discrete Fourier Transform (DFT), briefly described here, is a computationally feasible method for finding sampled values of the Truncated DCFT at a prescribed set of

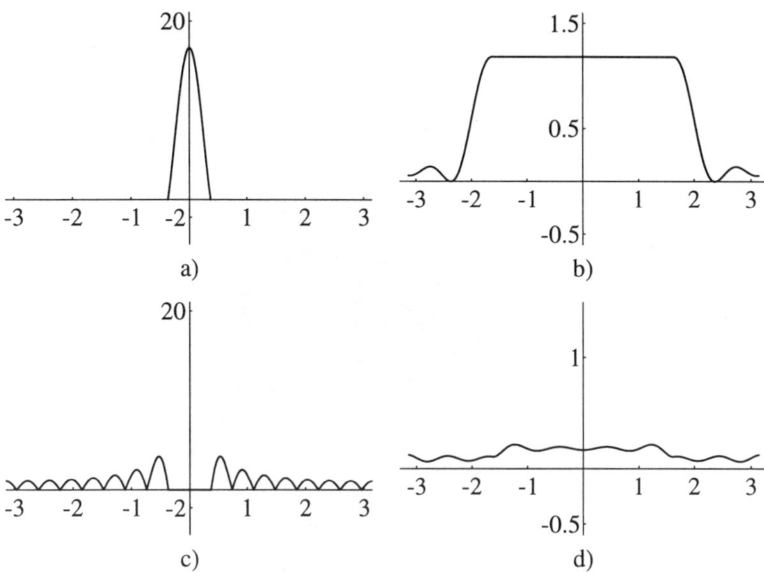

Figure 9: Moduli of: a) main lobe of DCFT of (10); b) smoothed DCFT of (16); c) side lobes of DCFT of (10); d) rippled DCFT of (16).

discrete frequencies. A determination of the sampled values involves a *finite* number of complex operations only and therefore is amenable to computer implementation. Moreover, there exists a computationally efficient algorithm for evaluating DFT, called the Fast Fourier Transform (FFT).

The procedure for evaluating the Truncated DCFT is based on the computation of its values at a *finite* set of multi-dimensional frequencies. If the frequencies are spaced close enough, a reasonable approximation of the DCFT is obtained. Denote the discrete sets of frequencies by

$$\omega_i^l = \frac{2\pi l_i}{2K_i + 1} \quad \text{for} \quad l_i = -K_i, \ldots, K_i \quad \text{and} \quad i = 1, \ldots, N. \tag{22}$$

Note that the numbers of discrete frequencies in these sets, i.e., $2K_i + 1$, exactly correspond to the extent of the data in each variable used in the definition (7) of the Truncated DCFT. These uniformly spaced frequencies cover one period of the DCFT, i.e., $-\pi < \omega_i \leq \pi$ for $i = 1, \ldots, N$. The evaluation of the Truncated DCFT on the discrete multi-dimensional frequency set gives rise to the Discrete Fourier Transform.

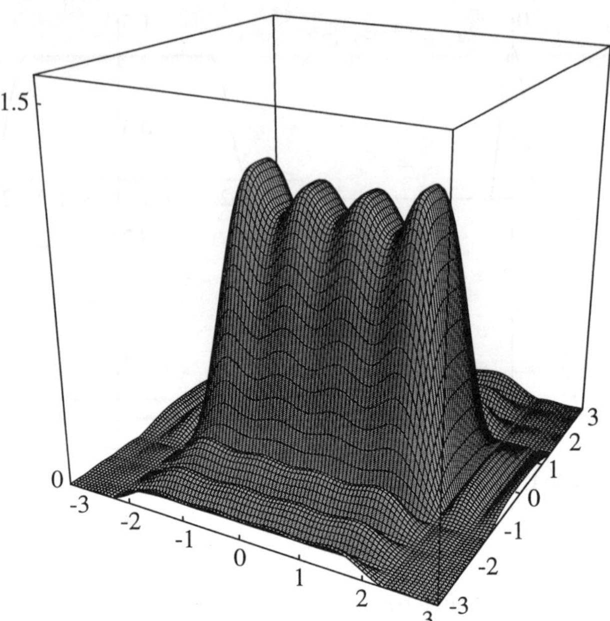

Figure 10: Modulus of Truncated DCFT of 2-D function (17).

Consider the Truncated DCFT computed at the discrete frequencies ω^l

$$F_K(e^{j\omega^l}) = \sum_{k=-K}^{K} f(k) e^{-j2\pi l \cdot k/(2K+1)} \tag{23}$$

where $\omega^l = (\omega_1^l, \ldots, \omega_N^l)$, with ω_i^l as in (22), and $l = (l_1, \ldots, l_N)$, while $k/(2K+1) = (k_1/(2K_1+1), \ldots, k_N/(2K_N+1))$. Computation of discrete frequency samples of the Truncated DCFT according to (23) is called the *Discrete Fourier Transform (DFT)* and gives a computer programmable method of its characterisation. The preferred compact definition of DFT is the simple modification of (23)

$$F_K(l) \stackrel{\text{def}}{=} F_K(e^{j\omega^l}), \tag{24}$$

convenient for computer implementation. Thus $F_K(l)$ denotes the Truncated DCFT's values at the discrete frequencies $\omega_i^l = 2\pi l_i/(2K_i + 1)$, $l_i = -K_i, \ldots, K_i$ and $i = 1, \ldots, N$. For a given l the term $F_K(l)$ is called the lth DFT coefficient and the expression (24) is the $(2K_1 + 1) \cdot \ldots \cdot (2K_N + 1)$-point N-dimensional DFT.

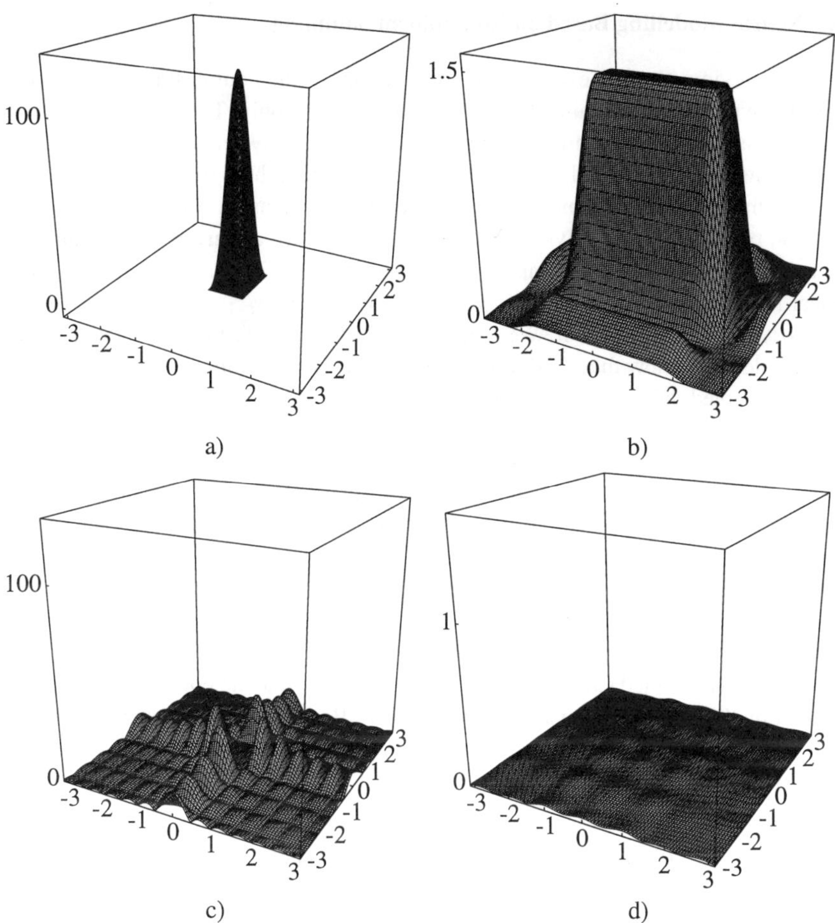

Figure 11: Moduli of: a) main lobe of 2-D version of (10); b) smoothed 2-D DCFT of (17); c) side lobes of 2-D version of (10); d) rippled 2-D DCFT of (17).

DFT provides a feasible method of the Fourier transform computer implementation, but is relatively slow from the computational point of view. A revolutionary change happened in signal processing after the introduction of very efficient algorithms, collectively known as the *Fast Fourier Transform (FFT)*. Thus in practical application DFT usually means its famous implementation, FFT.

3. Neural Modelling Based on Nonuniform Sampling

The previous Section explained approximations of the Fourier transform, which have to be made due to availability of finite discrete data only. This sets the stage for our new method of neural modelling of NARMA control systems (1). We are given past inputs and outputs and we want to reconstruct from them the unknown f. We do it by finding an approximation to its space-limited extension $\tilde{f}$ given by (2).

The problem we are solving is stated as follows. We are given a finite number of nonuniformly [2,3] spread samples $\tilde{f}(\lambda_k)$, where $\lambda_k = (\lambda_{k_1}, \ldots, \lambda_{k_N})$, of the nonlinear function $\tilde{f} = \tilde{f}(x)$, where $x = (x_1, \ldots, x_N) = (y_t, y_{t-1}, \ldots, y_{t-n+1}, u_t, u_{t-1}, \ldots, u_{t-m+1})$, i.e., $N = m + n$. We want to find the function $\tilde{f}$. From the statement of our problem we know that $\tilde{f}$ is of bounded support (space-limited). We assume that *all* the sampled values $\tilde{f}(\lambda_k)$ are given, which means that we deal with an off-line problem.

The main idea of our method is based on the transformation of the nonuniform sampling problem in the space domain to a uniform problem in the Fourier Transform domain, similarly to the Kim & Bose [9] method. However, taking advantage of the specific features of the problem of reconstruction of f in (1), we propose a simpler approach. Since $\tilde{f}$ is of bounded support, in order to reconstruct it properly we need its Continuous Fourier Transform (CFT), $\tilde{F}$. Therefore, our solution consists of two basic steps. First, we find an approximation of the CFT on the basis of given nonuniformly sampled values of $\tilde{f}$, i.e., $\tilde{f}(\lambda_k)$. Then we find the Fourier inverse of the approximation of $\tilde{F}$ to get an approximation of $\tilde{f}$. Both steps involve approximations and we will point out the sources of them in the sequel.

Note that CFT for the space-limited function $\tilde{f}$ is given by the following pair of relations

$$\tilde{F}(\omega) = \int_{\mathbb{R}^N} \tilde{f}(x) e^{-j\omega \cdot x} dx = \int_D \tilde{f}(x) e^{-j\omega \cdot x} dx, \qquad (25)$$

$$\tilde{f}(x) = \left(\frac{1}{2\pi}\right)^N \int_{\mathbb{R}^N} \tilde{F}(\omega) e^{j\omega \cdot x} d\omega, \qquad (26)$$

where $\int_D$ denotes the multiple definite integral over $D = [a, b]^n \times [c, d]^m$, $N = n+m$. Integration limits a, b correspond to the y components and c, d to the u components of vector x, respectively.

In the first step of our algorithm we approximate (25) by a finite Riemann sum of the form:

$$\tilde{F}_K(\omega) = \sum_{k=0}^{K-1} \tilde{f}(\lambda_k) e^{-j\omega \cdot \lambda_k} \Delta_k. \qquad (27)$$

Here $\sum_{k=0}^{K-1}$ stands for the multiple summation $\sum_{k_1=0}^{K_1-1} \sum_{k_2=0}^{K_2-1} \cdots \sum_{k_N=0}^{K_N-1}$ with $k = (k_1, \ldots, k_N)$ and $K = (K_1, \ldots, K_N) \in \mathbb{N}^N$; the K_i are related to the observation horizon time of system (1). Also, $\lambda_k = (\lambda_{k_1}, \ldots, \lambda_{k_N})$ is the vector of the kth sample from $D \subset \mathbb{R}^N$ (see (1) and [2]) and $\Delta_k = \prod_{i=1}^N \Delta_{k_i} = \Delta_{k_1} \cdot \Delta_{k_2} \cdot \ldots \cdot \Delta_{k_N}$, where $\Delta_{k_i} = \lambda_{k_i} - \lambda_{k_i-1}$ or $\Delta_k = \lambda_k - \lambda_{k-1}$. Also $\lambda_{k_i} > \lambda_{k_i-1}$ for all $k_i = 0, 1, \ldots, K_i - 1$ and $i = 1, 2, \ldots, n, n+1 \ldots, N$, where

$$\lambda_{K_i} = \begin{cases} b, & \text{for } i = 1, \ldots, n; \\ d, & \text{for } i = n+1, \ldots, N. \end{cases}$$

Furthermore $\omega = (\omega_1, \ldots, \omega_N) \in \mathbb{R}^N$ and, finally, $\omega \cdot \lambda_k$ is the Euclidean inner product $\sum_{i=1}^N \omega_i \lambda_{k_i}$. Let us note that (27) bears a close resemblance to the DCFT of f, given by (6). In the case of space-limited $\tilde{f}$ such a DCFT is naturally truncated, i.e., equals to the Truncated DCFT; cf. (7).

The core issue of the proposed algorithm is how to use (27) to get the best approximation of $\tilde{f}$. There seem to be several natural moves possible. However, some of them do not yield satisfactory answer to our problem. We cannot invert (27) directly, because this way we get the sampled values $\tilde{f}(\lambda_k)$ and not the function of continuous arguments we are looking for. Another problem is how to define such an inverse transform in this case. Yet another possibility is to convolve (27) with some other function of ω, say $G = G(\omega)$ and then perform an inverse transform. In this case we also face the same problem of inverse transform definition. Moreover, since $\tilde{f}$ is not band-limited, this inverse would not result in the reconstruction of $\tilde{f}$. This problem would remain even if we knew how to construct G for band-limited $\tilde{f}$.

Therefore we propose to sample $\tilde{F}_K$ of (27) and use these sampled values as a basis for further approximation of $\tilde{f}$. The crucial observation is that we may choose the sample locations arbitrarily. The simplest choice is to sample $\tilde{F}_K$ *uniformly*. The sampling intervals on all N axes are provided by the N-D Shannon Theorem [10].

Theorem 1 (N-Dimensional Sampling Theorem) *Let $h: \mathbb{R}^N \to \mathbb{R}$ be such that both its N-dimensional direct, H, and inverse Fourier transforms are well-defined. If the spectrum $H = H(\omega_1, \ldots, \omega_N)$ vanishes outside a bounded subset of N-dimensional space, then h can be everywhere reconstructed from its samples $\{h(x_k)\}_{k \in \mathbb{Z}^N}$, taken over a lattice of points $\{k_1 v_1 + k_2 v_2 + \ldots + k_N v_N\}$, $k_i \in \mathbb{Z}, i = 1, \ldots, N$, provided that the vectors $\{v_1, v_2, \ldots, v_N\}$, where $\{v_i\} \in \mathbb{R}^N$, $i = 1, \ldots, N$, are small enough to ensure non-overlapping of the spectrum $H(\omega_1, \ldots, \omega_N)$ with its periodic images on the lattice defined by the vectors $\{w_1, w_2, \ldots, w_N\}$, where $\{w_j\} \in \mathbb{R}^N$, $j = 1, \ldots, N$, with*

$$v_i \cdot w_j = 2\pi \delta_{ij} \tag{28}$$

where δ_{ij} is the Kronecker's symbol and $\cdot$ the inner product.

Condition (28) is a multi-dimensional generalisation of the Shannon condition [11].

Given the specific shape of the support D of $\tilde{f}$ we are able to calculate the appropriate sampling intervals for $\tilde{F}$ from (28), where the w_j's represent the space extent of $\tilde{f}$ and the v_i's give the required sampling intervals of $\tilde{F}$. Choosing the hyperrectangular sampling geometry for $\tilde{F}$ we obtain unique values for sampling intervals assuming the same geometry for the locations of repetitive parts of D.

Let us discuss this issue in detail for a simple 2-D example.

Example 1 In this case $m = n = 1$, $N = m + n = 2$ and NARMA model has the form

$$y_{t+1} = f(y_t, u_t),$$

where $y \in [a, b]$ and $u \in [c, d]$. Put $x = (x_1, x_2) = (y_t, u_t)$.

For the rectangular sampling geometry we obtain from (28)

$$v_1 \cdot w_1 = 2\pi \quad \text{and} \quad v_2 \cdot w_2 = 2\pi,$$

where $v_1 = [T_1, 0]^T$, $v_2 = [0, T_2]^T$, $w_1 = [\max\{|a|, |b|\}, 0]^T$ and $w_2 = [0, \max\{|c|, |d|\}]^T$. This way we get the desired sampling intervals T_1 and T_2 in the form

$$T_1 = \frac{2\pi}{\max\{|a|, |b|\}} \quad \text{and} \quad T_2 = \frac{2\pi}{\max\{|c|, |d|\}}.$$

This is not the case for any sampling geometry. If either of the geometries is not hyperrectangular then our result would not be unique. ■

The result can be generalised to the $N = m + n$ dimensions

$$T_i = \begin{cases} \frac{2\pi}{\max\{|a|,|b|\}}, & \text{for } i = 1, \ldots, n; \\ \frac{2\pi}{\max\{|c|,|d|\}}, & \text{for } i = n+1, \ldots, N. \end{cases} \tag{29}$$

Sampling the generalised DCFT (27) gives us the following representation of $\tilde{F}_K$

$$\tilde{F}_K(\omega) = \sum_{l \in \mathbb{Z}^N} \tilde{F}_K(\omega_S^l) \text{sinc}\,(\omega - \omega_S^l), \tag{30}$$

where $\omega = (\omega_1, \omega_2, \ldots, \omega_N)$, $\omega_S^l = (l_1\omega_{S_1}, l_2\omega_{S_2}, \ldots, l_N\omega_{S_N}) = (2\pi l_1/T_1, 2\pi l_2/T_2, \ldots, 2\pi l_N/T_N)$, $l = (l_1, \ldots, l_N)$ and $\text{sinc}\,(\omega - \omega_S^l) = \prod_{i=1}^N \text{sinc}\,(\omega_i - l_i\omega_{S_i}) = \prod_{i=1}^N \frac{\sin(\omega_i - l_i\omega_{S_i})}{\omega_i - l_i\omega_{S_i}}$.

Note that sampled version of (27) is equivalent to the following generalised Discrete Fourier Transform

$$\tilde{F}_K(\omega_S^l) = \sum_{k=0}^{K-1} f(\lambda_k) \mathrm{e}^{-j2\pi(\omega_S^l/K)\cdot\lambda_k}, \tag{31}$$

where $\omega_S^l/K = (l_1\omega_{S_1}/K_1, \ldots, l_N\omega_{S_N}/K_N)$. The inverse Fourier Transform of (30) would give us the sought $\tilde{f}$ in the form of an infinite linear combination of rectangular windowing functions. The theoretical background for such a reconstruction is provided by Theorem 1.

However, from the practical point of view we are not able to use the infinite number of samples. Moreover, this approximation is useless for control applications, as rectangular windowing functions result in an awkward representation of $\tilde{f}$. Our problem is to reconstruct $\tilde{F}_K$ in its entirety out of a given *finite* number of values $\tilde{F}_K(\omega_S^l)$ and obtain a useful representation of the dynamic system. Therefore, we have to introduce further approximation steps. Firstly, we will approximate sinc functions in (30) by some other functions of similar (but smooth) Fourier inverses. Gaussian functions seem to be suitable in this context. Secondly, we use only a finite number of terms in the approximation of (30). The adequacy of acting this way is ensured by the so called "$2X\Omega$ Theorem" (see [12,13] for details). This result allows us to use a finite linear combination of non-linear functions φ_k in order to approximate the given function f with a desired accuracy

$$\int_{-X/2}^{X/2}\Big[f(x)-\sum_{k=0}^{K-1}a_k\varphi_k(x)\Big]^2dx<\varepsilon, \tag{32}$$

where number of terms in (32) K depends on the accuracy ε. The functions used in (32) have to be Fourier transformable, i.e., belong to $L^1(\mathbb{R}^{m+n})\cap L^2(\mathbb{R}^{m+n})$. Taking this into account we end up with the following approximate representation of $\tilde{F}_K$ of (27)

$$\tilde{F}_K^a(\omega)=\sum_{l=-K/2}^{K/2-1}\tilde{F}_K(\omega_S^l)\mathrm{e}^{\frac{\|\omega-\omega_S^l-c_l\|^2}{\sigma_l^2}} \tag{33}$$

valid for K even, where $\frac{\|\omega-\omega_S^l-c_l\|^2}{\sigma_l^2}=\frac{|\omega_1-l_1\omega_{S_1}-c_{l_1}|^2}{\sigma_{l_1}^2}+\ldots+\frac{|\omega_N-l_N\omega_{S_N}-c_{l_N}|^2}{\sigma_{l_N}^2}$. For K odd the summation indices in (33) need a straightforward reformulation. Taking the inverse Fourier Transform of (33) we obtain the following approximate representation of the $\tilde{f}$

$$\begin{aligned}
\tilde{f}^a(x) &= \left(\frac{1}{2\pi}\right)^N\int_{\mathbb{R}^N}\tilde{F}_K^a(\omega)\mathrm{e}^{j\omega\cdot x}d\omega\\
&= \left(\frac{1}{2\pi}\right)^N\int_{\mathbb{R}^N}\sum_{l=-K/2}^{K/2-1}\tilde{F}_K(\omega_S^l)\mathrm{e}^{\frac{\|\omega-\omega_S^l-c_l\|^2}{\sigma_l^2}}\mathrm{e}^{j\omega\cdot x}d\omega\\
&= \left(\frac{1}{2\pi}\right)^N\sum_{l=-K/2}^{K/2-1}\tilde{F}_K(\omega_S^l)\int_{\mathbb{R}^N}\mathrm{e}^{\frac{\|\omega-\omega_S^l-c_l\|^2}{\sigma_l^2}+j\omega\cdot x}d\omega\\
&= \left(\frac{1}{2\sqrt{\pi}}\right)^N\sum_{l=-K/2}^{K/2-1}\tilde{F}_K(\omega_S^l)\sigma\mathrm{e}^{j(\omega_S^l+c_l)}\mathrm{e}^{-\frac{x^2\sigma_l^2}{4}},
\end{aligned} \tag{34}$$

where $\sigma_l=(\sigma_{l_1},\sigma_{l_2},\ldots,\sigma_{l_N})$, $\sigma=\sigma_{l_1}\cdot\sigma_{l_2}\cdot\ldots\cdot\sigma_{l_N}$, $x^2\sigma_l^2=x_1^2\sigma_{l_1}^2+\ldots+x_N^2\sigma_{l_N}^2$ and $c_l=(c_1,c_2,\ldots,c_N)$.

Combining this with existence results from the theory of neural approximation [14], we note that (34) gives a neural network architecture with gaussian RBFs as network nodes. However, the function reconstructed from (34) may take complex values. This means that we have to implement an additional step of 'phase discrimination' in our procedure. When the phase value of the complex number (reconstructed function value) lies in the interval $(-\pi/2, \pi/2)$ we assume phase to be equal 0 (i.e., value of the complex number is equal to its modulus); when it lies in the interval $(\pi, 3/2\pi)$ we assume phase to be equal π (value of the complex number is equal to its modulus with negative sign).

3.1. *Algorithm of the Method*

The summary of the algorithm is as follows:

1. from *nonuniform* samples λ_k, $\tilde{f}(\lambda_k)$, $0 \leq k_i \leq K_i - 1$, $K_i \in \mathbb{N}$, $i = 1, \ldots, N$, compute the approximate Fourier Transform representation: formula (27)
2. sample $\tilde{F}_K(\omega)$ *uniformly* with the Nyquist frequency: formula (29)
3. reconstruct an approximation of $\tilde{F}_K(\omega)$ out of given finite number of values of $\tilde{F}_K(\omega_S^l)$ using (33) (to replace sinc functions used in (30) use RBF's or other functions of similar spectral properties)
4. invert (33) according to (34) and discriminate the phase.

3.2. *Advantages of the Method*

The main advantage of this algorithm is its relative simplicity. By an appropriate choice of basis functions in the reconstruction formula we obtain a neural network approximation of NARMA model (1). This serves as a starting point for further control applications of this method. An example of such an application is given in [2].

Let us now summarise the advantages of the proposed algorithm:

1. *Tailored to the real-world data from input-output discrete-time models.* The data entering the algorithm are exactly the same data we measure as inputs and outputs of the plant,
2. *Computationally simple.* In comparison with many function approximation and reconstruction algorithms it requires only simple transformations and computations. As opposed to other approaches, this algorithm neither involves matrix inversion (the Kim & Bose [9] method), nor iterations (the Sandberg [15] method),

3. *Wide applicability—mild assumptions.* The only assumption on the given function is that both its direct and inverse Fourier transform exist,

4. *Flexible neural implementation.* In order to approximate the interpolation filter of (30) we may use several neural architectures. The use of gaussian RBFs is justified by their 'nice' properties. However, other basis functions may be considered,

5. *Usefulness for control purposes.* Appropriate choice of basis functions in (30) ensures the control applicability of the reconstructed function and especially its smoothness and accuracy.

3.3. *Comparison with the Sanner & Slotine Approach*

We now focus on basic differences between our and the Sanner & Slotine [4] approach which was a primary inspiration for our research.

- Our approach is based on the assumption of availability of the sampled values of function f spread nonuniformly, while Sanner & Slotine adopted the simplified view of uniform samples locations;
- We consider a discrete-time input-output models, while they dealt with a continuous-time setting;
- In our approach the function to be reconstructed is assumed to be space-limited, which results in very mild assumptions on f. Sanner & Slotine required f to be band-limited, which means very high smoothness;
- In our method the reconstruction is done in the frequency domain, while in theirs in the spatial domain.

3.4. *Extensions and Further Research*

The method presented seems to be a promising tool in the area of non-linear control systems modelling. The version discussed in this paper is based on the assumption of availability of all samples of the function f. This means it is *off-line* in character. This assumption has a direct influence on the definition of Riemann sums in (27). To define the intervals Δ_k we need all the λ_k's. Note that the order in which this values appear in the model does not correspond to the order in which they are summed up. This means we re-number the data in comparison with their natural indexing. While this is not a problem in the off-line approach, it may cause some difficulties when trying to construct an on-line version of the method.

However, it is possible to reformulate this method in a recursive manner, i.e., adapt it to account for incoming data *on-line*. The core issue is to decide how to define the summation pattern similar to the one given by (27). The most natural way is to subdivide the region in which $\tilde{f}$ is non-zero into an increasing number of intervals along with incoming data. Thus we start with 2^N subregions obtained from the initial values $\lambda_0 = (y_{t_0}, y_{t_0-1}, \ldots, y_{t_0-n+1}, u_{t_0}, u_{t_0-1}, \ldots, u_{t_0-m+1})$; then, after the arrival of the first measured value of the function $\tilde{f}$, the number of subdivisions increases to 3^N etc. At each step we are able to evaluate the approximate value of (27) in the form

$$\tilde{F}_i(\omega) = \sum_{k=0}^{i} \tilde{f}(\lambda_k) e^{-j\omega\cdot\lambda_k} \Delta_k, \quad \text{with} \quad i = 0, 1, \ldots, K-1, \tag{35}$$

where Δ_k are computed on-line. In the first instance these are only two subintervals along each axis obtained from λ_0. They are of the form $\{[a, y_{t_0}], [y_{t_0}, b]\}$, $\{[a, y_{t_0-1}], [y_{t_0-1}, b]\}, \ldots, \{[a, y_{t_0-n+1}], [y_{t_0-n+1}, b]\}$, $\{[c, u_{t_0}], [u_{t_0}, d]\}$, $\{[c, u_{t_0-1}], [u_{t_0-1}, d]\}, \ldots, \{[c, u_{t_0-m+1}], [u_{t_0-m+1}, d]\}$. Next, in each of these pairs of subintervals one of them (it depends whether the value of the incoming sample lies in the first or second subinterval of the pair) is further subdivided giving the triple of subintervals along each axis and so on. This proceeds so long as new measurements arrive. Finally, after all K samples have arrived, we obtain the same $\tilde{F}_K(\omega)$ as with (27). The on-line method based on the above approach is the subject of research in progress.

Some other interesting details of the method are also being investigated. The problem of determining a function of bounded support from the values of its Fourier transform (which is a Paley-Wiener function, see Section 4.) on a finite set is ill-posed in the sense of Hadamard and its solution is not unique. However, it is possible to find an approximate solution with a small error by the universal methods of the Tikhonov regularisation (see [5,6]).

4. Entire Functions and Paley-Wiener Theory

This Section presents a tutorial survey of entire functions and the Paley-Wiener theory. The Fourier transform of a space-limited function is entire (in fact, it is a Paley-Wiener function), which motivates our interest in the mathematical apparatus. Its results can be further used for the ill-posed problem [5,6] of reconstruction of the Fourier transform $\tilde{F}$ of (2) from its samples on a finite set.

We start with the one-dimensional case in Section 4.1.. The setting is valuable for explaining motivations for the notions introduced and allows their clear illustrations. The tutorial exposition concentrates here, setting the stage for the more important multi-dimensional case in Section 4.2.

The basic concepts are those of entire functions, their growth and regularity of growth, leading to the central notion of Paley-Wiener functions. The main result of Section 4.1. is the celebrated Paley-Wiener theorem (Theorem 3); its important complement is Theorem 4. In the multi-dimensional case (Section 4.2.) the fundamental development consists in the Plancherel-Pólya theorem (Theorem 9). However, the most important result, directly relevant to our control problem, is Theorem 10 due to Wiegerinck.

According to the earlier notation $\tilde{f}$ is the space-limited extension of f and $\tilde{F}$, F are the corresponding Fourier transforms. In this Section we drop the tilde for simplicity, while constantly alluding to space-limitedness.

4.1. One Complex Variable

Complex differentiability is a much stronger property than in the real variable case (see Chapter 3 §2 and Chapter 4 §8 in [16]). In an open and connected neighbourhood of a point it is equivalent to analyticity, i.e., the function has the (complex) Taylor expansion in the neighbourhood. The larger the domain[a] of analyticity, the more well-behaved the function, which motivates the following definition.

Definition 1 *A function* $F\colon \mathbb{C} \to \mathbb{C}$ *is called* entire *if and only if it is analytic everywhere, i.e.,* $\forall z_0 \in \mathbb{C}\ \ \forall z \in \mathbb{C}\ \ F(z) = \sum_{k=0}^{\infty} a_k (z - z_0)^k$.

Since this holds everywhere in $\mathbb{C}$, we can assume $z_0 = 0$ without loss of generality.

Entire functions [17] are a generalisation of polynomials.[b] Polynomials are, up to a constant, determined by their *complex* zeros, hence the complex plane setting here. Entire functions of one variable possess isolated zeros and can be factored into *infinite* products (see Chapter XII §57 in [18]). Many mathematical facts about these factorisations are available (see Chapter 2 in [17]) and—to a limited extent—can be used in an engineering (Signal Processing) context [19,20]. The main difficulty is finding the (complex) zeros and the approach cannot be usefully generalised to higher dimensions (see Section 4.2.).

However, it is possible to give an alternative characterisation of polynomials. By the Fundamental Theorem of Algebra a polynomial of order n, $P(z) = \sum_{k=0}^{n} a_k z^k$, has n complex zeros: $\zeta_1, \ldots, \zeta_n$. On the other hand, its modulus, $|P(z)|$, behaves as $|z|^n$ for large $|z|$ and thus asymptotic properties of P, or its *growth*, say something important about P without the need of finding ζ_i, $i = 1, \ldots, n$. This line of thought can be effectively pursued for entire functions of one variable (see Section 4.1.1. below) and usefully generalised for several variables (see Section 4.2.2.).

The Maximum Modulus Theorem characterises ‘monotonicity’ of analytic functions and makes possible analysis of their growth. Before formulating the result,

---

[a] Domain in the complex plane $\mathbb{C}$ is an open and connected subset of $\mathbb{C}$.

[b] This statement should not be taken too literally; see Chapter 4 §4 in [16].

we recall that a domain $\Omega \subset \mathbb{C}$ does not contain its boundary $\partial\Omega$, but its closure $\overline{\Omega} = \Omega \cup \partial\Omega$ does and is a closed set.

Theorem 2 (Maximum Modulus Theorem) *Let $\Omega \subset \mathbb{C}$ be a bounded domain and $F : \overline{\Omega} \to \mathbb{C}$ a function continuous in $\overline{\Omega}$ and analytic in Ω. Then*

$$\max_{z \in \overline{\Omega}} |F(z)| = \max_{z \in \partial\Omega} |F(z)|.$$

In other words, if Ω is a *bounded* domain, then the modulus of F attains its maximum on the boundary $\partial\Omega$. If we take for Ω an open disc, centered at the origin, then $|F(z)|$ is largest somewhere on the bounding circle. Thus, for an entire F its *maximum* modulus is a (real) monotonic increasing function of the radius of the circles.

Definition 2 *Let $F : \mathbb{C} \to \mathbb{C}$ be an entire function. The function $M : (0, \infty) \to [0, \infty)$ defined by*

$$M(r) = \max_{|z|=r} |F(z)|$$

is called the maximum modulus *of F.*

4.1.1. Growth of Entire Functions of One Complex Variable

In the light of Theorem 2 and Definition 2, to characterise the growth of an entire function is to investigate its maximum modulus $M(r)$. Thus, the asymptotic (for $r \to \infty$) behaviour of $M(r)$ is compared with that of the exponential function, whose exponent behaves like a polynomial μr^{λ}. We seek the lowest values of λ and μ for which $\exp(\mu r^{\lambda})$ captures the growth of $M(r)$.

We look for the greatest lower bound of the set of numbers $\lambda \geq 0$ (λ nonnegative for r^{λ} characterises a *polynomial*) such that $M(r) < \exp(r^{\lambda})$, as $r \to \infty$. Concisely (remembering that $r > 0$)

$$\inf\{\lambda \geq 0 |\ M(r) < \mathrm{e}^{r^{\lambda}},\ r \to \infty\} = \inf\{\lambda \geq 0 |\ \frac{\log\log M(r)}{\log r} < \lambda,\ r \to \infty\}. \tag{36}$$

Recall that the upper limit of a (real) function $h(r)$ at ∞ is $\limsup_{r\to\infty} h(r) = \inf_{R\in\mathbb{R}}(\sup_{r>R} h(r))$. Taking $h(r) = \log\log M(r)/\log r$, we see that lim sup is exactly what we need, with $\sup_{r>R} h(r)$ taking care of the defining inequality in (36), while $\inf_{R\in\mathbb{R}}$ getting the required greatest lower bound. Hence the following definition.

Definition 3 *The entire function F is of order ρ if and only if*

$$\limsup_{r\to\infty} \frac{\log\log M(r)}{\log r} = \rho, \tag{37}$$

where the values $0 \leq \rho \leq \infty$ are allowed.

The nonnegativity restriction on ρ in Definition 3 means that, for example, for an F having $M(r) \to \exp(1/r^2)$, as $r \to \infty$, order is not defined, although the limit (37) exists (and is equal to -2).

The reasoning for λ can be repeated for μ and is summarised below.

Definition 4 *The entire function F of positive order ρ is of type τ if and only if*

$$\limsup_{r\to\infty} \frac{\log M(r)}{r^\rho} = \tau, \tag{38}$$

where the values $0 \le \tau \le \infty$ are allowed.

Note that Definition 4 insists on $\rho > 0$, so that type is undefined for $\rho = 0$. If, however, ρ is positive, then F is of (i) minimum type if $\tau = 0$, (ii) normal type for $0 < \tau < \infty$ and (iii) maximum type when $\tau = \infty$.

Example 2 The canonical entire function for the notions of order and type is

$$F(z) = \mathrm{e}^{cz^n} = \sum_{k=0}^{\infty} \frac{c^k}{k!} z^{nk} \qquad \text{with} \quad c > 0. \tag{39}$$

Putting $z = |z|\mathrm{e}^{j\phi}$, $-\pi < \phi \le \pi$, the maximum modulus is $M(r) = \mathrm{e}^{cr^n}$. Hence the order is $\rho(F) = n$ and the type $\tau(F) = c$, as expected. ■

The defining property of entire functions is their representation everywhere by the power series $F(z) = \sum_{k=0}^{\infty} a_k z^k$. It is not straightforward to obtain $M(r)$ from that, hence the need to express order and type in terms of the coefficients a_k. For functions of finite order this is done as follows (Theorems 2.2.2/p. 9 and 2.2.10/p. 11 in [17]):

$$\rho = \limsup_{k\to\infty} \frac{k \log k}{\log(1/|a_k|)}, \qquad \tau = \frac{1}{\mathrm{e}\rho} \limsup_{k\to\infty} \left(k |a_k|^{\frac{\rho}{k}} \right). \tag{40}$$

If $a_k = 0$, then the quotient in the formula for ρ is to be taken as 0.

Example 3 Find order and type of the entire functions:

$$F_1(z) = \sum_{k=1}^{\infty} \left(\frac{\log k}{k} \right)^{\frac{k}{c}} z^k, \qquad F_2(z) = \sum_{k=2}^{\infty} \left(\frac{1}{k \log k} \right)^{\frac{k}{c}} z^k, \qquad c > 0.$$

Using formulae (40) we easily get $\rho(F_1) = \rho(F_2) = c$ with $\tau(F_1) = \infty$ and $\tau(F_2) = 0$ (imagine finding $M(r)$ here). Incidentally, F_1 and F_2 are examples of entire functions of maximal and minimal type, respectively, not intuitively obvious cases in the light of the canonical function of Example 2. ■

Example 4 We can use formulae (40) to cross-check results of Example 2. The Taylor coefficients are:

$$a_k = \begin{cases} c^l/l!, & \text{if } k = ln \text{ with } l \in \mathbb{Z}_+; \\ 0, & \text{otherwise.} \end{cases}$$

We now make an explicit use of the upper limit in the definition of order and compute

$$\rho(\mathrm{e}^{cz^n}) = \limsup_{l\to\infty} \frac{(ln)\log(ln)}{\log(l!/c^l)} = n \cdot \limsup_{l\to\infty} \frac{\log(ln)}{\log \sqrt[l]{l!} - \log c} = n \cdot \limsup_{l\to\infty} Q(l) = n,$$

since $\lim_{l\to\infty} Q(l) = \lim_{l\to\infty}(\log l / \log \sqrt[l]{l!}) = 1$ (as $\sqrt[l]{l!} \sim l/\mathrm{e}$, see page 73 in [18]). Similarly,

$$\tau(\mathrm{e}^{cz^n}) = \frac{c}{\mathrm{e}} \limsup_{l\to\infty} \frac{l}{\sqrt[l]{l!}} = c,$$

as expected. ■

4.1.2. Entire Functions and Fourier Transform

As mentioned before, the relevance of entire functions for us is that they arise in the study of the Fourier transform of functions with bounded support.[c]

Definition 5 *An entire function of order* 1 *and finite (normal) type is called an* entire function of exponential type *(EFET).*

The chief motivation for our interest in EFETs is the following fundamental result (see pp. 12–13 in [22] and pp. 170–171 in [23]).

Theorem 3 (Paley-Wiener) *The function* $F\colon \mathbb{C} \to \mathbb{C}$ *is entire of exponential type and belongs to* L^2 *on the real axis, or* F *is a* Paley-Wiener function, *if and only if*

$$F(z) = \int_a^b f(x)\mathrm{e}^{-jzx}dx, \tag{41}$$

where $f\colon \mathbb{R} \to \mathbb{R}$ *is a square integrable function of bounded support and the smallest closed interval* $[a, b]$ *containing its support is given by*

$$a = -\limsup_{y\to\infty} \frac{\log|F(-jy)|}{y}, \qquad b = \limsup_{y\to\infty} \frac{\log|F(jy)|}{y}, \tag{42}$$

where $y = \operatorname{Im} z$.

Let us explain the meaning of Theorem 3, relevant to our context.[d] If f is a real function of bounded support, then its Fourier transform $F\colon \mathbb{R} \to \mathbb{C}$ is $F(\omega) =$

[c] Recall [21] that for a real function $f\colon \mathbb{R} \to \mathbb{R}$ its support is defined as $\operatorname{supp} f = \overline{\{x \in \mathbb{R} |\ f(x) \neq 0\}}$, i.e., it is the closure of the set of points for which f is non-zero. If f is continuous, then the set $\{x \in \mathbb{R} |\ f(x) \neq 0\}$ is open and hence the closure to get the *smallest* closed set containing it. Then x is not in the support iff x has a neighbourhood on which f vanishes identically (as the complement of $\operatorname{supp} f$ is open). Thus, $\operatorname{supp} f$ has *all* the points at which continuous f is non-zero or is about to become so. Since $\operatorname{supp} f$ is closed, when it is bounded it is compact.

[d] It is a mathematical tradition not to have minus in the exponent in (41), i.e.,

$$G(z) = \int_a^b g(x)\mathrm{e}^{jzx}dx. \tag{41a}$$

But if (41a) is true, then substituting $g(x) = f(-x)$ we recover (41) and vice versa (see page 108 in [24]).

$\int_a^b f(x)\exp(-j\omega x)dx$ and: 1) F is square (Lebesgue) integrable $\int_{-\infty}^{\infty}|F(\omega)|^2 d\omega < \infty$; 2) the extension of F to the whole of complex plane[e] results in an entire function of exponential type. Thus the Fourier transform of a function with bounded support, considered in the complex domain, $F: \mathbb{C} \to \mathbb{C}$, is a Paley-Wiener function.

A remarkable feature of Theorem 3 is that it is a necessary *and* sufficient condition for the relation between f of bounded support and the extension of its Fourier transform. Hence our study of EFETs here.

Example 5 A familiar example of a Paley-Wiener function is the sinc function:

$$F(z) = \frac{\sin z}{z} = \frac{\mathrm{e}^{jz} - \mathrm{e}^{-jz}}{2jz} = \sum_{k=0}^{\infty} \frac{(-1)^k}{(2k+1)!} z^{2k}. \tag{43}$$

We have

$$|F(z)| = \frac{\sqrt{\mathrm{e}^{2|z|\sin\phi} + \mathrm{e}^{-2|z|\sin\phi} - 2\cos[2(|z|\cos\phi)]}}{2|z|}. \tag{44}$$

For a given $|z|$, $|F(z)|$ has (as a function of ϕ) two equal maxima: at $\phi = -\pi/2$ and $\phi = \pi/2$ (see Figure 12a). Hence $M(r) = \sqrt{\mathrm{e}^{2r} + \mathrm{e}^{-2r} - 2}/2r$ and for r large enough there exist $0 < a < 1$ and $b > 1$, such that

$$A(r) = \frac{\log\log(a\mathrm{e}^r/r)}{\log r} \leq \frac{\log\log M(r)}{\log r} \leq \frac{\log\log(b\mathrm{e}^r/r)}{\log r} = B(r)$$

and thus $\rho(F) = 1$, since $\lim_{r\to\infty} A(r) = \lim_{r\to\infty} B(r) = 1$. Similarly, $\tau(F) = 1$.

The result is intuitively clear: the term $\exp(jz)$ should asymptotically dominate the expression $[\exp(jz) - \exp(-jz)]/2jz$. Note that (44) reduces to the real function $|\sin\omega/\omega|$ for $\phi = 0$ ($\omega > 0$) and $\phi = \pi$ ($\omega < 0$), as expected (see Figure 12b). We also have

$$\int_{-\infty}^{\infty} \left|\frac{\sin\omega}{\omega}\right|^2 d\omega = 2 \cdot \left(\int_0^1 \frac{\sin^2\omega}{\omega^2} d\omega + \int_1^{\infty} \frac{\sin^2\omega}{\omega^2} d\omega\right) < 4$$

and thus $\sin z/z$ is L^2 along the real axis, as stipulated in Theorem 3.

Finally, the corresponding function of bounded support is

$$f(x) = \begin{cases} 1, & \text{if } -1 < x < 1; \\ 0, & \text{elsewhere} \end{cases} \tag{45}$$

and its support coincides with the interval $[a, b] = [-1, 1]$, according to (42). ■

[e] The extension from $F(\omega)$ to $F(z)$ should not be confused with the relation between Fourier and Laplace transforms (note that $\omega = \mathrm{Re}\, z$).

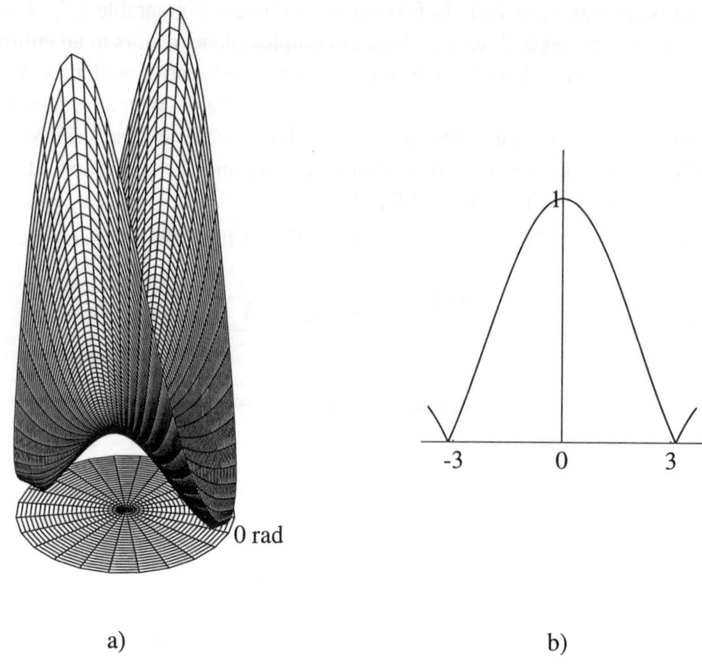

Figure 12: Growth of $F(z) = \sin z/z$: a) $|F(z)|$ in the polar coordinates $(|z|, \phi)$; the maximum modulus is attained for $\phi = \pm\pi/2$ for all $|z|$, b) the real function $|\sin \omega/\omega|$ obtained from $|\sin z/z|$ for $\phi = 0$ $(\omega > 0)$ and $\phi = \pi$ $(\omega < 0)$.

4.1.3. Regular Growth of Entire Functions of Exponential Type

For EFETs there is a more subtle asymptotic measure of growth than type (38). While τ is a global parameter, the notion of the *indicator function* (Chapter 5 in [17] and Chapter I §§ 15–16 and 19–20 in [25]) gives information about growth in different directions, along rays $\phi = \text{const}$.

Definition 6 *Let* $F: \mathbb{C} \to \mathbb{C}$ *be an entire function of exponential type. The* indicator function *of* F *is defined as*

$$h(\phi) = \limsup_{r\to\infty} \frac{\log |F(re^{j\phi})|}{r}. \tag{46}$$

Roughly speaking, (46) is (38) as a function of direction $\phi = \text{const}$ in the complex plane. In fact, $\tau = \max_{-\pi<\phi\leq\pi} h(\phi)$ (see Theorem 5.4.1/p. 75 in [17]).

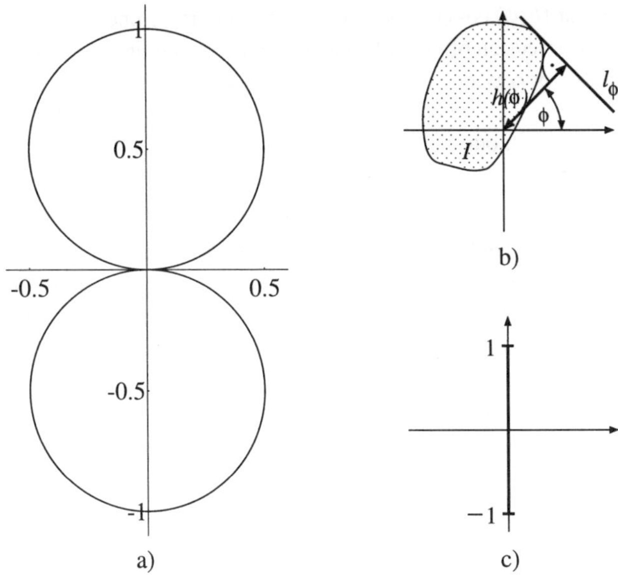

Figure 13: Illustrations of the indicator function h of $F(z) = \sin z/z$: a) the indicator function $h(\phi) = |\sin\phi|$ in the polar coordinates $(h(\phi), \phi)$; b) general idea of indicator diagram I of an EFET: h is the supporting function of the closed convex region I obtained from the intersection of half-planes defined by the lines l_ϕ, $-\pi < \phi \leq \pi$; c) the indicator diagram of $F(z) = \sin z/z$ is the interval $[-1, 1]$ on the imaginary axis, coinciding with the largest closed interval containing the support of $f(x) = \mathfrak{F}^{-1}(F(\omega))$, see Equation (45).

Example 6 The indicator function of $\sin z/z$ from Example 5 is

$$h(\phi) = \limsup_{r\to\infty} \frac{\sqrt{e^{2r\sin\phi} + e^{-2r\sin\phi} - 2\cos[2(r\cos\phi)]}}{2r} = |\sin\phi|. \qquad (47)$$

The indicator function is illustrated in Figure 13a in the polar coordinates $(h(\phi), \phi)$. Comparison with Figure 12a shows that both (asymptotically) agree: $|\sin z/z|$ tends to 0 along the real axis ($\phi = 0$ or $\phi = \pi$), has the most rapid growth along the imaginary axis ($\phi = \pm\pi/2$) and its increase is varying in between. ■

It turns out that h is a *supporting function* (Chapter I §19 in [25]) of a bounded, closed convex set I in the complex plane, called the *indicator diagram*. That is, for a given $\phi \in (-\pi, \pi]$, $h(\phi)$ is the length of the segment in the plane with an end-point in (0, 0) and lying on the ray defined by ϕ (see Figure13b). The line l_ϕ, perpendicular to the segment at $(h(\phi)\cos\phi, h(\phi)\sin\phi)$, is a supporting line[f] of the convex set I. The set is the intersection of all half-planes defined by the lines l_ϕ, $-\pi < \phi \leq \pi$.

The equation for l_ϕ is $x\cos\phi + y\sin\phi - h(\phi) = 0$, where $x = \operatorname{Re} z$ and $y = \operatorname{Im} z$, but this—in general—does not allow construction of the indicator diagram I from a known h. However, in our context, things simplify considerably (see page 108 in [17] and page 387 in [25]).

Theorem 4 *Let F be a Paley-Wiener function, as in Theorem 3. Then its indicator diagram I is the segment $[ja, jb]$ of the imaginary axis.*

Recall from Theorem 3 that $[a, b]$ is the smallest closed interval containing the support of $f(x) = \mathfrak{F}^{-1}(F(\omega))$, where $\mathfrak{F}^{-1}$ is the inverse Fourier transform. For (45) of Example 5 this is illustrated in Figure 13c and follows immediately from (47) of Example 6, because $h(0) = h(\pi) = 0$ and $h(\pi/2) = h(-\pi/2) = 1$.

As mentioned in Section 4.1., growth is a convenient alternative to zeros of entire functions. However, the order ρ and type τ are substitutes only for the degree of a polynomial. Information about the distribution of zeros is given by the indicator function (Chapter II §1 in [25]).

Definition 7 *The set of zeros ζ_k of an entire function of exponential type is called* regular *if and only if*

1. *the limit*
$$\lim_{r\to\infty} \frac{n(r, \phi_1, \phi_2)}{r} \tag{48}$$
exists for all $-\pi < \phi_1, \phi_2 \leq \pi$ (except possibly for a denumerable set of them), where $n(r, \phi_1, \phi_2)$ is the number of zeros in the sector $\{z \in \mathbb{C} \mid \phi_1 < \operatorname{Arg} z < \phi_2 \text{ and } |z| < r\}$;

2. *the limit*
$$\lim_{r\to\infty} \Big(c + \sum_{|\zeta_k|\leq r} \frac{1}{\zeta_k}\Big) \tag{49}$$
exists, where c is the highest coefficient in the polynomial $P(z)$ in the Hadamard expansion of f: $f(z) = z^m e^{P(z)} \prod_{k=1}^{\infty} G(z/\zeta_k, p)$.

Condition (49) is technical and we shall ignore it, but (48) is clear about distribution of zeros: the number of zeros in every sector should be (asymptotically) proportional to its radius. The rigmarole of Definition 7 can be replaced by the clean wording of growth in Definition 8 and Theorem 5 (see page 158 in [25]).

[f] A supporting line l of a bounded, closed plane convex set I is a line such that 1) I lies entirely in a half-plane defined by l and 2) the intersection of l and I is either a point or a segment. Clearly, it is a generalisation of a tangent line, coinciding with it at points of smoothness of ∂I, and being well defined at non-differentiable points (take I to be a triangle and consider its vertices).

Definition 8 *An entire function of exponential type F is said to be of* regular growth *if and only if the limit*

$$\lim_{r\to\infty} \frac{\log|F(re^{j\phi})|}{r} \tag{50}$$

exists for all $-\pi < \phi \le \pi$, except for a negligible set of pairs (r, ϕ).

Note that in (50) we have, unlike in (46), the limit (as opposed to lim sup) and thus we have to exclude the points for which $|F(re^{j\phi})| = 0$. Hence the negligible set (see pages 139 and 96 in [25]), not arising if lim sup is used, as in Definition 6.

Theorem 5 *An entire function of exponential type is of regular growth if and only if its set of zeros is regular.*

4.2. Several Complex Variables

Theory of entire functions of several complex variables is more involved, due to non-trivial geometry of $\mathbb{C}^N$, which is that of $\mathbb{R}^{2N}$ with the asymmetry [26] introduced by real and imaginary parts of $(z_1, \ldots, z_N) = z = x + jy = (x_1, \ldots, x_N) + j(y_1, \ldots, y_N)$; here $x \in \mathbb{R}^N$ and $y \in \mathbb{R}^N$.

The concept of complex differentiability in a neighbourhood of a point in $\mathbb{C}^N$ (holomorphy) is a direct analogue (Chapter I §2.3 in [26]) of the $N = 1$ case.

Definition 9 *A function $F: \mathbb{C}^N \to \mathbb{C}$ is called* entire *if and only if it is analytic everywhere, i.e., $\forall z_0 \in \mathbb{C}^N\ \forall z \in \mathbb{C}^N\ F(z) = \sum_{k\in\mathbb{Z}_+^N} a_k(z - z_0)^k$, where $k = (k_1, \ldots, k_N)$ and $(z - z_0)^k = (z_1 - z_1^0)^{k_1} \cdot \ldots \cdot (z_N - z_N^0)^{k_N}$.*

The coefficients $a_k \in \mathbb{C}^N$ of the Taylor expansion are given by

$$a_k = a_{(k_1,\ldots,k_N)} = \frac{1}{k_1! \cdot \ldots \cdot k_N!} \left.\frac{\partial^{k_1+\ldots+k_N} F(z)}{\partial z_1^{k_1} \cdots \partial z_1^{k_1}}\right|_{z=z_0}.$$

Again, since this holds everywhere in $\mathbb{C}^N$, we can assume $z_0 = 0$.

Characterisation of entire functions of several complex variables in terms of their zero sets, i.e., those $z \in \mathbb{C}^N$ for which $F(z) = 0$, is very difficult. The zero sets are not isolated points, but—in general—analytic varieties (see Chapter 2 in [27] and Chapter 4 in [23]), a special generalisation of curves and surfaces. This leads to quite involved analysis with far weaker results than in the one-dimensional case. First of all, only very few [28] polynomials in $\mathbb{C}^N$ admit factorisations. For general entire functions, representation in terms of the Osgood product (see Part II (Zweites Kapitel) §§4–8 in [29]) or via the Weierstrass Preparation Theorem (Chapter II §8.23 in [26]) are of not too much practical value (Chapter 10 §3 in [30]). Hence growth properties offer an attractive alternative and can be effectively used after introduction of little new machinery, needed because of geometry of $\mathbb{C}^N$, $N > 1$.

There is a direct analogue (see pp. 21–22 in [26]) of Theorem 2 for $\mathbb{C}^N$.

Theorem 6 *Let $\Omega \subset \mathbb{C}^N$ be a bounded domain and $F: \overline{\Omega} \to \mathbb{C}$ a function continuous in $\overline{\Omega}$ and analytic in Ω. Then*

$$\max_{z \in \overline{\Omega}} |F(z)| = \max_{z \in \partial\Omega} |F(z)|.$$

While $|F(z)|$ attains its maximum on the boundary $\partial\Omega$, it is not straightforward to take for Ω an 'open disc' to obtain an analogue of Definition 2. We discuss the issue in Section 4.2.2. and now give a version (see page 130 in [23]) convenient in the several complex variables setting.

Definition 10 *Let $F: \mathbb{C}^N \to \mathbb{C}$ be an entire function. The function $M: \mathbb{R}_+^N - \{0\} \to \mathbb{R}_+ = [0, \infty)$ defined by*

$$M(r) = \max_{\substack{|z_k| = r_k \\ 1 \le k \le N}} |F(z)|,$$

where $r = (r_1, \ldots, r_N)$, is called the maximum modulus *of F.*

Another possibility (Chapter 1 §3 in [27]) is $M^p: (0, \infty) \to [0, \infty)$ given by $M^p(r) = \max_{p(z) \le r} |F(z)|$, where p is a (real-valued) norm on $\mathbb{C}^N$ and r a positive *scalar* variable. While it looks very much like Definition 2, with $p(z) = |z|$ for $z \in \mathbb{C}$, it turns out somewhat awkward in computations. For example, the function of Example 2 is defined in $\mathbb{C}^N$ as $\exp(cz^n) = \exp(c(z_1^n + \ldots + z_N^n))$ to take advantage of the separation of variables; similarly, the N-dimensional sinc (Example 5) is $\operatorname{sinc} z = \operatorname{sinc} z_1 \cdot \ldots \cdot \operatorname{sinc} z_N$, where $\operatorname{sinc} z_k = \sin z_k / z_k$, $k = 1, \ldots, N$. Calculation of M is immediate in both cases, while M^p is involved for the Euclidean norm $p(z) = [\sum_{k=1}^N (x_k^2 + y_k^2)]^{1/2}$ and still non-trivial for $p(z) = \max_{1 \le k \le N} |z_k|$.

Before studying growth of entire functions in $\mathbb{C}^N$ in Section 4.2.2., we need a few facts about harmonic and subharmonic functions. This apparatus is very helpful in investigations of regularity of growth in several complex variables. Recall that this means (for one variable) relating the distribution of zeros of an entire function to its growth. For $\mathbb{C}^N$, $N > 1$, the underlying zero set may be intricate, so some additional tractable tools are needed. These are briefly described in Section 4.2.1. below.

4.2.1. Harmonic and Subharmonic Functions

There are close links between real functions satisfying the Laplace partial differential equation (PDE) and entire functions. For $N = 1$, the methods of complex variable are used to solve problems related to the PDE. For $N > 1$, the reverse is advantageous and therefore the one-dimensional case below is given as a motivating introduction to the important results for several complex variables.

One Complex Variable A function $F: \mathbb{C} \to \mathbb{C}$, given by $F: x + jy \mapsto u(x, y) + jv(x, y)$, is entire if the functions $u: \mathbb{R}^2 \to \mathbb{R}$ and $v: \mathbb{R}^2 \to \mathbb{R}$ satisfy the Cauchy-Riemann equations, $u_x = v_y$, $u_y = -v_x$. They also are harmonic, i.e., each satisfies the Laplace equation $\Delta h = h_{xx} + h_{yy} = 0$ (see Chapter X in [16]). Thus entire functions give rise to a subclass of harmonic functions, viz., those obeying the Cauchy-Riemann conditions. This is an important connection, allowing application of the insights in the Dirichlet Problem[g] to the investigation of entire functions.

Another definition of the harmonic function h is via the Mean Value Property.

Definition 11 *A continuous function* $h: \Omega \to \mathbb{R}$, *where* $\Omega \subset \mathbb{C}$ *is a domain, has the* Mean Value Property *if and only if*

$$h(z_0) = \frac{1}{2\pi} \int_0^{2\pi} h(z_0 + re^{j\phi}) d\phi \tag{51}$$

for any closed disc $\overline{B}(z_0; r) = \{z \in \mathbb{C} | \; |z - z_0| \leq r\} \subset \Omega$.

Thus the value of h at z_0 is the mean of its values on the circle bounding any disc (in Ω) with centre z_0. The result of far-reaching consequences is that h is harmonic in Ω if and only if it has the Mean Value Property there.

Solution of the Dirichlet Problem requires a generalisation of harmonic functions, derived from the Mean Value Property.

Definition 12 *A continuous function* $h: \Omega \to \mathbb{R}$, *where* $\Omega \subset \mathbb{C}$ *is a domain, is called* subharmonic *if and only if*

$$h(z_0) \leq \frac{1}{2\pi} \int_0^{2\pi} h(z_0 + re^{j\phi}) d\phi \tag{52}$$

for any closed disc $\overline{B}(z_0; r) = \{z \in \mathbb{C} | \; |z - z_0| \leq r\} \subset \Omega$.

Note that equality (51) is changed to inequality (52) and thus h is harmonic if and only if both h and $-h$ are subharmonic. For technical reasons it is convenient to replace continuity with the weaker requirement of upper semi-continuity[h] in Definition 12 and then interpret (52) as the Lebesgue integral.

The usefulness of subharmonic functions consists in the following result (see pp. 266–267 in [16]).

Proposition 1 *Let a continuous function* $H: \partial\Omega \to \mathbb{R}$, *where* $\Omega \subset \mathbb{C}$ *is a domain, be given together with the family* $\mathcal{P}(H, \Omega)$ *of all subharmonic functions* $\chi: \Omega \to \mathbb{R}$, *such that* $\limsup_{z \to z_0} \chi(z) \leq H(z_0)$ *for all* $z_0 \in \partial\Omega$. *Then*

$$h(z) = \sup\{\chi(z) | \; \chi \in \mathcal{P}(H, \Omega)\} \tag{53}$$

is a harmonic function on Ω.

[g] The Dirichlet Problem in the plane is to determine all domains $\Omega \subset \mathbb{C}$ such that for any continuous function $H: \partial\Omega \to \mathbb{R}$ there exists a function $h: \overline{\Omega} \to \mathbb{R}$, continuous in $\overline{\Omega} = \Omega \cup \partial\Omega$, so that $h(z) = H(z)$ on $\partial\Omega$ and h is harmonic in Ω (for unbounded Ω, ∞ is adjoined to $\mathbb{C}$, see pp. 8–9 in [16]).

[h] Recall (see page 83 in [31]) that the function $h: \Omega \to \mathbb{R}$ is called *upper semi-continuous* at $z_0 \in \Omega$ if $h(z_0) \geq \limsup_{z \to z_0} h(z)$. If h is upper semi-continuous in $\overline{\Omega}$, then h is bounded above and attains its supremum.

With an appropriate (see Chapter X §4 in [16]) strengthening of assumptions on Ω, it is also true that $\lim_{z\to z_0} h(z) = H(z_0)$ for all $z_0 \in \partial\Omega$. Hence the Dirichlet Problem can be solved by means of the harmonic function h majorising the subharmonic functions χ, according to (53).

Several Complex Variables We consider entire functions $F\colon \mathbb{C}^N \to \mathbb{C}$ with $F\colon x + jy \mapsto u(x, y) + jv(x, y)$, where $x = (x_1, \ldots, x_N) \in \mathbb{R}^N$, $y = (y_1, \ldots, y_N) \in \mathbb{R}^N$ and $u\colon \mathbb{R}^{2N} \to \mathbb{R}$, $v\colon \mathbb{R}^{2N} \to \mathbb{R}$. We say that $h = h(x, y)$ is harmonic if it satisfies the $2N$-dimensional Laplace equation $\sum_{k=1}^{N}(h_{x_k x_k} + h_{y_k y_k}) = 0$. However, $u = \operatorname{Re} F$ and $v = \operatorname{Im} F$ obey the Cauchy-Riemann ($2N$ real) equations $u_{x_k} = v_{y_k}$, $u_{y_k} = -v_{x_k}$, $k = 1, \ldots, N$ and thus give rise to a special class of harmonic functions, called *pluriharmonic* functions (see Chapter I §2.4 in [26]).

The following class of functions is of special interest for the study of growth of entire functions in $\mathbb{C}^N$ (see Chapter 2 in [23] and Appendix I in [27]).

Definition 13 *An upper semi-continuous function $h\colon \Omega \to \mathbb{R}$, where $\Omega \subset \mathbb{C}^N$ is a domain, is called* plurisubharmonic *if and only if*

$$h(z_0) \leq \frac{1}{2\pi}\int_0^{2\pi} h(z_0 + re^{j\phi}w)\,d\phi \tag{54}$$

for any line segment $\overline{L}(z_0; r; w) = \{z_0 + \tau w \in \mathbb{C}^N \mid |\tau| \leq r, \tau \in \mathbb{C}\} \subset \Omega$.

A plurisubharmonic function is the complex analogue of a convex function. A function $g\colon \Omega \to \mathbb{R}$, where $\Omega \subset \mathbb{R}^N$ is a domain, is called convex if restricted to the segments in Ω of the straight lines $x = x_0 + tv$ (here $x_0, v \in \mathbb{R}^N$ and t is a real parameter) is a convex function of t. Similarly, $h\colon \Omega \to \mathbb{R}$, where $\Omega \subset \mathbb{C}^N$ is a domain, is plurisubharmonic if for each complex line $z = z_0 + \tau w$ ($z_0, w \in \mathbb{C}^N$ and τ is a complex parameter) its restriction to any $\overline{L}(z_0; r; w) \subset \Omega$ is a subharmonic function of τ (compare Definitions 12 and 13).

The plurisubharmonic (PSH) functions play the same role for several complex variables as the subharmonic functions for one variable. If both h and $-h$ are plurisubharmonic, then h is pluriharmonic and when h is twice differentiable, then h is PSH iff $\Delta h \geq 0$, where Δ is the Laplacian. The most important fact (see page 105 in [23] and page 230 in [27]) in our context of entire functions $F\colon \mathbb{C}^N \to \mathbb{C}$ ($\Omega = \mathbb{C}^N$) is that M (also $|F|$) of Definition 10 and $\log M$ (also $\log|F|$) are PSH (and thus $\log\log M$ is).

Before we study the growth of F we have to sort out the last technicality. If $\{h_k\}_{k=1}^{s}$ is a finite family of PSH functions, then $H(z) = \sup_{1\leq k\leq s} h_k(z)$ is also PSH. This is not necessarily true for an infinite PSH sequence $\{h_k\}_{k\in S}$ and thus we want to find the smallest PSH majorant of $\{h_k\}_{k\in S}$. This is done by *regularisation* (see Chapter 1 §7 in [27]).

Definition 14 *Let $h\colon \Omega \to \mathbb{R}$, where $\Omega \subset \mathbb{C}^N$ is a domain, be locally bounded from above. Then*

$$h^*(z_0) = \limsup_{\Omega \ni z \to z_0} h(z) \qquad \forall\, z_0 \in \Omega$$

is called the regularisation *of h.*
Local boundedness of h means that for any $z_0 \in \Omega$ there exists a neighbourhood of z_0 in which $h(z_0)$ is bounded. By definition, h^* is the smallest upper semi-continuous majorant of h. We can now state the main result.
Theorem 7 *Let $\{h_k\}_{k \in S}$ be a family of PSH functions defined on a domain $\Omega \subset \mathbb{C}^N$ and locally bounded from above and $H(z) = \sup_{k \in S} h_k(z)$. Then the regularisation H^* of H is also PSH on Ω.*

4.2.2. Growth of Entire Functions of Several Complex Variables

In this Section we describe the N-dimensional analogues of the notions and results of Sections 4.1.1.–4.1.3.. We take for granted the motivations and illustrations presented there, and therefore give a more concise account.

The last question of the background material is an N-dimensional analogue of an open disc in $\mathbb{C}$, signalled when Definition 10 was introduced. Geometry of $\mathbb{C}^N$, $N > 1$, allows many generalisations and we settle on the *multi-circular* domains (see Introduction §6 in [23] and Chapter I §1.2.4 in [26]), also known as Reinhardt or N-circular domains.
Definition 15 *A domain $\Omega \subset \mathbb{C}^N$ is called* multi-circular with centre at z_0 *if and only if for every its point z', it contains also the circles, centred at z_0 and passing through z', i.e., for every $z' \in \Omega$ the set*

$$C(z_0; z') = \{z \in \mathbb{C}^N \mid |z_k - z_k^0| = |z_k' - z_k^0|,\ k = 1, \dots, N\} \tag{55}$$

is contained in Ω.
The set $C(z_0; z')$ is a multi-circle, because, for each k, $|z_k - z_k^0| = |z_k' - z_k^0|$ determines (in $\mathbb{C}$) the circle with centre at z_k^0 and radius $|z_k'|$. A multi-circular domain $\Omega \subset \mathbb{C}^N$ is said to be *complete* if, together with each point $z' \in \Omega$, it contains the entire polydisc $D(z_0; z') = \{z \in \mathbb{C}^N \mid |z_k - z_k^0| \le |z_k' - z_k^0|,\ k = 1, \dots, N\}$. Balls $B(z_0; r) = \{z \in \mathbb{C}^N \mid |z - z_0| < r\}$ are complete multi-circular domains (thus also the disc in $\mathbb{C}$), while every multi-annulus $\{z \in \mathbb{C}^N \mid r < |z_k - z_k^0| < R,\ k = 1, \dots, N\}$ is multi-circular, but incomplete.

Upon translation, z_0 in (55) can be made 0 and then $C(z_0; z')$ contains all points with the same $|z_k'|$, $k = 1, \dots, N$. Thus, putting $r_k = |z_k - z_k^0|$, we define the mapping $\alpha\colon \Omega \to \mathbb{R}_+^N$ given by $z \mapsto \alpha(z) = (r_1, \dots, r_N)$. This motivates defining a complete domain G in $\mathbb{R}_+^N$ by requiring that G is the largest open subset of $\mathbb{R}_+^N$ such that for every $r' = (r_1', \dots, r_N')$ in G, all $r = (r_1, \dots, r_N)$, with $0 \le r_k \le r_k'$

$(k = 1, \ldots, N)$, are also in G. This is useful, because: (i) the image under α of a complete multi-circular domain $\Omega \subset \mathbb{C}^N$ is a complete domain $G \subset \mathbb{R}^N$, (ii) the dimension is reduced by half (from $\mathbb{C}^N \sim \mathbb{R}^{2N}$ to $\mathbb{R}^N$).

We are ready to study growth of entire functions in $\mathbb{C}^N$.

Definition 16 *Let G be a complete domain in $\mathbb{R}^N_+$ and, for $r = (r_1, \ldots, r_N) \in G$, define $tr = (tr_1, \ldots, tr_N)$, $t > 0$. The entire function $F\colon \mathbb{C}^N \to \mathbb{C}$ is of order ρ if and only if*

$$\limsup_{t\to\infty} \frac{\log\log M(tr)}{\log t} = \rho, \tag{56}$$

where values $0 \le \rho \le \infty$ are allowed.

Order (56) does not depend (see Chapter 2 §6 and Chapter 3 §1 in [23]) on the choice of G through which M grows. This is not so with type and hence G in Definitions 16 and 17. The issue does not arise for $N = 1$, where G is naturally implied in Definition 2.

Definition 17 *Let G be a complete domain in $\mathbb{R}^N_+$ and, for $r = (r_1, \ldots, r_N) \in G$, define $tr = (tr_1, \ldots, tr_N)$, $t > 0$. The entire function $F\colon \mathbb{C}^N \to \mathbb{C}$ of positive order ρ is of type τ_G if and only if*

$$\limsup_{t\to\infty} \frac{\log M(tr)}{t^\rho} = \tau_G, \tag{57}$$

where values $0 \le \tau_G \le \infty$ are allowed.

Formulae (40) have the following N-dimensional analogues (see page 131 in [23]).

Theorem 8 (Gol'dberg) *Let G be a complete domain in $\mathbb{R}^N_+$. The order ρ and type τ_G of the entire function $F\colon \mathbb{C}^N \to \mathbb{C}$*

$$F(z) = \sum_{k\in\mathbb{Z}^N_+} a_k z^k = \sum_{\|k\|=0}^{\infty} a_k z^k,$$

where $\|k\| = k_1 + \ldots + k_N$, are related to the coefficients of its Taylor expansion by

$$\rho = \limsup_{\|k\|\to\infty} \frac{\|k\| \log \|k\|}{\log(1/|a_k|)}, \qquad \tau_G = \frac{1}{e\rho} \limsup_{\|k\|\to\infty} \left(\|k\| \left[|a_k| d_k(G) \right]^{\frac{\rho}{\|k\|}} \right),$$

with $d_k(G) = \max_{r\in G} r^k$, where $r^k = r_1^{k_1} \cdot \ldots \cdot r_N^{k_N}$.

If $a_k = 0$, then the quotient in the formula for ρ is to be taken as 0. Note that $d_k(G)$ explicitly takes account for the dependence of τ_G on the domain. For domains

$$G_1 = \{r \in \mathbb{R}^N_+ \mid (\textstyle\sum_{k=1}^N r_k^2)^{1/2} < 1\} \quad \text{and} \quad G_2 = \{r \in \mathbb{R}^N_+ \mid \textstyle\sum_{k=1}^N r_k < 1\}$$

we have (page 134 in [23])

$$d_k(G_1) = \sqrt{k_N^{k_1} \cdot \ldots \cdot k_N^{k_N} / \|k\|^{\|k\|}} \quad \text{and} \quad d_k(G_2) = k_N^{k_1} \cdot \ldots \cdot k_N^{k_N} / \|k\|^{\|k\|}$$

with $\|k\| = k_1 + \ldots + k_N$, as before.

It should be emphasised that F is defined on the *whole* of $\mathbb{C}^N$ and the domain $G \subset \mathbb{R}_+^N$ arises only when $M(r)$ of Definition 10 is investigated. In the plane r naturally grows radially and it is a non-issue, while in the $N > 1$ case $r = (r_1, \ldots, r_N)$ may tend to infinity in many ways and hence it matters *how* it traverses $\mathbb{R}_+^N$.

As in the one-dimensional case, an entire function is said to be of exponential type if $\rho = 1$ and τ_G is finite. We now introduce a generalisation of Definition 6 with a view of stating an analogue of Theorems 3 and 4.

Definition 18 *Let $F\colon \mathbb{C}^N \to \mathbb{C}$ be an entire function of exponential type. The P-*indicator *h of F is defined as*

$$h(\Phi) = \sup_{x \in \mathbb{R}^N} \left(\limsup_{t \to \infty} \frac{\log |F(x_1 + jt\Phi_1, \ldots, x_N + jt\Phi_N)|}{t} \right), \tag{58}$$

where $\Phi = (\Phi_1, \ldots, \Phi_N)$ is a point on the unit sphere in $\mathbb{R}^N$.

In Definition 6 the argument of h was the angle ϕ. For the P-indicator the direction is chosen in the imaginary subspace of $\mathbb{C}^N$, i.e., $\mathbb{R}^N$ corresponding to $y_k = \operatorname{Im} z_k$, $k = 1, \ldots, N$. The direction is determined by the point Φ on the sphere in this $\mathbb{R}^N$.

The following result extends Theorems 3 and 4 to several complex variables (see pp. 171–175 in [23]) and applies to (25).

Theorem 9 (Plancherel-Pólya) *The function $F\colon \mathbb{C}^N \to \mathbb{C}$ is entire of exponential type and $F(\operatorname{Re} z)$ belongs to $L^2(\mathbb{R}^N)$, or F is a* Paley-Wiener function, *if and only if*

$$F(z) = \int_D f(x) e^{-jz \cdot x} dx,$$

where $f\colon \mathbb{R}^N \to \mathbb{R}$ is a square integrable function of bounded support D. Moreover, the supporting function $H(\Phi)$ of the smallest convex domain Ω off which $f \equiv 0$ coincides for any Φ with the P-indicator $h(\Phi)$ of F.

The supporting function $H\colon \mathbb{R}^N \to \mathbb{R}$ of a convex set $\Omega \subset \mathbb{R}^N$ is $H(\Phi) = \sup_{x \in \Omega}(\Phi_1 x_1 + \ldots + \Phi_N x_N)$. If the length of Φ is 1, then $\Phi \cdot x = \sum_{k=1}^N \Phi_k x_k$ is the projection of vector x on the direction Φ and then $H(\Phi)$ is the distance from the origin to the nearest supporting hyperplane of Ω (cf. comments after Example 6).

The connection between $\Omega \supset D$ and the P-indicator in Theorem 9 is as between the interval $[a, b]$ of Theorem 3 and the indicator diagram of Theorem 4.

We now proceed to formulate an N-dimensional analogue of Theorem 5. It turns out that a more powerful result, directly relevant to our problem, is available [32].

To define regular growth of entire functions of exponential type (EFET) in $\mathbb{C}^N$ (see Chapter 4 in [27]), we start with a generalisation of Definition 6.

Definition 19 *Let $F\colon \mathbb{C}^N \to \mathbb{C}$ be an entire function of exponential type. The* radial indicator *of F is defined as*

$$h(z) = \limsup_{t \to \infty} \frac{\log |F(tz)|}{t}. \tag{59}$$

We also consider its regularisation (cf. Definition 14), h^*, which is plurisubharmonic (cf. Definition 13) and positively homogeneous, i.e., $h^*(tz) = th^*(z)$ for $t > 0$.

As for $N = 1$, we could define regularity of growth by requiring the existence of the limit of h^* outside a suitable negligible set (cf. Definition 8). However, for $N > 1$ it is possible for F to have irregular growth on a small set of rays without affecting the global asymptotic behaviour of the function or its zero set. For example, $z_2 \exp z_1$ has regular growth except for $z_2 = 0$, when it is identically 0. The reason is that h and h^* may be different on a small set of rays, so it is plausible to study behaviour of $\log |F|$ on ever smaller neighbourhoods of a ray, than on the ray itself. This motivates the following definition, having the advantage of the absence of exceptional sets (see page 96 in [27] and page 97 in [32]).

Definition 20 *Let $F: \mathbb{C}^N \to \mathbb{C}$ be an entire function of exponential type and $B(z_0; R)$ an open ball in $\mathbb{C}^N$, centre $z_0 \in \mathbb{C}^N$ and radius $R > 0$, with $V(R)$ being its volume. For $z \in \mathbb{C}^N$ put*

$$I(z; \delta, t) = \frac{1}{V(\delta t)} \int_{B(tz;\delta t)} \frac{\log |F(w)|}{t} dw, \tag{60}$$

where δ, t are positive scalars, $\delta > 0$ and $t > 0$. Then F has regular growth in the direction *$z \in \mathbb{C}^N - \{0\}$ if and only if for every $\varepsilon > 0$ and $\delta_0 > 0$ there exist δ with $0 < \delta < \delta_0$ and $T = T(\varepsilon, \delta) > 0$ such that for $t > T$:*

$$|I(z; \delta, t) - h^*(z)| < \varepsilon, \tag{61}$$

where h^* is the regularisation of the radial indicator h of F. Also, F has regular growth on a set *Ω if and only if it has regular growth in all directions $z \in \Omega$, $z \neq 0$.*
The direction $z = 0 \in \mathbb{C}^N$ is excluded to avoid triviality of tz. Note that w in (60) is taken over $B(tz; \delta t)$ and hence the dependence of I on z.

Comparison of (59) with (60) and (61) reveals how $\log |F|$ can be studied on ever smaller neighbourhoods of the ray tz. For a given $\varepsilon > 0$ the closeness in (61) means that for large enough t and small enough δ regularity of growth of F is discernible from its properties on the ball $B(tz; \delta t)$, 'sitting' on the the ray tz of the direction z. The ball is centred at tz and defines a neighbourhood of radius δt swept by the (Lebesgue) integral (60), so that all irregular behaviour occurring on exceptional sets of (Lebesgue) measure zero is eliminated.

Thus defined regular growth can be related to the 'distribution' of the zero set ζ of F, i.e., relation between the maximum modulus M and the area of ζ (see Chapter 3 in [27] and page 98 in [32]). This is not pursued here.

With Definition 20 explained, we are ready to state the result (Theorem 6 in [32]) directly relevant to our control problem.

Theorem 10 (Wiegerinck) *Let $f \in L^2(\mathbb{R}^N)$ have bounded support $D \subset \mathbb{R}^N$ and the* *convex hull of D be a polyhedron. Then* $F\colon \mathbb{C}^N \to \mathbb{C}$ *given by*

$$F(z) = \int_D f(x)\mathrm{e}^{-jz\cdot x}dx$$

is a Paley-Wiener function of regular growth on $\mathbb{C}^N$.

Recall (cf. (2) and (25)) that in our context the support of the space-limited extension of f, i.e., $\tilde{f}$, is $D = [a,b]^n \times [c,d]^m \subset \mathbb{R}^N$, $N = n + m$, so it is equal to its convex hull (i.e., the smallest convex set containing D). Thus, according to Theorem 10, not only is $\tilde{F}(z) = \int_D \tilde{f}(x)\mathrm{e}^{-j(z\cdot x)}dx$ a Paley-Wiener function, but it has regular growth on the whole of $\mathbb{C}^N$ (in terms of Definition 20 $\Omega = \mathbb{C}^N$).

5. Conclusions

This Chapter described a new approach to modelling of NARMA control systems with feedforward neural networks. The method is based on Fourier analysis and nonuniform multi-dimensional sampling and aims at reconstruction of the unknown right-hand side of the NARMA model, given past inputs and outputs. The approach has the following advantages: realistic engineering setting, computational simplicity, wide applicability (mild assumptions), flexible neural implementation and usefulness for control purposes.

The algorithm uses an approximation of the Fourier transform of the right-hand side of the NARMA model. Therefore ways of doing so and their motivations were described.

Real-world engineering constraints lead to the consideration of Fourier transforms which are Paley-Wiener functions. A tutorial survey of the relevant mathematical theory was presented and its usefulness for the ill-posed problems setting mentioned. This is a subject of further research.

The new method seems to be a promising technique for neurocontrol. Interesting areas of further research are, among others, aspects of the on-line algorithm and the ill-posed problems approach to reconstruction.

Acknowledgements

Work supported by ESPRIT III Basic Research Project No. 8039 *Neural Adaptive Control Technology*.

6. References

1. S. Chen and S. A. Billings, "Representation of non-linear systems: the NARMAX model," *International Journal of Control*, vol. 49, pp. 1013–1032, 1989.

2. A. Dzieliński and R. Żbikowski, "Feedforward neural networks: n-D systems theoretic aspects," in *Proceedings of the Third European Control Conference, September 5–8, 1995, Rome, Italy* (A. Isidori, S. Bittanti, E. Mosca, A. D. Luca, M. D. D. Benedetto, and G. Oriolo, eds.), vol. 2, pp. 1595–1600, European Union Control Association, 1995.

3. A. Dzieliński and R. Żbikowski, "N-dimensional sampling as a tool for nonlinear neurocontrol," in *Proceedings of the Second International Symposium on Methods and Models in Automation and Robotics, MMAR '95, Miedzyzdroje, Poland, August 30 – September 2 1995* (S. Bańka, S. Domek, and Z. Emirsajłow, eds.), (Szczecin, Poland), pp. 69–74, Szczecin University of Technology Press, 1995.

4. R. M. Sanner and J.-J. E. Slotine, "Gaussian networks for direct adaptive control," *IEEE Transactions on Neural Networks*, vol. 3, no. 6, pp. 837–863, 1992.

5. A. N. Tikhonov and V. Y. Arsenin, *Solution of Ill-posed Problems*. New York: John Wiley & Sons, 1977.

6. M. M. Lavrentiev, *Some Improperly Posed Problems of Mathematical Physics*. Berlin, Heidelberg and New York: Springer Verlag, 1967.

7. J. A. Cadzow, *Foundations of Digital Signal Processing and Data Analysis*. New York, NY: Macmillan Publishing Company, 1987.

8. A. V. Oppenheim and R. W. Schafer, *Digital Signal Processing*. Englewood Cliffs, NJ: Prentice-Hall, 1975.

9. S. P. Kim and N. K. Bose, "Reconstruction of 2-D bandlimited discrete signals from nonuniform samples," *IEE Proceedings, Part F*, vol. 137, pp. 197–204, 1990.

10. D. P. Petersen and D. Middleton, "Sampling and reconstruction of wave-number-limited functions in n-dimensional euclidean spaces," *Information and Control*, vol. 5, pp. 279–323, 1962.

11. C. E. Shannon, "Communication in the presence of noise," *Proceedings of the IRE*, vol. 37, pp. 10–21, 1949.

12. D. Slepian, "On bandwidth," *Proceedings of the IEEE*, vol. 64, pp. 292–300, March 1976.

13. A. Dzieliński and R. Żbikowski, "Irregular sampling approach to neurocontrol with feedforward neural networks," *Annals of Mathematics and Artificial Intelligence*, 1996. (to appear).

14. R. Żbikowski and A. Dzieliński, "Neural approximation: A control perspective," in *Advances in Neural Networks for Control Systems* (K. J. Hunt, G. R. Irwin, and K. Warwick, eds.), Advances in Industrial Control, pp. 1–25, Berlin, New York: Springer-Verlag, 1995.

15. I. W. Sandberg, "The reconstruction of band-limited signals from nonuniformly spaced samples," *IEEE Transactions on Circuits and Systems–I: Fundamental Theory and Applications*, vol. 41, pp. 64–66, 1994.

16. J. B. Conway, *Functions of One Complex Variable*. New York: Springer-Verlag, Second ed., 1978.

17. R. P. Boas, *Entire Functions*. New York: Academic Press, 1954.

18. K. Knopp, *Theory and Application of Infinite Series*. New York: Dover Publications, 1990.

19. A. A. G. Requicha, "The zeros of entire functions: Theory and engineering applications," *Proceedings of the IEEE*, vol. 68, no. 3, pp. 308–328, 1980.

20. F. A. Marvasti, *A Unified Approach to Zero-Crossings and Nonuniform Sampling*. Oak Park, Illinois, USA: Nonuniform, 1987.

21. J. Dugundji, *Topology*. Boston: Allyn and Bacon, 1966.

22. R. E. A. C. Paley and N. Wiener, *Fourier Transforms in the Complex Domain*. Colloquium Publications, Volume 19, Providence, RI: American Mathematical Society, 1934.

23. L. I. Ronkin, *Introduction to the Theory of Entire Functions of Several Variables*. Translations of Mathematical Monographs, Volume 44, Providence, RI: American Mathematical Society, 1974.

24. E. M. Stein and G. Weiss, *Introduction to Fourier Analysis on Euclidean Spaces*. Princeton, NJ: Princeton University Press, 1971.

25. B. J. Levin, *Distribution of Zeros of Entire Functions*. Translations of Mathematical Monographs, Volume 5, Providence, RI: American Mathematical Society, 1964.

26. B. V. Shabat, *Introduction to Complex Analysis. Part II: Functions of Several Variables*. Translations of Mathematical Monographs, Volume 110, Providence, RI: American Mathematical Society, 1992.

27. P. Lelong and L. Gruman, *Entire Functions of Several Complex Variables*. Berlin: Springer-Verlag, 1986.

28. M. H. Hayes and J. H. McClellan, "Reducible polynomials in two or more variables," *Proceedings of the IEEE*, vol. 70, no. 2, pp. 197–198, 1982.

29. W. F. Osgood, *Lehrbuch der Funktionentheorie. Zweiter Band (Erste Lieferung)*. Leipzig, Berlin: Teubner, 1924. (In German).

30. N. E. Hurt, *Phase Retrieval and Zero Crossings. Mathematical Methods in Image Reconstruction*. Dordrecht, Boston and London: Kluwer Academic Publications, 1989.

31. J. C. Burkill and H. Burkill, *A Second Course in Mathematical Analysis*. Cambridge, England: Cambridge University Press, 1970.

32. J. J. O. O. Wiegerinck, "Growth properties of Paley-Wiener functions on $\mathbb{C}^n$," *Proceedings of the Koninklijke Nederlands Akademie van Wetenschappen, Series A: Mathematical Sciences*, vol. 87, pp. 95–112, 1984.

Part II

Nonlinear Control Fundamentals for Neural Networks

GEOMETRIC METHODS IN NONLINEAR CONTROL THEORY: A SURVEY

Witold Respondek*
Institut National des Sciences Appliquées de Rouen
Département Génie Mathématique
B.P. 08 Place Émile Blondel
76131 Mont Saint Aignan Cedex
France
E-mail: wresp@lmi.insa-rouen.fr

ABSTRACT

This Chapter is a survey of some geometric methods used in nonlinear control theory. After recalling main tools from Differential Geometry the following basic control theoretic concepts are discussed in the nonlinear context: controllability, observability, equivalence of systems, linearization, and decoupling. The presented results are illustrated by some engineering examples.

1. Introduction

The aim of these notes is to give a survey of some methods and results of geometric nonlinear control theory. This theory deals with control problems and attacks them, and solves some of them, using methods of Differential Geometry. Therefore problems, motivations, and intuitions are coming from physical and engineering considerations, while powerful and elegant tools from geometry. Among many reasons to use geometric methods in nonlinear control we would like to emphasize two. Firstly, in many physical and engineering problems the natural state-space is not a linear space but has a structure of a differentiable manifold, i.e., a multi-dimensional analog of a curve or surface. For example, the natural state space of a pendulum is the surface of cylinder, while for conservative systems the surface of ellipsoid corresponding to a constant energy level. Thus differential geometric tools appear very naturally. Secondly, even if we consider local aspects of a control problem only (and thus we can assume that the state-space is an open subset of a cartesian space), the application of geometric methods is very useful, because they describe invariant properties of the system which are independent of any particular coordinates in which we represent the system.

Although classical results of Chow and Rashevskii have a clear interpretation in terms of nonlinear controllability, the idea of using methods and tools from Differential Geometry appeared in the sixties (see Hermann [1]) and geometric methods have been extensively used from the beginning of the seventies in works of Brockett, Hermes, Hirschorn, Krener, Jurdjevic, Lobry, Sussmann, and others. The first prob-

*on leave from Institute of Mathematics, Polish Academy of Sciences

lem studied via geometric methods was that of nonlinear controllability, followed by those of nonlinear observability and nonlinear realization. Starting from the seventies differential geometric methods have also been successfully used in optimal control problems. In the eighties geometric methods have been applied (besides the above-mentioned fundamental system theory concepts of controllability and observability) to compensate nonlinearities (linearization problem) and to synthesize controls with decoupling properties (disturbance decoupling and input-output decoupling problems). Afterwards, the nonlinear tool-kit has been rapidly growing and many fundamental control problems have been successfully studied and solved using differential geometric methods and their combinations with (differential) algebraic methods. To mention just a few: nonlinear stabilizability, synthesis of nonlinear observers, output tracking, inversion, use of dynamic feedback, and many others.

The survey is organized as follows. We start with some basic tools and notations from Differential Geometry. We discuss the fundamental concept of the Lie bracket, give its equivalent definitions, and illustrate it by examples. In Section 3. we discuss the problem of nonlinear controllability. We introduce the concepts of accessibility and strong accessibility and provide their geometric characterizations. Then we state two results on complete controllability. Finally, we introduce the notion of orbit and discuss its relations with the integrability of distributions. In Section 4. we briefly discuss nonlinear observability and provide a sufficient (and almost necessary) condition for local observability. The next Section is devoted to the problem of equivalence of nonlinear control systems. We discuss state-space equivalence and feedback equivalence. We also give necessary and sufficient conditions for state-space equivalence to a linear system. Section 6. is devoted to the feedback linearization problem. We give necessary and sufficient conditions for a nonlinear system to be feedback linearizable. We also discuss restricted feedback linearization, global aspects of the problem, and, finally, partial linearization. Section 7. is devoted to the decoupling problem. We discuss disturbance decoupling and input-output decoupling, and provide geometric conditions for their solvability: for the former problem they are based on the notion of controlled invariant distribution and for the latter on the notion of characteristic index. Throughout the paper we give many examples to illustrate the introduced concepts and given results, and to show their possible applications in engineering problems.

As clearly follows from the description of the contents, our survey is far from complete. We concentrate on basic system theoretic concepts (controllability, observability, equivalence, linearization, and decoupling) and we do not touch such important and well understood (due to geometric methods) problems and concepts as stabilizability, observers, zero dynamics, output tracking, dynamic feedback, disturbance attenuation, and many others. Fortunately, the interested reader may find detailed study of all those problems, and many others, in two very nice monographs on nonlinear geometric control theory: A. Isidori [2] and H. Nijmeijer and A. J. van der Schaft [3]. Our survey can be treated as a "kindergarten" of geometric control theory and we strongly encourage the interested reader to proceed from the kindergarten to a geometric control school provided by the two books. There he/she will find detailed study of the problems and examples we discuss here, as well as analysis of many others.

2. Basic Tools and Notations

We shall be dealing with two classes of nonlinear control systems, the class of general systems

$$\Pi: \quad \dot{x} = f(x, u), \quad x(t) \in X, \quad u(t) \in U, \tag{1}$$

and the class of control-affine systems

$$\Sigma: \quad \dot{x} = f(x) + \sum_{i=1}^{m} u_i g_i(x), \quad x(t) \in X, \quad u(t) = (u_1(t), \ldots, u_m(t))^T \in U, \tag{2}$$

where X is an open subset of $\mathbb{R}^n$ (or a differentiable manifold of dimension n) and U is an open subset of $\mathbb{R}^m$ (often $U = \mathbb{R}^m$). Control functions u belong to a class of admissible controls $\mathcal{U}$ which is assumed to contain all piecewise constant controls $\mathcal{PC}$ and to be contained in the class $\mathcal{M}$ of all measurable controls.

All considered objects: functions, vector fields, differential forms etc. are of class C^∞. We will denote smooth functions and smooth vector fields by $C^\infty(X)$ and $V^\infty(X)$, respectively. The blanket assumption of smoothness is very convenient when the apparatus of Differential Geometry is used. In most cases, the assumption of infinite differentiability is not essential, but introduced only to avoid having to count the degree of differentiability in a particular argument.

A vector field f on X is a mapping $X \ni p \mapsto f(p) \in T_pX$ which assigns a tangent vector $f(p)$ at p to any point p in X. Here T_pX is the tangent space of X at p, i.e., the linear space spanned by all tangent vectors at p. For example, if X is the sphere $S^2 = \{(x_1, x_2, x_3) \in \mathbb{R}^3 \mid x_1^2 + x_2^2 + x_3^3 = 1\}$, then T_pX is the tangent plane to the sphere at $p \in S^2 = X$. However, the definition of T_pX is independent of the coordinate system. Note that controlled differential equation (1) can be viewed as a of family of vector fields $\{f(x, u) \mid u \in U\}$ on X, corresponding to constant controls.

Consider a diffeomorphism $\Phi \colon X \longrightarrow X$, i.e., a smooth one-to-one mapping (bijection) with smooth inverse. It defines the following transformation of a vector field f (change of coordinates)

$$(\Phi_* f)(p) = d\Phi(q) f(q), \quad q - \Phi^{-1}(p),$$

where $d\Phi$ denotes the tangent map of Φ (the Jacobian mapping of Φ). We have

$$(\Phi \circ \Psi)_* f = \Phi_* \Psi_* f,$$

where $\circ$ denotes composition of maps.

All the notions are (differential) geometric in nature, i.e., their definition (and thus 'objective existence') is independent of the coordinate system used. The setting allows discovering their *intrinsic* properties, or properties *invariant* under changes of coordinates. Of course, concrete computations must be done in a (conveniently chosen) coordinate system.

In given local[a] coordinates $x = (x_1, \ldots, x_n)^T$ we write $f = (f_1, \ldots, f_n)^T$, where $f_i = f_i(x)$. For a vector field f let γ_t^f, or simply γ_t, denote the flow of f (the solution of the differential equation defined by f), i.e., $\frac{d}{dt}\gamma_t = f(\gamma_t)$, $\gamma_0 = x_0$.

Any vector field $f \in V^\infty(X)$ defines a linear operator in the space $C^\infty(X)$ by assigning to any $\phi \in C^\infty(X)$

$$(L_f\phi)(p) = \left.\frac{\partial}{\partial t}\right|_{t=0} \phi(\gamma_t(p)) = \sum_{i=1}^{n} f_i(p)\frac{\partial}{\partial x_i}\phi(p), \tag{3}$$

which is called the Lie derivative of ϕ along f. Observe that the second expression in the above formula gives an invariant definition while the third one provides a formula in local coordinates $x = (x_1, \ldots, x_n)^T$. One often identifies $f = L_f = \sum_{i=1}^n f_i \partial/\partial x_i$.

2.1. *Lie Bracket*

One of the crucial objects in nonlinear geometric control theory is that of the Lie bracket. A nonlinear control system can be considered as a collection of dynamical systems (vector fields) parametrized by a parameter called control. The Lie bracket allows studying interconnections between different dynamical systems in a coordinate independent way. The Lie bracket of two vector fields is another vector field which measures non-commutativeness of the flows of both vector fields. We give three equivalent definitions of the Lie bracket, each of them being useful for us.

We start with the easiest (but coordinate dependent) definition in $\mathbb{R}^n$. Let $X \subset \mathbb{R}^n$ and let $f, g \in V^\infty(X)$. The *Lie bracket* of f and g is another vector field on X

$$[f, g](p) = dg(p)f(p) - df(p)g(p), \tag{4}$$

where df and dg denote the Jacobian maps of f and g.

[a] Often the state-space of a nonlinear control system cannot be described by a single (global) coordinate system. For example, it is impossible to define a global coordinate system on the sphere S^2 in $\mathbb{R}^3$. One has to patch at least two local coordinate systems, e.g., stereographic projections from the northern and southern poles. Hence the notion of *local* coordinates.

Example 1 Let $f = (1, 0)^T$ and $g = (0, x_1)^T$ on $\mathbb{R}^2$ then $[f, g] = (0, 1)^T$. Note that the Lie bracket of f and g adds a new direction to the space spanned by f and g at the origin. ■

Example 2 Let $g = b$ be a constant vector field and $f = Ax$ be a linear vector field. Then $[f, g] = [Ax, b] = 0 - Ab = -Ab$. ■

For any $p \in X$ the Lie bracket $[f, g](p)$ is an element of T_pX (i.e., it is a tangent vector at p) but is $[f, g]$ a vector field which transforms with coordinate changes like a vector field? The following geometric definition of the Lie bracket

$$[f, g](p) = \frac{\partial}{\partial t}\bigg|_{t=0} d\gamma^f_{-t}(\gamma^f_t(p))\, g(\gamma^f_t(p)) = \frac{\partial}{\partial t}\bigg|_{t=0} ((\gamma^f_{-t})_* g)(p).$$

coincides with the previous one and thus gives an affirmative answer to the question.

Some basic properties of the Lie bracket are given by the following four statements.

Proposition 1 *If Φ is a diffeomorphism of X then $[\Phi_* f, \Phi_* g] = \Phi_*[f, g]$.*

This fundamental property implies that the Lie bracket has invariant (geometric) meaning: if we compute it in the 'old' coordinates and then transform the result to the 'new' ones (the term on the right) we get the same vector field as when making computations in the 'new' coordinates (the term on the left).

Proposition 2 *The Lie bracket of vector fields f and g is identically equal to zero if and only if their flows commute, i.e.,*

$$[f, g] \equiv 0 \Longleftrightarrow \gamma^f_t \circ \gamma^g_s(p) = \gamma^g_s \circ \gamma^f_t(p) \quad \forall s, t \in \mathbb{R}, \forall p \in X.$$

Proposition 3 *Let us fix a $p \in X$ and consider the curve*

$$\alpha(t) = \gamma^g_{-t} \circ \gamma^f_{-t} \circ \gamma^g_t \circ \gamma^f_t(p). \tag{5}$$

Then $\alpha'(0) = 0$ and $\alpha''(0) = 2[f, g](p)$.

Thus the tangent vector to the curve $\alpha(\cdot)$ at zero is equal to $2[f, g](p)$ which gives the third definition of the Lie bracket. It is extremely important in (nonlinear) Control Theory, because it implies that the points attainable from p by means of the vector fields f and g lie not only in the 'directions' $f(p)$ and $g(p)$, but also in the 'direction' of the Lie bracket $[f, g](p)$. Indeed, if the control system $\dot{x} = uf(x) + vg(x)$ is subject to the control $(u(t), v(t))$ equal successively to $(1, 0)$, $(0, 1)$, $(-1, 0)$ and $(0, -1)$, each on time interval of length t, then for $t \to 0$ we reach points lying "infinitesimally" in the direction of $[f, g](p)$.

Example 3 Consider linear vector fields in $\mathbb{R}^2$: $f(x) = (x_2, -x_1)^T$ and $g(x) = (-x_1, -2x_2)^T$. Their flows, i.e., solutions of $\dot{x} = g(x)$ and $\dot{x} = f(x)$ are:

$$\gamma^f_t(p) = \begin{pmatrix} p_1 \cos t + p_2 \sin t \\ -p_1 \sin t + p_2 \cos t \end{pmatrix} \qquad \gamma^g_t(p) = \begin{pmatrix} p_1 e^{-t} \\ p_2 e^{-2t} \end{pmatrix},$$

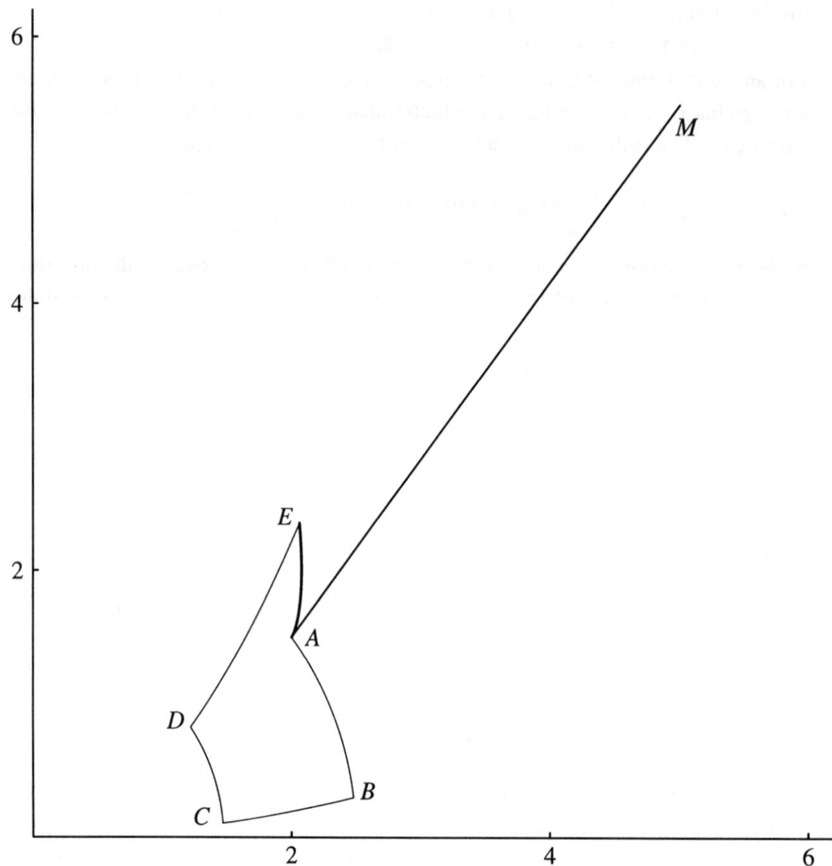

Figure 1: Illustration of Proposition 3 via simulations of the flows f, g of Example 3. Point A is the initial condition $p = (2, 1.5)^T$, arc AB corresponds to $\gamma_s^f(A)$, $0 \le s \le t$, $B = \gamma_t^f(A)$, arc BC to $\gamma_s^g(B)$, $0 \le s \le t$, $C = \gamma_t^g \circ \gamma_t^f(A)$, arc CD to $\gamma_s^{-f}(C)$, $0 \le s \le t$, $D = \gamma_t^{-f} \circ \gamma_t^g \circ \gamma_t^f(A)$, arc DE to $\gamma_s^{-g}(D)$, $0 \le s \le t$; finally, $E = \alpha(t) = \gamma_{-t}^g \circ \gamma_{-t}^f \circ \gamma_t^g \circ \gamma_t^f(p)$. The vector $\overrightarrow{AM} = \alpha''(0) = 2[f, g](p)$ is tangent to the curve $\gamma_{-s}^g \circ \gamma_{-s}^f \circ \gamma_s^g \circ \gamma_s^f(p)$, $0 \le s \le t$, given by the arc AE; see formula (5).

where p is the initial condition.

We have $[f, g] = (x_2, x_1)^T$ and

$$\gamma_t^{[f,g]}(p) = \begin{pmatrix} p_1 \cosh t + p_2 \sinh t \\ p_1 \sinh t + p_2 \cosh t \end{pmatrix}.$$

The interpretation of the Lie bracket according to (5) is shown in Figure 1.

Finally, for $\lambda_1 = 1$ and $\lambda_2 = 2$, we have

$$\gamma_t^{\lambda_2 g + \lambda_1 f}(p) = \begin{pmatrix} p_1(1+t)\mathrm{e}^{-3t} + p_2 t \mathrm{e}^{-3t} \\ -p_1 t \mathrm{e}^{-3t} + p_2(1-t)\mathrm{e}^{-3t} \end{pmatrix}. \quad \blacksquare$$

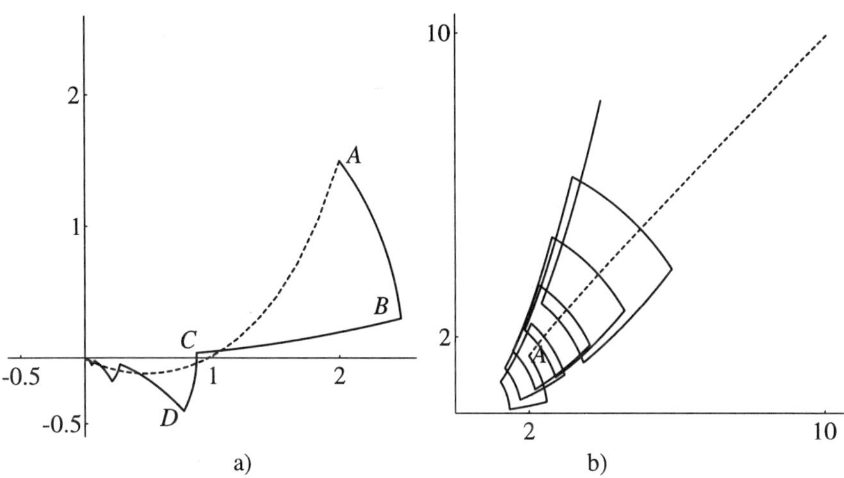

Figure 2: Illustration of Proposition 4 (vector fields f, g of Example 3): a) succession $AB \leftrightarrow \gamma^f_{\lambda_1 t}$, $BC \leftrightarrow \gamma^g_{\lambda_2 t}$, $CD \leftrightarrow \gamma^f_{\lambda_1 t}, \ldots$ of curves defined by ϕ_t with $\lambda_1 = 1$ and $\lambda_2 = 2$; the dotted line, starting at $A = p = (2, 1.5)^T$, is the arc of $\gamma_t^{\lambda_2 g + \lambda_1 f}(p)$; b) sequence of curves defined by ψ_t and the asymptotic trajectory $\gamma_{t^2}^{[f,g]}(p)$.

Proposition 4 *Let $f, g \in V^\infty(X)$, $p \in X$ and let $\lambda_1, \lambda_2 \in \mathbb{R}$. Put*

$$\phi_t = \gamma^g_{\lambda_2 t} \circ \gamma^f_{\lambda_1 t}, \qquad \psi_t = \gamma^g_{-t} \circ \gamma^f_{-t} \circ \gamma^g_t \circ \gamma^f_t.$$

Then for the curves

$$\begin{aligned} \alpha_k(t) &= \phi_{t/k} \circ \cdots \circ \phi_{t/k}(p), \quad k \text{ times}, \\ \beta_k(t) &= \psi_{\sqrt{t}/k} \circ \cdots \circ \psi_{\sqrt{t}/k}(p), \quad k^2 \text{ times}, \end{aligned}$$

we have the convergence

$$\alpha_k(t) \longrightarrow \gamma_t^{\lambda_2 g + \lambda_1 f}(p), \quad \text{and} \quad \beta_k(t) \longrightarrow \gamma_{t^2}^{[f,g]}(p) \quad \text{as } k \longrightarrow \infty.$$

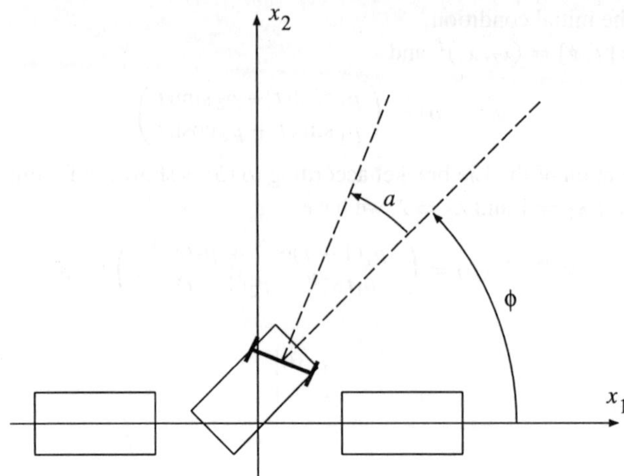

Figure 3: Illustration of Example 4: Car parking.

Example 4 (Car parking) We want to repark a car blocked by two others, see Figure 3. The kinematic movements of the car in coordinates (x, y, ϕ) can be described by the following two vector fields on $\mathbb{R}^2 \times (-\pi, \pi) \subset \mathbb{R}^3$

$$f = (r\cos(\phi + a), r\sin(\phi + a), b)^T, \quad g = (r\cos(\phi - a), r\sin(\phi - a), -b)^T.$$

Here a and r are positive constants depending on the geometry of the car: a is the angle between the axis of the car and the instantaneous direction of the movement of the geometric center of the car. We have

$$[f, g] = br(-\sin(\phi - a) - \sin(\phi + a), \cos(\phi - a) + \cos(\phi + a), 0)^T.$$

In particular, at $\phi = 0$ we have

$$[f, g] = (0, 2br\cos(a), 0)^T.$$

Therefore if we use the strategy given by β_k of Proposition 4, then infinitesimally we are able to move perpendicularly to the axis of the street. ■

Throughout this paper we will use the following notation for the 'iterated' Lie brackets: $\mathrm{ad}_f g = [f, g]$ and inductively $\mathrm{ad}_f^0 g = g$ and, for $i \geq 1$, $\mathrm{ad}_f^i g = [f, \mathrm{ad}_f^{i-1} g] = \underbrace{\mathrm{ad}_f \cdots \mathrm{ad}_f}_{i \text{ times}} g$.

3. Controllability

The aim of this Section is to discuss some nonlinear controllability results, both local and global. We also discuss relations of nonlinear controllability with the problem of integrability of distributions.

3.1. Accessibility

We consider two classes of control systems: the general systems Π given by (1) and the control-affine systems Σ given by (2). For controlled dynamics $f(x, u)$ we put $\gamma_t^u = \gamma_t^{f_u}$, where $f_u(\cdot) = f(\cdot, u)$.

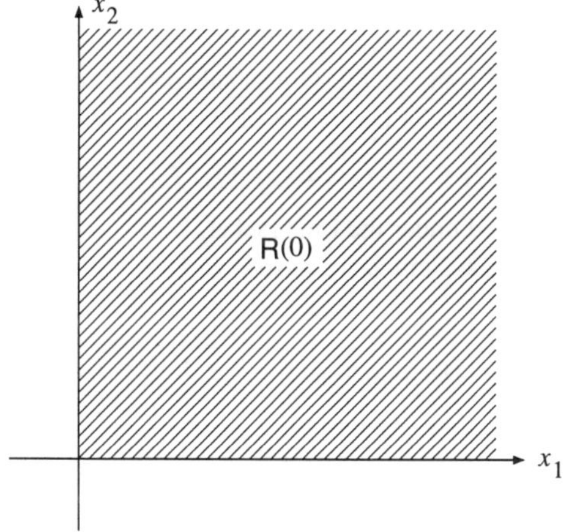

Figure 4: Reachable set R(0) from 0 for the system $\dot{x}_1 = u_1^2,\ \dot{x}_2 = u_2^2$ in $\mathbb{R}^2$.

Definition 1 *The reachable set from $x_0 \in X$ for system Π for the class of piecewise constant controls is the set*

$$\mathrm{R}(x_0) = \{\gamma_{t_k}^{u_k} \circ \cdots \circ \gamma_{t_1}^{u_1} x_0 \in X \mid \quad k \geq 1,\ u_1, \ldots, u_k \in U,\ t_1, \ldots, t_k \geq 0\}.$$

The set of above points with $t_1 + \cdots + t_k = t$ will be called the reachable set at time t from x_0 and denoted by $\mathrm{R}_t(x_0)$.

It is unreasonable to expect that the reachable set of a nonlinear control system will have, in general, a simple structure. Almost never will it be a linear subspace, even if $X = \mathbb{R}^n$ and $U = \mathbb{R}^m$. For example, for the system in the plane

$$\dot{x}_1 - u_1^2, \quad \dot{x}_2 - u_2^2$$

with $U = \mathbb{R}^2$ the reachable set from the origin is the positive orthant, see Figure 4. Therefore, our aim will be to establish qualitative properties of the reachable sets.

Definition 2 *The system* Π *is* accessible from x_0 *if* $\operatorname{int} \mathsf{R}(x_0) \neq \varnothing$. *The system* Π *is* strongly accessible from x_0 *if* $\operatorname{int} \mathsf{R}_t(x_0) \neq \varnothing$ *for any* $t > 0$.

Here int R means the interior of the set R.

We now define the following two families of vector fields: $\mathcal{F} = \{f_u\}_{u \in U}$ and $\mathcal{G} = \{f_u - f_v \mid u, v \in U\}$, where $f_u(\cdot) = f(\cdot, u)$. Observe that $\mathcal{F}$ gives the collection of all vector fields corresponding to all constant controls. The *Lie algebra of system* Π is the smallest linear space $\mathcal{L}$ of $V^\infty(X)$ which contains $\mathcal{F}$ and is closed under the Lie bracket:

$$f_1, f_2 \in \mathcal{L} \Rightarrow [f_1, f_2] \in \mathcal{L},$$

or equivalently

$$f_1 \in \mathcal{F}, \ f_2 \in \mathcal{L} \Rightarrow [f_1, f_2] \in \mathcal{L}.$$

The *Lie ideal of system* Π is the smallest linear subspace $\mathcal{L}_0$ of $V^\infty(X)$ which contains the family $\mathcal{G}$ and is closed under taking the Lie brackets with the elements of $\mathcal{F}$:

$$f_1 \in \mathcal{F}, \ f_2 \in \mathcal{L}_0 \Rightarrow [f_1, f_2] \in \mathcal{L}_0.$$

It follows that $\mathcal{L}_0$ is closed under the Lie bracket.

For the control-affine system Σ we have

$$\begin{aligned} \mathcal{L} &= \operatorname{Lie}\{f, g_1, \ldots, g_m\} \\ &= \operatorname{span}\{[g_{i_1}, \ldots, [g_{i_{k-1}}, g_{i_k}], \ldots] \mid \ k \geq 1, \ 0 \leq i_1, \ldots, i_k \leq m\}, \\ \mathcal{L}_0 &= \operatorname{span}\{[g_{i_1}, \ldots, [g_{i_{k-1}}, g_{i_k}], \ldots] \mid \ k \geq 1, \ 0 \leq i_1, \ldots, i_k \leq m, \ i_k \neq 0\}, \end{aligned}$$

where $g_0 = f$. Here Lie $\mathcal{A}$ stands for the Lie algebra generated by a family of vector fields $\mathcal{A}$, i.e., the smallest linear subspace containing $\mathcal{A}$ and closed under the Lie bracket.

Example 5 Consider the linear system

$$\Lambda: \qquad \dot{x} = Ax + Bu = Ax + \sum_{i=1}^{m} u_i b_i.$$

We have $g_1 = b_1, \ldots, g_m = b_m$ with $f = g_0 = Ax$ and

$$[Ax, b_i] = -Ab_i, \quad [Ax, [Ax, b_i]] = [Ax, -Ab_i] = A^2 b_i, \ldots,$$

$$\operatorname{ad}_{Ax} \cdots \operatorname{ad}_{Ax} b_i = \operatorname{ad}_{Ax}^j b_i = (-1)^j A^j b_i.$$

Therefore the Lie ideal $\mathcal{L}_0$ consists of constant vector fields only and we have

$$\begin{aligned} \mathcal{L}_0 &= \operatorname{span}\{A^j b_i \mid \ j \geq 0, \ 1 \leq i \leq m\} \\ &= \operatorname{span}\{A^j b_i \mid \ 0 \leq j \leq n-1, \ 1 \leq i \leq m\}. \end{aligned}$$

Moreover, $\mathcal{L} = \operatorname{span}\{Ax, \mathcal{L}_0\}$. ■

For any family of vector fields $\mathcal{H} \subset V^\infty(X)$ we put $\mathcal{H}(x) = \text{span}\,\{h(x)|\ h \in \mathcal{H}\}$. The two following results of Sussmann & Jurdjevic [4] and Krener [5] give fundamental accessibility properties.

Theorem 1 *Consider a smooth system* Π.

(i) If it satisfies $\dim \mathcal{L}(x_0) = n$, *then it is accessible from* x_0.

(ii) If Π *is analytic and* $\dim \mathcal{L}(x_0) < n$, *then it is not accessible from* x_0.

The analyticity assumption in statement (ii) cannot be dropped.

Example 6 Consider the system with the scalar control $u \in U = \mathbb{R}$

$$\dot{x}_1 = u, \quad \dot{x}_2 = x_1^k, \qquad k \geq 2.$$

The Taylor linearization of this system, at the equilibrium $x_0 = 0$ and $u_0 = 0$, is not controllable. We have $f = (0, x_1^k)^T$ and $g = (1, 0)^T$. Thus

$$[g, f] = (0, kx_1^{k-1})^T, \quad [g, [g, f]] = (0, k(k-1)x_1^{k-2})^T, \quad \text{ad}_g^k f = (0, k!)^T,$$

and so $\dim \mathcal{L}_0(x) = \dim \mathcal{L}(x) = 2$ for all x, in particular the system is strongly accessible from the origin. ■

There is an analogous relation between the Lie ideal $\mathcal{L}_0$ and the attainable set at time t.

Theorem 2 *Consider a smooth system* Π.

(i) If it satisfies $\dim \mathcal{L}_0(x_0) = n$ *then it is strongly accessible from* x_0, *i.e.,* $\text{int}\, \mathsf{R}_t(x_0) \neq \varnothing$ *for any* $t > 0$.

(ii) If Π *is analytic and* $\dim \mathcal{L}_0(x_0) < n$, *then* $\text{int}\, \mathsf{R}_t(x_0) = \varnothing$ *for any* $t > 0$.

Example 7 Consider the system on $\mathbb{R}^2$

$$\dot{x}_1 = 1, \quad \dot{x}_2 = u\, x_1^2,$$

and take $x_0 = (0, 0)$, and $U = \mathbb{R}$. We have

$$\mathcal{F} = \{(1, u\, x_1^2)^T |\quad u \in \mathbb{R}\}, \quad \mathcal{G} = \text{span}\,\{(0, x_1^2)^T\}.$$

The Lie algebra $\mathcal{L}$ contains the vector fields

$$f_1 = (1, 0)^T, \ \ f_2 = (1, x_1^2)^T, \ \ f_3 = [f_1, f_2] = (0, 2x_1)^T, \ \ [f_1, f_3] = (0, 2)^T.$$

Therefore $\dim \mathcal{L}(x_0) = 2$ and so the system is accessible from x_0. On the other hand $\mathcal{L}_0(x_0) = \text{span}\,\{(0, 1)^T\}$ and so the interior of the attainable set at time t, $t > 0$, is empty. In fact, $\mathsf{R}(x_0)$ is equal to the open right half plane including the origin and $\mathsf{R}_t(x_0) = \{\, x_1 = t, x_2 \in \mathbb{R}\}$. ■

Example 8 In this example we discuss accessibility of linear systems without constraints. Consider the linear system Λ of the form $\dot{x} = Ax + \sum_{i=1}^{m} u_i b_i = Ax + Bu$. We have (cf. Example 5) $\mathcal{L}_0(x) = \mathrm{Im}\,[B, AB, \ldots, A^{n-1}B]$ and $\mathcal{L}(x) = \mathrm{span}\,\{Ax, \mathcal{L}_0(x)\}$. Therefore the unconstrained linear system is strongly accessible from x if and only if the controllability matrix $[B, AB, \ldots, A^{n-1}B]$ is of rank n which means, according to the Kalman controllability rank condition, that the linear system is controllable.

Note, however, that an uncontrollable linear system may be accessible from x and this happens when $\dim \mathcal{L}(x) = n$ but $Ax \notin \mathrm{Im}\,[B, AB, \ldots, A^{n-1}B]$. ■

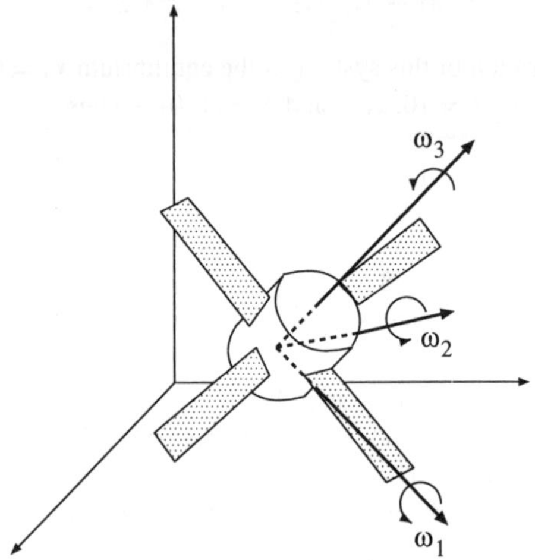

Figure 5: Illustration of Example 9: a spacecraft.

Example 9 Consider a spacecraft (see Figure 5) with two pairs of jets placed so that their angular momenta are parallel to principal axes of the spacecraft. Then the equations of motion for the angular velocities take the form

$$\begin{aligned} \dot{\omega}_1 &= a_1\omega_2\omega_3 + u_1, \\ \dot{\omega}_2 &= a_2\omega_3\omega_1 + u_2, \\ \dot{\omega}_3 &= a_3\omega_1\omega_2. \end{aligned}$$

We have

$$f = (a_1\omega_2\omega_3, a_2\omega_3\omega_1, a_3\omega_1\omega_2)^T, \quad g_1 = (1,0,0)^T, \quad g_2 = (0,1,0)^T$$

from which it follows that

$$[f, g_1] = -(0, a_2\omega_3, a_3\omega_2)^T, \quad [f, g_2] = -(a_1\omega_3, 0, a_3\omega_1)^T,$$

$$[g_1, [g_2, f]] = (0, 0, a_3)^T.$$

Therefore

$$\dim \mathcal{L}_0(x) = 3 \Longleftrightarrow \dim \mathcal{L}(x) = 3 \Longleftrightarrow a_3 \neq 0$$

for any $x = (\omega_1, \omega_2, \omega_3)$. Here $a_3 = (I_1 - I_2)/I_3$, where I_1, I_2, I_3 are the momenta of inertia along the principal axes. The system is accessible (equivalently, strongly accessible) if and only if the momenta of inertia of the space-craft along the axes with two pairs of jets are different. ■

3.2. *Complete Controllability*

In the previous subsection we considered accessibility which is a qualitative property of controllability meaning that the reachable set has nonempty interior. Of course, the 'ideal' situation occurs when the reachable set coincides with the whole state space. A control system is called *completely controllable* if $\mathsf{R}(x_0) = X$ for any $x_0 \in X$.

Of course, if a linear control system has the strong accessibility property then it satisfies the Kalman controllability condition (see Example 8) and thus it is completely controllable. For general nonlinear systems the property of complete controllability is quite rare. We show two classes of systems possessing it. The first is described by the following classical result of Chow and Rashevskii.

Theorem 3 *Let $\mathcal{F} = \{f_u\}_{u \in U}$ be a family of vector fields on X. If* $\dim \mathcal{L}(x) = n$ *for any $x \in X$ and the family is symmetric, i.e., for any $u \in U$ there exists $v \in U$ such that $f_u = -f_v$, then any two points of X can be joined piecewise by trajectories of the vector fields in $\mathcal{F}$.*

Example 10 Consider a control-linear system of the form

$$\Gamma\colon \quad \dot{x} = \sum_{i=1}^{m} u_i g_i(x), \qquad x(t) \in X, \quad u(t) = (u_1(t), \ldots, u_m(t))^T \in \mathbb{R}^m.$$

Of course, the family $\mathcal{G} = \{\sum_{i=1}^m u_i g_i | \quad u_i \in \mathbb{R}\}$ is symmetric and thus the control system Γ is completely controllable if $\dim \mathcal{L}(x) = n$, where $\mathcal{L}$ is the Lie algebra generated by the g_i's. Such control-linear systems (or equivalently distributions, as we will see in Section 3.3. below) are called completely non-holonomic, because of their interpretation in mechanics. ■

Recall that a point x is called *Poisson stable* for a vector field f on X if for any neighborhood V of this point and any $t > 0$ there is a $T > t$ such that $\gamma_T^f(x) \in V$. Periodic and almost periodic vector fields are examples of vector fields possessing Poisson stable points. The following result has been obtained by Bonnard and Crouch.[6,7]

Theorem 4 *If the set of Poisson stable points for the drift f of a control-affine system Σ is dense in X and the system satisfies* $\dim \mathcal{L}(x) = n$, *for any x in X then it is completely controllable.*

Example 11 (continuation of Example 9) In the case of a spacecraft with two pairs of jets placed along the principal axes of inertia the drift term has a dense set of Poisson stable points x and thus the spacecraft is completely controllable (for details, see [6,7]). ■

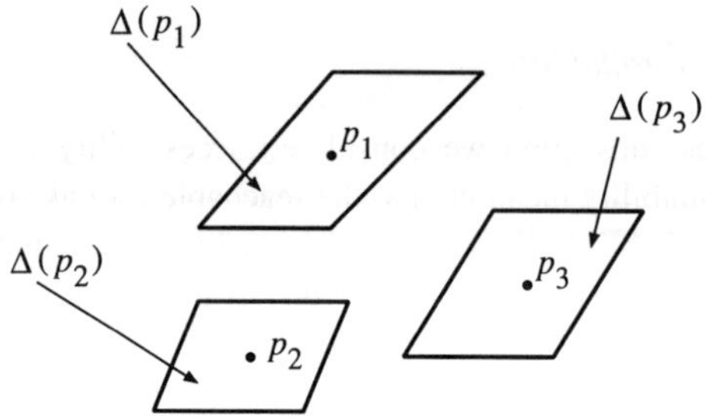

$\Delta(p_1)$ p_1 $\Delta(p_3)$ $\Delta(p_2)$ p_3 p_2

Figure 6: A two-dimensional distribution in $\mathbb{R}^3$, i.e., a field of planes in $\mathbb{R}^3$.

3.3. *Orbits*

In this Section we discuss the concept of orbit of a nonlinear control system and its relations with integrability of distributions. To start with recall that a *distribution on X* is a map Δ

$$p \mapsto \Delta(p) \subset T_pX$$

which assigns to each point p in X a subspace of the space of tangent vectors at this point, see Figure 6.

We say that $f \in \Delta$ if $f(p) \in \Delta(p)$ for all p in X. We will say that a distribution Δ is smooth if there exist smooth vector fields f_α, $\alpha \in A$, such that $\Delta(p) = \text{span}\,\{f_\alpha |\ \alpha \in A\}$. We will consider smooth distributions only.

A distribution Δ is called *involutive* if for any vector fields $f, g \in \Delta$ the Lie bracket is also in Δ, i.e., $[f, g] \in \Delta$.

The following is a local version of the classical theorem of Frobenius.

Theorem 5 *If Δ is an involutive distribution of dimension k on X, then, locally around any point in X, there exists a smooth change of coordinates which transforms Δ to* span $\{e_1, ..., e_k\}$, *where $e_1, ..., e_k$ are the constant vector fields* $e_i = (0, .., 0, 1, 0, .., 0)$, *with* 1 *at the i-th place.*

Definition 3 *The* orbit of a point $p \in X$ *of a family of vector fields* $\mathcal{F} = \{f_u\}_{u \in U}$ on *X is the set of points of X reachable from p piecewise by trajectories of vector fields in the family, i.e.,*

$$\text{Orb}\,(p) = \{\gamma_{t_k}^{u_k} \circ \cdots \circ \gamma_{t_1}^{u_1} | \quad k \geq 1,\ u_1, \ldots, u_k \in U,\ t_1, \ldots, t_k \in \mathbb{R}\}.$$

Observe that we allow time instants t_i to be positive as well as negative and thus the orbit of p consists of points 'reachable' from p going forward *and* backward in time.

The relation "q belongs to the orbit of p" is an equivalence relation on the space X which is a disjoint union of orbits (equivalence classes).

Let Γ be the smallest distribution on X which contains the family $\mathcal{F}$ (i.e., $f_u(p) \in \Gamma(p)$ for all $u \in U$) and is invariant under any flow γ_t^u, $u \in U$, that is,

$$g \in \Gamma \Longrightarrow (\gamma_t^u)_* g \in \Gamma, \quad \text{for any } u \in U \text{ and } t \in \mathbb{R}.$$

The following theorem was proved independently by Sussmann and Stefan.

Theorem 6 (Orbit Theorem) *Each orbit* $S = \text{Orb}\,(p)$ *of* $\mathcal{F} = \{f_u\}_{u \in U}$ *is an immersed submanifold. Moreover,* $T_pS = \Gamma(p)$ *for all* $p \in X$.

Corollary 1 *If the vector fields* f_u *are analytic, then* $T_pS = \mathcal{L}(p) = \{g(p) | \ g \in \mathcal{L}\}$. *In the smooth case the following inclusion holds* $\mathcal{L}(p) \subset \Gamma(p)$.

This result has the following very important consequence. If we fix an initial condition $p \in X$ then all trajectories starting from p and all trajectories ending at p are completely contained in the orbit Orb (p). Since Orb (p) has the structure of a submanifold we can always restrict our system to it. In the analytic case this restriction will possess the accessibility property from any point of the orbit.

In order to understand how the collection of all orbits fulfills the state space we need the following concept. A *foliation* $\{S_\alpha\}_{\alpha \in A}$ of dimension k of X is a partition $X = \cup_{\alpha \in A} S_\alpha$ of X into disjoint connected (immersed) submanifolds S_α, called *leaves*, which has the following property. For any $x \in X$ there exists a neighborhood U of x and a diffeomorphism $\Phi : U \longrightarrow V \subset \mathbb{R}^n$ onto an open subset V such that

$$\Phi((U \cap S_\alpha)_{cc}) = \{x = (x_1, \ldots, x_n) \in V | \quad x_{k+1} = c_\alpha^{k+1}, \ldots, x_n = c_\alpha^n\},$$

where P_{cc} denotes a connected component of the set P, see Figure 7.

A distribution Δ of constant dimension is *integrable* if there exists a foliation on X such that for any $p \in X$ we have $T_pS = \Delta(p)$, where S is the leaf passing through p. Finding the foliation is called integrating of the distribution, while the foliation and its leaves are called integral foliation and integral (sub)manifolds of the distribution.

We can now state a global version of the Frobenius Theorem.

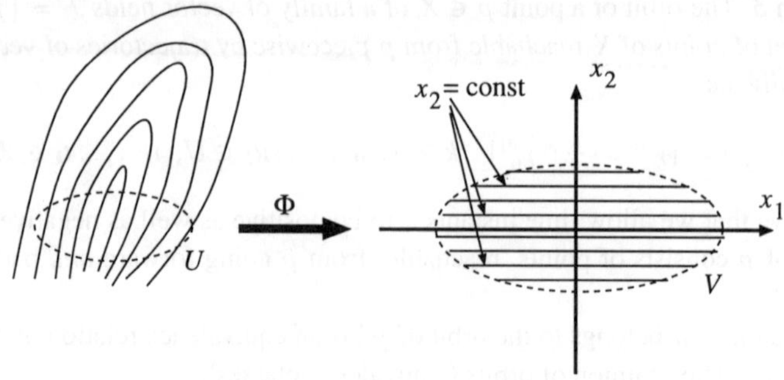

Figure 7: The idea of foliation.

Theorem 7 *A smooth distribution Δ of constant dimension is integrable if and only if it is involutive. The integral foliation of Δ is a partition of X into orbits of the family of vector fields $\{g \mid g \in \Delta\}$.*

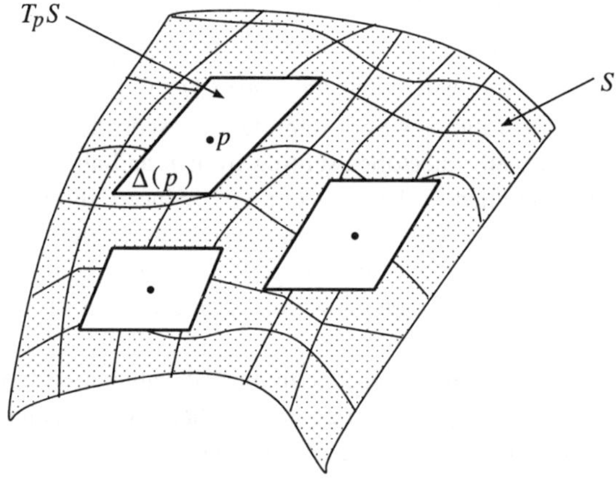

Figure 8: Integrable distribution; compare with Figure 6.

The global Frobenius Theorem implies that if all orbits are of constant dimension (in the analytic case if $\dim \mathcal{L}(x)$ is constant) then they form a foliation of the state space.

Example 12 Consider a spacecraft with one pair of jets:

$$\dot{\omega}_1 = a_1\omega_2\omega_3 + u$$

$$\begin{aligned}\dot{\omega}_2 &= a_2\omega_3\omega_1\\ \dot{\omega}_3 &= a_3\omega_1\omega_2.\end{aligned}$$

We have

$$f = (a_1\omega_2\omega_3, a_2\omega_3\omega_1, a_3\omega_1\omega_2)^T, \quad g = (1,0,0)^T,$$

$$[f,g] = -(0, a_2\omega_3, a_3\omega_2)^T = -(\omega_1)^{-1}f + (\omega_1)^{-1}a_1\omega_2\omega_3 g$$

and for the higher order Lie brackets:

$$[g,[f,g]] = 0, \quad [f,[f,g]] = (\phi,0,0)^T = \phi g,$$

where ϕ is a function.

Thus $\mathcal{L}(x) = \text{span}\,\{g(x), [f,g](x)\}$ and the system is never accessible: there is one 1-dimensional orbit and a continuum of 2-dimensional orbits. ■

4. Observability

In this Section we will consider very briefly the concept of nonlinear observability. As it is well known (e.g., see [8]) in the linear case the notions of controllability and observability are dual. We will show how that duality extends to the nonlinear case and will give a rank condition, which generalizes the Kalman observability condition, and which can be considered as a dual of the Lie rank condition for accessibility considered in Theorem 1.

Consider the class of nonlinear systems with outputs (measurements)

$$\Sigma_0: \quad \begin{aligned}\dot{x} &= f(x,u), & x(t)\in X,\ u(t)\in U,\\ y &= h(x), & y(t)\in Y.\end{aligned}$$

Here X and U are open subsets of $\mathbb{R}^n$ and $\mathbb{R}^m$ (or differentiable manifolds), respectively, $Y \subset \mathbb{R}^p$ and $h_i \in C^\infty(X)$, $i = 1,\ldots,p$, where $h = (h_1,\ldots,h_p)^T$. The class of admissible controls $\mathcal{U}$ is fixed and $\mathcal{PC} \subset \mathcal{U} \subset \mathcal{M}$, where $\mathcal{PC}$ denotes the class of piecewise constant controls and $\mathcal{M}$ the class of measurable controls.

For a fixed control $u \in \mathcal{U}$ and an initial condition $x(0) = p$ we denote by $y_{p,u}(t)$ the value of the output $y(\cdot)$ of system Σ_0 at time t (corresponding to the initial point p and the control u).

We call the system Σ_0 *observable* if for any two points $p_1, p_2 \in X$ there exists an admissible control $u \in \mathcal{U}$ and a time $t \geq 0$ such that $y_{p_1,u}(t) \neq y_{p_2,u}(t)$. The system is called *locally observable* at a point $p \in X$ if there exists a neighborhood V of p such that Σ_0 restricted to V is observable.

We would like to emphasize two features of nonlinear observability. Firstly, it is (like accessibility) a local concept. Indeed, local observability means that one distinguishes neighboring points by considering trajectories which stay close to the initial condition. Secondly, unlike in the linear case, controls play an important role. In general, there may exist controls which do not distinguish points; nevertheless the system can be observable if other controls distinguish, see Example 13 below.

Example 13 Consider the bilinear system

$$\begin{aligned} \dot{x}_1 &= x_2 u, \\ \dot{x}_2 &= 0, \\ y &= x_1 \end{aligned} \tag{6}$$

where $(x_1, x_2)^T \in \mathbb{R}^2$. This system is observable, because if we put $u(t) = 1$ we get an observable linear system. Notice, however, that the control $u(t) = 0$ does not distinguish x and $\bar{x}$ such that $x_1 = \bar{x}_1$ and $x_2 \neq \bar{x}_2$. ■

The *observation space* of Σ_0 is defined as

$$\mathsf{O} = \text{span}\,\{f_{u_k} \cdots f_{u_1} h_i | \quad 1 \leq i \leq p, k \geq 0, u_1, \ldots, u_k \in U\}$$

and its rank at q as

$$\dim \mathrm{d}\mathsf{O}(q) = \dim\,\{\mathrm{d}\phi(q), \phi \in \mathsf{O}\}.$$

The following result of Hermann and Krener [9] gives a dual of Theorem 1.

Theorem 8 *Assume that* Σ_0 *satisfies* $\dim \mathrm{d}\mathsf{O}(q) = n$, *then the system is locally observable at* q.

Example 14 Consider a linear control system with outputs of the form

$$\dot{x} = Ax + Bu, \quad y = Cx,$$

where $x \in \mathbb{R}^n$, $y \in \mathbb{R}^p$. We have

$$\mathrm{d}\mathsf{O}(x) = \mathrm{Im}\,[C^T, (CA)^T, \ldots, (CA^{n-1})^T].$$

Therefore the linear system is observable if the matrix

$$[C^T, (CA)^T, \ldots, (CA^{n-1})^T]$$

is of rank n. Observe that the rank depends neither on controls, nor on the initial condition. ■

Example 15 (continuation of Example 13) For the system (6) we have

$$\mathsf{O} = \text{span}\,\{x_1, x_2 u | \quad u \in \mathbb{R}\}$$

and thus $\dim\ \mathrm{d}\mathsf{O}(x) = 2$ for any $x \in \mathbb{R}^2$. ■

Proposition 5 *Assume that the control system* $\dot{x} = f(x, u)$ *is accessible on* X *(e.g., it satisfies the Lie algebra rank condition on* X*). Then* $\dim \mathrm{d}\mathsf{O}(q) = \text{const}$ *on* X*. Moreover, the system is locally observable at* q *if and only if* $\dim\ \mathrm{d}\mathsf{O}(q) = n$.

5. Equivalence of Control Systems

In this Section we consider equivalence of control systems. We start with state-space equivalence and then we define feedback equivalence.

5.1. *State-space Equivalence*

Two systems are state-space equivalent if they are related by a diffeomorphism (and then also their trajectories, corresponding to the same controls, are related by that diffeomorphism). A question of particular interest is when a nonlinear system is equivalent to a linear one. If this is the case, the nonlinearities of the considered system are not intrinsic, they appear because of a "wrong" choice of coordinates, and the nonlinear system shares all properties of its linear equivalent.

Consider a smooth nonlinear control system of the form

$$\Pi: \qquad \dot{x} = f(x, u), \quad x(t) \in X,\ u(t) \in U,$$

where X is an open subset of $\mathbb{R}^n$ (or an n-dimensional manifold) and U is an open subset of $\mathbb{R}^m$. $\mathcal{U}$ is the class of admissible controls (contained in the space of all U-valued measurable functions).

Consider another control system of the same form with the same control space U and the same class of admissible controls $\mathcal{U}$

$$\tilde{\Pi}: \qquad \dot{\tilde{x}} = \tilde{f}(\tilde{x}, u), \quad \tilde{x}(t) \in \tilde{X},\ u(t) \in U,$$

where $\tilde{X}$ is an open subset of $\mathbb{R}^n$ (or an n-dimensional manifold). Analogously to the transformation $\Phi_* g$ of a vector field $g(\cdot)$ by a diffeomorphism Φ, we define the transformation of $f(\cdot, u)$ by Φ. Put

$$(\Phi_* f)(\tilde{p}, u) = d\Phi(\Phi^{-1}(\tilde{p}))f(\Phi^{-1}(\tilde{p}), u).$$

We say that control systems Π and $\tilde{\Pi}$ are *state-space equivalent* (*locally state-space equivalent at points p and* $\tilde{p}$) if there exists a diffeomorphism $\Phi : X \to \tilde{X}$ (a *local* diffeomorphism $\Phi : X_0 \to \tilde{X}$, $\Phi(p) = \tilde{p}$, where X_0 is a neighborhood of p) such that $\Phi_* f = \tilde{f}$.

Recall that

$$\mathcal{F} = \{f_u |\ u \in U\} \quad \text{and} \quad \tilde{\mathcal{F}} = \{\tilde{f}_u |\ u \in U\},$$

where $f_u = f(\cdot, u)$ and $\tilde{f}_u = \tilde{f}(\cdot, u)$. (Local) state-space equivalence of Π and $\tilde{\Pi}$ means simply that $\Phi_* f_u = \tilde{f}_u$ for any $u \in U$, i.e., that Φ establishes correspondence between vector fields defined by constant controls. Assume $\dim \mathcal{L}(p) = \dim \tilde{\mathcal{L}}(\tilde{p}) = n$, i.e., Π and $\tilde{\Pi}$ are accessible from p and $\tilde{p}$, respectively. The following observation shows that (local) state-space equivalence is very natural.

Proposition 6 *Π and $\tilde{\Pi}$ are (locally) state-space equivalent if and only if there exists a (local) diffeomorphism Φ which (locally, in neighborhoods of p and $\tilde{p}$) preserves trajectories corresponding to the same controls $u \in \mathcal{U}$, i.e.,*

$$\Phi(\gamma_t^u(p)) = \tilde{\gamma}_t^u(\tilde{p}) \quad \text{for any } u \in \mathcal{U} ,$$

where $\gamma_t^u(p)$ (respectively $\tilde{\gamma}_t^u(\tilde{p})$) denotes the trajectory of Π (respectively $\tilde{\Pi}$) corresponding to the control function $u \in \mathcal{U}$.

Introduce the following notation for the left iterated Lie brackets

$$f_{[u_1 u_2 \ldots u_k]} = [f_{u_1}, [f_{u_2}, \ldots, [f_{u_{k-1}}, f_{u_k}] \ldots]]$$

and analogous for the tilded family. In particular $f_{[u_1]} = f_{u_1}$.

The following result result was established by Krener [10] (see also Sussmann [11]).

Theorem 9 *Assume that the systems Π and $\tilde{\Pi}$ are analytic and that* $\dim \mathcal{L}(p) = \dim \tilde{\mathcal{L}}(\tilde{p}) = n$. *Then the following holds.*

(i) The systems are locally state-space equivalent at p and $\tilde{p}$ if and only if there exists a linear isomorphism of the tangent spaces $F\colon T_pX \to T_{\tilde{p}}\tilde{X}$ such that

$$F f_{[u_1 u_2 \ldots u_k]}(p) = \tilde{f}_{[u_1 u_2 \ldots u_k]}(\tilde{p}) \tag{7}$$

for any $k \geq 1$ and any $u_1, \ldots u_k \in U$.

(ii) Assume, moreover, that X and $\tilde{X}$ are simply connected and that the Lie algebras $\mathcal{L}$ and $\tilde{\mathcal{L}}$ of Π and $\tilde{\Pi}$, respectively, consist of complete vector fields and satisfy the Lie rank condition everywhere. If there exist points $p \in X$ and $\tilde{p} \in \tilde{X}$ and a linear isomorphism $F\colon\ T_pX \to T_{\tilde{p}}\tilde{X}$ satisfying (7) then Π and $\tilde{\Pi}$ are state-space equivalent.

This theorem shows that all information concerning (local) behavior is contained in the values at the initial condition of the Lie brackets from $\mathcal{L}$. In a sense the (iterative) Lie brackets form invariant (higher order) derivatives of the dynamics of the system and in the analytic case they completely determine its local properties as (higher order) derivatives do for analytic functions.

Put $g_0 = f$. Using this theorem one obtains the following linearization result (compare [11,12]).

Proposition 7 *(i) An analytic control-affine system Σ at $p \in X$ is locally state-space equivalent to a linear controllable system of the form*

$$\Lambda_c\colon \qquad \dot{x} = Ax + c + Bu = Ax + c + \sum_{i=1}^{m} u_i b_i, \quad x \in \mathbb{R}^n, \quad u \in \mathbb{R}^m \tag{8}$$

at $x_0 \in \mathbb{R}^n$ if and only if

(E1) $[g_{i_1}, [g_{i_2}, \ldots [g_{i_{k-1}}, g_{i_k}] \ldots](p) = 0$ *for any* $k \geq 2$ *and any* $0 \leq i_j \leq m$, $1 \leq j \leq k$, *provided that at least two* i_j*'s are different from zero*

and

(E2) $\dim \operatorname{span} \{\mathrm{ad}_f^j g_i(p) | \quad 1 \leq i \leq m \,,\ 0 \leq j \leq n-1\}(p) = n.$

(ii) An analytic control-affine system Σ *is locally state-space equivalent at* $p \in X$ *to a linear controllable system of the form*

$$\Lambda: \qquad \dot{x} = Ax + Bu = Ax + \sum_{i=1}^{m} u_i b_i, \quad x \in \mathbb{R}^n, \ u \in \mathbb{R}^m \tag{9}$$

at $0 \in \mathbb{R}^n$ *if and only if* Σ *satisfies* (E1), (E2) *and* $f(p) = 0$.

(iii) An analytic system Σ *is state-space equivalent to a controllable linear system* Λ *if and only if it satisfies* (E1), (E2), *there exists* $p \in X$ *such that* $f(p) = 0$, X *is simply connected, and moreover*

(E3) *the vector fields* f *and* $g_1, \ldots, g_m$ *are complete.*

Recall that a vector field f is complete if the solution γ_t^f exists for any $t \in \mathbb{R}$.

5.2. *Feedback Equivalence*

The role of the concept of feedback in control cannot be overestimated and is very well understood, both in the linear and nonlinear cases. We would like to consider it as a way to transform nonlinear systems in order to achieve desired properties. When considering state-space equivalence the controls remain unchanged. The idea of feedback equivalence is to enlarge state-space transformations by allowing to transform controls as well and to transform them in a way which depends on the state: thus feeding the state back to the system.

Consider two general control systems Π and $\tilde{\Pi}$ given respectively by $\dot{x} = f(x, u)$, $x \in X$, $u \in U$ and $\dot{\tilde{x}} = \tilde{f}(\tilde{x}, \tilde{u})$, $\tilde{x} \in \tilde{X}$, $\tilde{u} \in \tilde{U}$. Assume that U ($\tilde{U}$) is an open subset of $\mathbb{R}^m$ ($\mathbb{R}^{\tilde{m}}$). We say that Π and $\tilde{\Pi}$ are *feedback equivalent* if there exists a diffeomorphism $\chi: X \times U \to \tilde{X} \times \tilde{U}$ of the form

$$\chi(x, u) = (\Phi(x), \Psi(x, u))$$

which transforms the first system to the second, i.e.,

$$d\Phi(x) f(x, u) = \tilde{f}(\Phi(x), \Psi(x, u)).$$

Observe that Φ plays the role of a coordinate change in X and Ψ, called feedback transformation, changes coordinates in the control space in a way which is state dependent.

When studying dynamical control systems with parameters (cf. the bifurcation theory) the situation is opposite: coordinate changes in the parameters space are state-independent, while coordinate changes in the state-space may depend on the parameters.

For control-affine case, i.e., for systems Σ of the form

$$\dot{x} = f(x) + \sum_{i=1}^{m} u_i g_i(x) = f(x) + g(x)u,$$

we will restrict feedback transformations to the control-affine ones

$$\tilde{u} = \Psi(x, u) = \tilde{\alpha}(x) + \tilde{\beta}(x)u,$$

where $\tilde{\beta}(x)$ is an invertible $m \times m$ matrix (in order to preserve the control-affine form of Σ). Denote the inverse feedback transformation by $u = \alpha(x) + \beta(x)\tilde{u}$; then feedback equivalence means that

$$\tilde{f} = \Phi_*(f + g\alpha) \quad \text{and} \quad \tilde{g} = \Phi_*(g\beta).$$

For control linear systems Γ of the form $\dot{x} = g(x)u = \sum_{i=1}^{m} g_i(x)u_i$, (local) feedback equivalence coincides with (local) equivalence of distributions $\mathcal{G}$ spanned by the vector fields g_i's.

6. Feedback Linearization

Since feedback transformations change dynamical behavior of a system, they are used to achieve its desired properties. In Sections 7.2. and 7.3. we will show how feedback transformations are used to synthesize controls with decoupling properties. In this Section we will study the problem of when a nonlinear system can be transformed to a linear form via feedback. The interest in this problem is two-fold. Firstly, if one is able to compensate nonlinearities by feedback then the modified system possesses all control properties of its linear equivalent and linear control theory can be used in order to study it and/or to achieve the desired control properties. This shows possible engineering applications of the feedback linearization. From the mathematical (or system theory) viewpoint, if one would like to classify nonlinear systems under feedback transformations (which define a group action on the space of all systems), then one of the most natural problems is to characterize those nonlinear systems which are feedback equivalent to linear ones.

6.1. Static Feedback Linearization

A general nonlinear control system Π given by (1) is *(locally) feedback linearizable* if it is (locally) feedback equivalent to a controllable linear system Λ_c of the form

$$\dot{x} = Ax + c + Bu.$$

Recall that $\mathcal{F} = \{f_u |\ u \in \mathbb{R}^m\}$ and $\mathcal{G} = \{f_{u_1} - f_{u_2} |\ f_{u_i} \in \mathcal{F},\ i = 1, 2\}$.

The following distributions will play a fundamental role in solving the feedback linearization problem:

$$\begin{aligned}
\Delta_1(x) &= \text{span}\, \mathcal{G}(x) \\
\Delta_2(x) &= \text{span}\, [\mathcal{F}, \Delta_1](x) = \text{span}\, \{[f_u, g](x) |\ f_u \in \mathcal{F},\ g \in \mathcal{G}\} \\
&\ \vdots \\
\Delta_j(x) &= \text{span}\, [\mathcal{F}, \Delta_{j-1}](x) \quad \text{(inductively)}.
\end{aligned}$$

Remark 1 One can show that $\Delta_1(x) = \text{span}_u\{\text{Im}\, \partial f(x, u)/\partial u\}$. ■

Remark 2 For the linear system Λ_c of (8) we have $\Delta_1 = \text{Im}\, B$ and, for $j \geq 0$, $\Delta_j = \text{Im}\, [B, \ldots, A^{j-1}B]$. ■

The feedback linearization problem was solved by Jakubczyk and Respondek [13], and independently by Hunt and Su [14].

Theorem 10 *A system* Π *given by (1) is locally feedback linearizable at* x_0 *if and only if in a neighborhood of* x_0 *it satisfies the following conditions:*

(A1) dim $\Delta_j(x) = \text{const}, \quad j = 1, \ldots, n,$

(A2) Δ_j *are involutive,* $\quad j = 1, \ldots, n,$

(A3) dim $\Delta_n(x_0) = n$.

In applications one is usually interested in the points of equilibria.

Corollary 2 *A system* Π *given by (1) is locally feedback equivalent at* x_0 *to a controllable system* Λ *(given by (9)) at* 0 *if and only if it satisfies the conditions* (A1), (A2), (A3) *of Theorem 10 and* $f(x_0) \in \Delta_1(x_0)$.

We now consider feedback equivalence of linear controllable multi-input systems Λ of the form $\dot{x} = Ax + Bu$ (in this case diffeomorphism $\Phi(x)$ and feedback $\Psi(x, u)$ are taken to be linear with respect to the state and control). As shown by Brunovsky [15] the only feedback invariants are the dimensions m_j of Im M^j, where the map M^j is defined as $[B, \ldots, A^{j-1}B]$. Put $n_0 = 0$ and $n_j = m_j - m_{j-1},\ j = 1, \ldots n$. If we define

$$\kappa_j = \max\{n_i |\ n_i \geq j\}, \tag{10}$$

then $\kappa_1 \geq \ldots \geq \kappa_m$ and $\sum_{i=1}^m \kappa_i = n$. The integers κ_i, called controllability (or Brunovsky) indices, form another set of complete invariants of feedback equivalence of linear controllable systems.

Every controllable system Λ given by (9) with indices $\kappa_i, i = 1, \ldots, m$, is feedback equivalent to the system (canonical form)

$$\begin{aligned}
\dot{x}_{i,j} &= x_{i,j+1}, \quad j = 1, \ldots, \kappa_i - 1, \\
\dot{x}_{i,\kappa_i} &= u_i,
\end{aligned} \tag{11}$$

which consists of m independent κ_i-fold integrators.

Very often we deal with control-affine systems. To state a feedback linearization result for them define the following distributions

$$\begin{aligned}\mathcal{D}^1(x) &= \operatorname{span}\{g_i(x) \mid 1 \leq i \leq m\} \\ \mathcal{D}^j(x) &= \operatorname{span}\{\operatorname{ad}_f^{q-1} g_i(x) \mid 1 \leq q \leq j,\ 1 \leq i \leq m\}, \quad j = 2, 3, \ldots\end{aligned}$$

If the dimensions $d_j(x)$ of $\mathcal{D}^j$ are constant (see (A1)' and (B1) below) we denote them by d_j and we define indices ρ_j as follows. Define $d_0 = 0$ and put $r_j = d_j - d_{j-1},\ j = 1, \ldots, n$. Also, define (cf. (10)) $\rho_j = \max\{r_i \mid r_i \geq j\}$.

Distributions Δ_j are feedback invariant. If, moreover, they are involutive, then in the control-affine case $\Delta_j = \mathcal{D}^j,\ j \geq 1$ and, in particular, ρ_j's are feedback invariant. In this case, the indices ρ_j coincide with κ_j, the controllability indices of the linear equivalent of Σ.

The following result describes linearizable control-affine systems Σ of the form (2).

Theorem 11 *The following three conditions are equivalent:*
(i) Σ is locally feedback linearizable at $x_0 \in \mathbb{R}^n$.
(ii) Σ satisfies in a neighborhood of x_0:
(A1)' $\dim \mathcal{D}^j(x) = \text{const},\ j = 1, \ldots, n,$
(A2)' $\mathcal{D}^j$ *are involutive*, $j = 1, \ldots, n,$
(A3)' $\dim \mathcal{D}^n(x_0) = n.$
(iii) Σ satisfies in a neighborhood of x_0:
(B1) $\dim \mathcal{D}^j(x) = \text{const},\ j = 1, \ldots, n,$
(B2) $\mathcal{D}^{\rho_j - 1}$ *are involutive*, $j = 1, \ldots, m,$
(B3) $\dim \mathcal{D}^{\rho_1}(x_0) = n$, *where ρ_1 is the largest controllability index.*

Corollary 3 *A scalar input system Σ is feedback linearizable if and only if it satisfies:*
(C1) $g(x_0), \ldots, \operatorname{ad}_f^{n-1} g(x_0)$ *are independent,*
(C2) $\mathcal{D}^{n-1}$ *is involutive.*

Especially simple is the planar case with scalar-control, i.e., $n = 2, m = 1$.

Corollary 4 *A planar system with scalar control, i.e., $n = 2,\ m = 1$, is locally feedback linearizable at x_0 if and only if g and $\operatorname{ad}_f g$ are independent at x_0.*

Example 16 Consider a pendulum (a rigid one-link manipulator) consisting of a mass m with control torque u, see Figure 9. The evolution of the pendulum is described by the Euler-Lagrange equation with external force

$$ml^2\ddot{\theta} + mgl \sin\theta = u.$$

We rewrite it as

$$\Sigma_1 : \quad \begin{aligned}\dot{\theta} &= \omega \\ \dot{\omega} &= -(g/l)\sin\theta + u/ml^2.\end{aligned}$$

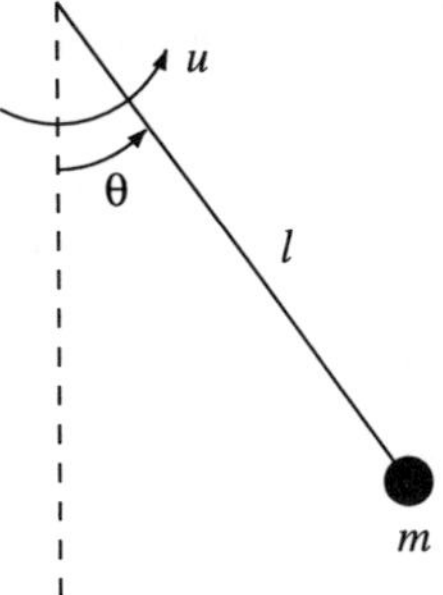

Figure 9: Illustration of Example 16: a one-link pendulum.

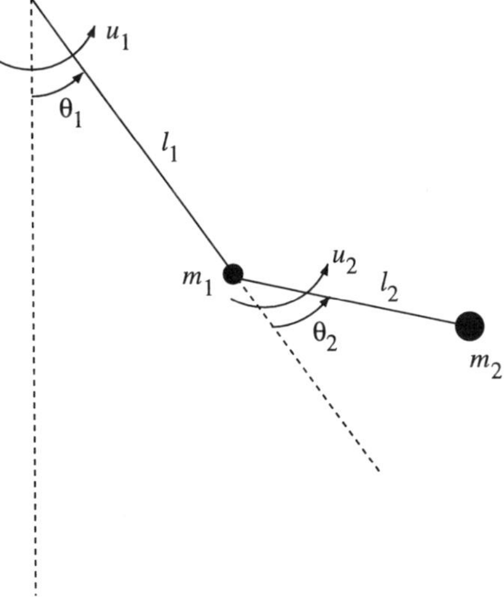

Figure 10: Illustration of Example 17: a two-link pendulum.

The natural state space of Σ_1 is $S^1 \times \mathbb{R}^1$, i.e., $(\theta, \omega) \in S^1 \times \mathbb{R}^1$, where S^1 stands for the unit circle.

Since g and $\mathrm{ad}_f g$ are independent everywhere, then Σ_1 is feedback linearizable: one can take, e.g., $u = mgl \sin\theta + ml^2\tilde{u}$ (no change of coordinates is needed) to get $\dot{\theta} = \omega, \ \dot{\omega} = \tilde{u}$. ■

Example 17 Consider the following *rigid two-link manipulator* (*double pendulum*)[16,3], see Figure 10 :

$$\begin{aligned} \dot{x}^1 &= x^2 \\ \dot{x}^2 &= -M^{-1}(x^1)(C(x^1, x^2) + k(x^1)) + M^{-1}(x^1)u, \end{aligned}$$

where $x^1 = (\theta_1, \theta_2), \ x^2 = (\dot{\theta}_1, \dot{\theta}_2), \ u = (u_1, u_2)$, and the positive definite symmetric matrix $M(x^1)$ is given by

$$\begin{pmatrix} m_1 l_1^2 + m_2 l_1^2 + m_2 l_2^2 + 2m_2 l_1 l_2 \cos\theta_2 & m_2 l_2^2 + m_2 l_1 l_2 \cos\theta_2 \\ m_2 l_2^2 + m_2 l_1 l_2 \cos\theta_2 & m_2 l_2^2 \end{pmatrix}.$$

The term $k(\theta)$ represents the gravitational force and the term $C(\theta, \dot{\theta})$ reflects the centripetal and Coriolis forces.

We see that $\mathcal{D}^1 = \mathrm{span}\,\{\partial/\partial x^2\} = \mathrm{span}\,\{\partial/\partial\dot{\theta}_1, \partial/\partial\dot{\theta}_2\}$ is involutive and also $\dim \mathcal{D}^2(x^1, x^2) = 4$; hence the doubled pendulum is feedback linearizable. A linearizing feedback is given, e.g., by $u = C(x^1, x^2) + k(x^1) + M(x^1)\tilde{u}$. ■

Example 18 Consider the following model of a permanent magnet stepper motor [17]

$$\begin{aligned} \dot{x}_1 &= -K_1 x_1 + K_2 x_3 \sin(K_5 x_4) + u_1 \\ \dot{x}_2 &= -K_1 x_2 + K_2 x_3 \cos(K_5 x_4) + u_2 \\ \dot{x}_3 &= -K_3 x_1 \sin(K_5 x_4) + K_3 x_2 \cos(K_5 x_4) - K_4 x_3 + K_6 \sin(4K_5 x_4) - \tau_L/J \\ \dot{x}_4 &= x_3, \end{aligned}$$

where x_1, x_2 denote currents, x_3 denotes the rotor speed and x_4 its position, J is the rotor inertia, and τ_L is the load torque. We see that if $x_{40} \neq 0$ the distributions $\mathcal{D}^1 = \mathrm{span}\,\{\partial/\partial x_1, \partial/\partial x_2\}$ and $\mathcal{D}^2 = \mathrm{span}\,\{\partial/\partial x_1, \partial/\partial x_2, \partial/\partial x_3\}$ are involutive and that $\dim \mathcal{D}^3(x) = 4$ and thus the system is feedback linearizable. ■

Example 19 Consider the following model of unicycle, i.e., a one-wheel vehicle driven by pedals (see Figure 11):

$$\begin{aligned} \dot{x}_1 &= u_1 \cos\theta \\ \dot{x}_2 &= u_2 \sin\theta \\ \dot{\theta} &= u_2, \end{aligned}$$

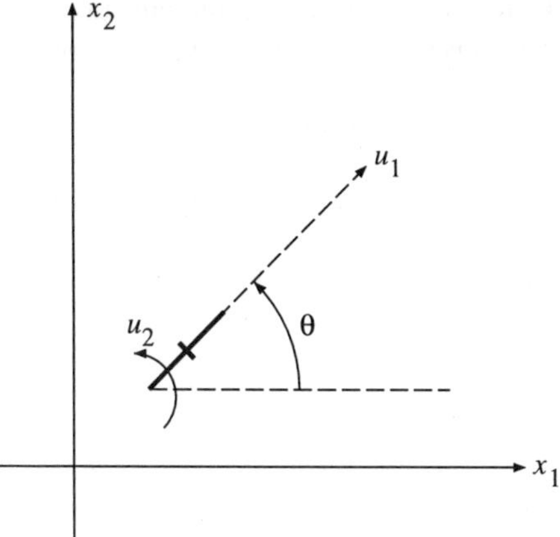

Figure 11: Illustration of Example 19: a unicycle.

where $(x_1, x_2, \theta) \in \mathbb{R}^2 \times S^1$. We have

$$g_1 = (\cos\theta, \sin\theta, 0)^T, \quad g_2 = (0, 0, 1)^T$$

thus $[g_1, g_2] = (\sin\theta, -\cos\theta, 0)^T$ and hence $\mathcal{D}^1$ is not involutive: the unicycle is not static feedback linearizable. ■

6.2. *Restricted Feedback Linearization*

Consider a control-affine system Σ and a feedback transformation $u = \alpha(x) + \beta(x)\tilde{u}$ which can be interpreted as an (affine) change of coordinates (depending on the state) in the input space. The term β allows to choose generators of the distribution $\mathcal{D}^1$, whereas the term α changes the drift f. Restricted feedback allows to transform the drift f only and keeps the g_i's unchanged. More precisely, two control affine systems Σ and $\tilde{\Sigma}$ are *restricted feedback equivalent* if there exists a diffeomorphism Φ between their state spaces and a restricted feedback of the form $u = \alpha(x) + \tilde{u}$ such that

$$\tilde{f} = \Phi_*(f + g\alpha) \quad \text{and} \quad \tilde{g}_i = \Phi_* g_i, \; i = 1, \ldots, m.$$

We will be interested in equivalence to linear controllable systems under such feedback and we will call it *restricted feedback linearization.*

The three main reasons to discuss restricted feedback linearization are as follows. Firstly, it was Brockett's restricted feedback linearization result [18] which began an increasing interest in various kinds of feedback linearization problems for nonlinear systems. Secondly, there is a nice stochastic interpretation of the restricted feedback linearization [18]. Thirdly, it is relatively easy, as we will show it, to proceed from local results to global ones.

We consider scalar input systems of the form

$$\Sigma: \quad \dot{x} = f(x) + g(x)u, \quad x(t) \in X, \ u(t) \in \mathbb{R},$$

and study their equivalence with linear scalar input systems of the form

$$\Lambda_c: \quad \dot{\tilde{x}} = A\tilde{x} + c + b\tilde{u}, \quad \tilde{x}(t) \in \mathbb{R}^n, \ \tilde{u}(t) \in \mathbb{R}.$$

We have the following result [18].

Theorem 12 *Σ is locally restricted feedback linearizable at x_0 if and only if it satisfies in a neighborhood of x_0 the following conditions:*

(RC1) $g(x_0), \ldots, \mathrm{ad}_f^{n-1} g(x_0)$ *are independent,*

(RC2) $[\mathrm{ad}_f^q g, \mathrm{ad}_f^r g] \subset \mathcal{D}^{n-2}$ *for any* $0 \le q, r \le n-1$.

Remark 3 As in the case of feedback linearization (see Corollary 2), Σ is restricted feedback equivalent at x_0 to Λ_c, with $c = 0$, at 0 if and only if it satisfies (RC1), (RC2) and $f(x_0) \in \mathcal{D}^1(x_0)$. ■

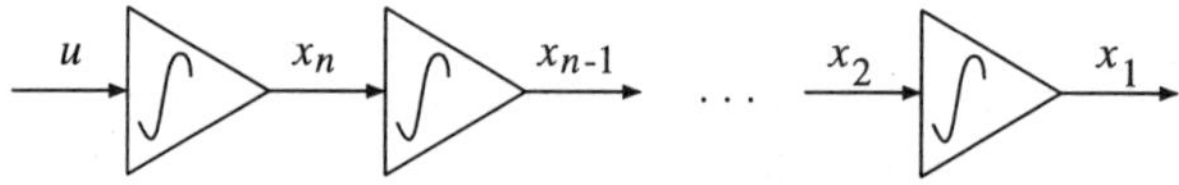

Figure 12: Canonical form for the single-input case.

In the single-input case all linearizable systems are equivalent to the canonical form (cf. (11))

$$\begin{aligned} \dot{x}_i &= x_{i+1}, \quad i = 1, \ldots, n-1, \\ \dot{x}_n &= u, \end{aligned} \tag{12}$$

illustrated in Figure 12.

If Σ is (locally) feedback linearizable, then there are many pairs (α, β) and many (local) diffeomorphisms which transform Σ into its canonical form. However, if we allow for restricted feedback only, then α transforming Σ into the canonical form is unique and is given by

$$\alpha = (-1)^{n-1} L_f^{n-1} \gamma_n, \tag{13}$$

where L_f stands for the Lie derivative along f and the smooth function γ_n is uniquely defined by $f = \sum_{i=1}^n \gamma_i \mathrm{ad}_f^{i-1} g$. This observation is crucial for establishing the following result on restricted feedback linearization [19,20].

Theorem 13 *Σ is restricted globally feedback linearizable if and only if it satisfies the conditions* (RC1), (RC2) *of Theorem 12 and*

(RC3) *the vector fields $\tilde{f}$ and g are complete, where $\tilde{f} = f + g\alpha$ and α is defined by (13),*

(RC4) *the state-space X is simply connected.*

Example 20 (Continuation of Example 16) We have

$$f = \omega\frac{\partial}{\partial\theta} - (g/l)\sin\theta\frac{\partial}{\partial\omega} \quad \text{and} \quad g = (1/ml^2)\frac{\partial}{\partial\omega}.$$

Therefore $[\mathrm{ad}_f g, g] = 0$ and since g and $\mathrm{ad}_f g$ are independent everywhere then the system is restricted feedback linearizable. Indeed, it is immediate to see that the feedback $u = mgl\sin\theta + \tilde{u}$ brings the system to a linear form (no action of diffeomorphism is needed). ■

Pendulum is globally equivalent to a linear system evolving not on $\mathbb{R}^2$ but on $S^1 \times \mathbb{R}^1$. If we enlarge the class of linear systems to include systems of the form $\dot{x} = Ax + Bu$, where every component x_i of x is either a global coordinate on $\mathbb{R}^1$ or a global coordinate (angle) on S^1, then Theorem 13 remains true if we drop the assumption (RC4). This includes many mechanical control systems.

6.3. *Partial Linearization*

The linearizability conditions are restrictive (except for the scalar input affine systems on the plane). Given a nonlinearizable system it is therefore natural to ask what is its largest linearizable subsystem. Consider a partially linear system Λ_p of the form

$$\begin{aligned} \dot{\tilde{x}}_1 &= A\tilde{x}_1 + c + \sum_{i=1}^{m}\tilde{u}_i b_i \\ \dot{\tilde{x}}_2 &= \tilde{f}_2(\tilde{x}_1, \tilde{x}_2) + \sum_{i=1}^{m}\tilde{u}_i\tilde{g}_{2,i}(\tilde{x}_1, \tilde{x}_2). \end{aligned}$$

Recall that the Lie ideal $\mathcal{L}_0$ is defined as $\mathcal{L}_0 = \mathrm{Lie}\,\{\mathrm{ad}_f^q g_i \mid 1 \le i \le m,\ j \ge 0\}$. With the help of $\mathcal{L}_0$ we define another Lie ideal by putting $\mathcal{L}^2 = [\mathcal{L}_0, \mathcal{L}_0]$. Denote by $\mathcal{E}$ the distribution spanned by $\mathcal{L}^2$. It is $\mathcal{L}^2$ which contains all intrinsic nonlinearities not removable by action of diffeomorphisms [21].

Theorem 14 *(i) If Σ is locally state-space equivalent at x_0 to a partially linear Λ_p, then $\dim\mathcal{E}(x) < n$ in a neighborhood of x_0.*

(ii) Assume that Σ satisfies $\dim\mathcal{L}_0(x_0) = n$ and that $\dim\mathcal{E}(x) = e = \mathrm{const}$ in a neighborhood of x_0. Then Σ is locally state-space equivalent to a partially linear Λ_p, such that $\dim\tilde{x}_1 = n - e$ and the linear subsystem is controllable.

Corollary 5 *Let an analytic system Σ satisfy $\dim \mathcal{L}_0(x_0) = n$. It is locally state-space equivalent at x_0 to a partially linear Λ_p if and only if $\dim \mathcal{E}(x_0) < n$. Moreover, there exists Λ_p, with an $(n-e)$-dimensional linear controllable subsystem, which is state-space equivalent to Σ.*

We now consider the problem of transforming a nonlinear system to a partially linear one via feedback. This problem has been studied and solved in the scalar-input case in [22] and in the multi-input case in [23] and [21]. Recall that for a smooth distribution $\mathcal{D}$ we denote by $\overline{\mathcal{D}}$ its involutive closure, i.e., the smallest distribution containing $\mathcal{D}$ and closed under the Lie bracket. Consider a scalar input system Σ.

Theorem 15 *(i) If Σ is locally feedback equivalent at x_0 to a partially linear Λ_p with ρ-dimensional linear controllable subsystem then Σ satisfies the following conditions*

(PC1) $g(x_0), \ldots, \mathrm{ad}_f^{\rho-1}(x_0)$ *are independent,*

(PC2) $\dim \overline{\mathcal{D}}^{\rho-1}(x) < n$ *in a neighborhood of* x_0,

(PC3) $\mathrm{ad}_f^{\rho-1} g(x_0) \notin \overline{\mathcal{D}}^{\rho-1}(x_0)$.

(ii) Assume that there exists ρ such that $\dim \overline{\mathcal{D}}^{\rho-1}(x) = \mathrm{const}$ and that (PC1), (PC2), (PC3) *are satisfied. Then Σ is locally feedback equivalent to a partially linear system Λ_p with a ρ-dimensional linear controllable subsystem. Moreover, the largest ρ satisfying the above conditions gives the largest dimension of the linear subsystem among all possible partial linearizations.*

Example 21 Consider a symmetric spacecraft (two inertia momenta are equal) with one pair of jets

$$\begin{aligned} \dot\omega_1 &= a\omega_2\omega_3 + e_1 u \\ \dot\omega_2 &= -a\omega_1\omega_3 + e_2 u \\ \dot\omega_3 &= e_3 u. \end{aligned}$$

We have

$$g = (e_1, e_2, e_3)^T \quad \mathrm{ad}_f g = a(e_2\omega_3 + e_3\omega_2, -e_1\omega_3 - e_3\omega_1, 0)^T$$

and $[g, \mathrm{ad}_f g] = 2ae_3(e_2, -e_1, 0) \in \mathcal{D}^2 = \mathrm{span}\,\{g, \mathrm{ad}_f g\}$ if and only if $e_3 = 0$ or $e_1 = e_2 = 0$, i.e., the symmetric spacecraft is controlled in a symmetric way: the angular momentum of the jet is parallel to the third principal axis. In all cases the system is not linearizable (either the orbit is not three-dimensional or $\mathcal{D}^2$ is not involutive) but it contains a two-dimensional linear subsystem for an open and dense set of initial conditions. ■

7. Decoupling

In this Section we show how static feedback allows to transform the dynamics of a nonlinear system in order to achieve the desired decoupling properties. In Section 7.2.

we consider disturbance decoupling while in Section 7.3. we deal with input-output decoupling. We start with the crucial concept of invariant distributions.

7.1. Invariant Distributions

Consider a smooth nonlinear control system with outputs of the form

$$\Sigma_0: \quad \begin{array}{rcl} \dot{x} & = & g_0(x) + \sum_{i=1}^m g_i(x) = g_0(x) + g(x)u \\ y & = & h(x), \end{array}$$

where $x(t) \in X$, $u(t) \in \mathbb{R}^m$, $y(t) \in \mathbb{R}^p$.

A distribution $\mathcal{D}$ is called *invariant* for Σ_0 if $[g_i, \mathcal{D}] \subset \mathcal{D},\ i = 0, \dots, m$. If a distribution is not invariant under Σ_0 it may become invariant under a suitable feedback modification. A distribution $\mathcal{D}$ is called *controlled invariant* if there exists an invertible feedback of the form

$$u = \alpha(x) + \beta(x)\tilde{u}, \quad \beta(\cdot) \text{ invertible},$$

such that $\mathcal{D}$ is invariant under the feedback modified dynamics $\dot{x} = \tilde{g}_0(x) + \sum_{i=1}^m \tilde{g}_i(x)$, i.e.,

$$[\tilde{g}_i, \mathcal{D}] \subset \mathcal{D}, \quad i = 0, \dots, m,$$

where $\tilde{g}_0 = g_0 + g\alpha$ and $\tilde{g} = g\beta$.

Example 22 In the case of a linear system of the form

$$\Lambda: \quad \dot{x} = Ax + Bx, \quad x \in \mathbb{R}^n,\ u \in \mathbb{R}^m,$$

a subspace $V \subset \mathbb{R}^n$ is said to be invariant under Λ if $AV \subset V$. We say that V is controlled invariant (or (A, B)-invariant) if there exists a linear feedback of the form $u = Fx + G\tilde{u}$ such that $(A + BF)V \subset V$. Observe that in the linear case (A, B)-invariance does not depend on G, so one can take $G = Id$ or G-noninvertible.

One can check by a direct calculation that (A, B)-invariance is equivalent to

$$AV \subset V + \operatorname{Im} B. \tag{14}$$

We refer to [8] for an extensive treatment of the concept of invariance in the linear case.
■

Put $\mathcal{G} = \text{span}\,\{g_1, \dots, g_m\}$. In the nonlinear case controlled invariance implies the following property of *local controlled invariance* (cf. (14)) $[g_i, \mathcal{D}] \subset \mathcal{D} + \mathcal{G},\ i = 0, \dots, m$. For involutive distributions the converse holds locally under regularity assumptions (see [24,25,26]).

Proposition 8 *Assume that the distributions $\mathcal{D}$, $\mathcal{G}$ and $\mathcal{D} \cap \mathcal{G}$ are nonsingular. If $\mathcal{D}$ is involutive and locally controlled invariant then locally it is controlled invariant.*

7.2. Disturbance Decoupling

In this Section we apply the concept of controlled invariant distributions to solve the nonlinear disturbance decoupling problem. Consider the following nonlinear system affected by disturbances d (which are assumed to be bounded functions of time)

$$\Sigma_d: \quad \begin{array}{rcl} \dot{x} &=& g_0(x) + \sum_{i=1}^{m} u_i g_i(x) + \sum_{i=1}^{k} d_i q_i(x) \\ y &=& h(x), \end{array}$$

where $x(t) \in X$, $u(t) \in \mathbb{R}^m$, $y(t) \in \mathbb{R}^p$ and $d(t) \in \mathbb{R}^k$. All data are smooth, i.e., $f, g_0, g_1, \ldots, g_m \in V^\infty(X)$, $h_i \in C^\infty(X)$, where $h = (h_1, \ldots, h_p)^T$, and the disturbance vector fields $q_1, \ldots, q_k \in V^\infty(X)$.

We say that the disturbance decoupling problem (DDP) is solvable, if there exists an invertible feedback of the form $u = \alpha(x) + \beta(x)\tilde{u}$ such that the output $y(t) = h(x(t))$ of the feedback modified system $\dot{x} = \tilde{g}_0(x) + \sum_{i=1}^{m} \tilde{u}_i \tilde{g}_i(x) + \sum_{i=1}^{k} d_i q_i(x)$ does not depend on the disturbances $d(t)$.

Put $\mathcal{Q} = \text{span}\,\{q_1, \ldots, q_k\}$. The following result has been proved by Isidori *et al* [25].

Theorem 16 *If DDP is solvable then there exists an involutive controlled invariant distribution $\mathcal{V}$ such that $\mathcal{Q} \subset \mathcal{V} \subset \operatorname{Ker} dh$.*

This result suggests the following approach to DDP. Look for the maximal controlled invariant distribution in Ker dh and check whether it contains $\mathcal{Q}$. However, in general such a maximal distribution may not exist. Moreover, even if it exists and contains disturbance vector fields, it is not necessarily true that DDP is solvable. On the other hand, there always exists the maximal locally controlled invariant distribution $\mathcal{V}^*$ in Ker dh and leads to the following solution of DDP (see [24] and [25]).

Theorem 17 *Assume that $\mathcal{V}^*$, $\mathcal{V}^* \cap \mathcal{G}$ and $\mathcal{G}$ are nonsingular. If $\mathcal{Q} \subset \mathcal{V}^*$ then DDP is solvable.*

Example 23 Consider the linear system with disturbances

$$\Lambda_d: \quad \begin{array}{rcl} \dot{x} &=& Ax + Bx + Ed \\ y &=& Cx, \end{array}$$

where $x \in \mathbb{R}^n$, $u \in \mathbb{R}^m$, $y \in \mathbb{R}^p$, $d \in \mathbb{R}^k$, and d denotes disturbances. DDP is solvable if and only if Im $E \subset V^*$, where V^* is the largest controlled invariant subspace in Ker C (see [8]). ■

Example 24 Consider a particle of unit mass moving on the surface of a cylinder according to a potential force given by the potential function V (see [3])

$$\begin{array}{rclrcl} \dot{q}_1 &=& p_1 & \dot{q}_2 &=& p_2 \\ \dot{p}_1 &=& -(\partial V/\partial q_1)(q_1, q_2) + u & \dot{p}_2 &=& -(\partial V/\partial q_2)(q_1, q_2) + d, \end{array}$$

where $(q_1, q_2, p_1, p_2) \in T(S^1 \times \mathbb{R})$. Let the output be given as $y = q_1$. One can see that $\mathcal{V}^* = \text{span}\,\{\partial/\partial q_2, \partial/\partial p_2\}$. Moreover, $\partial/\partial p_2 \in \mathcal{V}^*$ and DDP is solvable by the feedback $\tilde{u} = -(\partial V/\partial q_1)(q_1, q_2) + u$. ■

7.3. *Input-output Decoupling*

Consider a smooth nonlinear control system with outputs of the form

$$\Sigma_0: \quad \begin{aligned} \dot{x} &= f(x) + \textstyle\sum_{i=1}^m u_i g_i(x) \\ y &= h(x), \end{aligned}$$

where $x(t) \in X$, $u(t) \in \mathbb{R}^m$, $y(t) \in R^p$. We say that the input-output noninteracting problem is solvable for Σ_0 if there exists an invertible feedback of the form $u = \alpha(x) + \beta(x)\tilde{u}$ such that the feedback modified system $\dot{x} = \tilde{f}(x) + \sum_{i=1}^m \tilde{u}_i \tilde{g}_i(x),\ y = h(x)$, where $\tilde{f} = f + g\alpha,\ \tilde{g} = g\beta$, satisfies

$$y_i^{(k_i)} = \tilde{u}_i, \quad i = 1, \dots, p \tag{15}$$

for suitable integers k_i. Observe that we assume that the input-output map of the modified system is linear. Therefore there is no loss of generality in assuming the form (15), because if the transfer matrix of the input-output response is diagonal (which is the usual definition of noninteracting) we can always achieve (15) by applying a suitable linear feedback.

For each output channel define its *relative degree (characteristic number)* ρ_i to be the smallest integer such that for any neighborhood V of x_0 we have $L_{g_j} L_f^{\rho_i - 1} h_i(x) \neq 0$ for some $1 \leq j \leq m$ and for some $x \in V$. Denote by $D(x)$ the $(p \times m)$ decoupling matrix whose (i, j)-entry is $L_{g_j} L_f^{\rho_i - 1} h_i(x)$.

Theorem 18 *(i) The system Σ_0 is input-output decouplable at x_0 via an invertible feedback of the form $u = \alpha(x) + \beta(x)\tilde{u}$ if and only if* rank $D(x_0) = p$.

(ii) Moreover, the feedback which yields $y_i^{(k_i)} = \tilde{u}_i$, where $k_i = \rho_i,\ i = 1, \dots, p$, is given by $\tilde{u}_i = L_f^{\rho_i} h_i + \sum_{j=1}^m L_{g_j} L_f^{\rho_i - 1} h_i u_j$.

Remark 4 The above-defined feedback renders $\mathcal{V}^*$, the maximal controlled invariant distribution in Ker dh, invariant. ■

Example 25 Consider the following rigid two-link manipulator, or double pendulum, (see [3] and also Example 17)

$$\begin{aligned} \dot{x}^1 &= x^2 \\ \dot{x}^2 &= -M^{-1}(x^1)(C(x^1, x^2) + k(x^1)) + M^{-1}(x^1)u, \end{aligned}$$

where $x^1 = (\theta_1, \theta_2)$, $x^2 = (\dot{\theta}_1, \dot{\theta}_2)$, $u = (u_1, u_2)$, and the positive definite symmetric matrix $M(x^1)$ is given by

$$\begin{pmatrix} m_1 l_1^2 + m_2 l_1^2 + m_2 l_2^2 + 2m_2 l_1 l_2 \cos\theta_2 & m_2 l_2^2 + m_2 l_1 l_2 \cos\theta_2 \\ m_2 l_2^2 + m_2 l_1 l_2 \cos\theta_2 & m_2 l_2^2 \end{pmatrix}.$$

The term $k(\theta)$ represents the gravitational force and the term $C(\theta, \dot{\theta})$ reflects the centripetal and Coriolis force.

As the outputs we take the cartesian coordinates of the endpoint

$$\begin{aligned} y_1 &= h_1(\theta_1, \theta_2) &= l_1 \sin\theta_1 + l_2 \sin(\theta_1 + \theta_2) \\ y_2 &= h_2(\theta_1, \theta_2) &= l_1 \cos\theta_1 + l_2 \cos(\theta_1 + \theta_2). \end{aligned}$$

We have (by a direct computation) $\rho_1 = \rho_2 = 2$ and rank $D(x) = 2$ if and only if $l_1 l_2 \sin\theta_2 \neq 0$. Thus the system is input-output decouplable if $\theta \neq k\pi$ (a physical interpretation of these points which we have to exclude is clear). ■

Of course, Theorem 18 deals with the simplest version of the input-output decoupling problem. To illustrate variety and power of geometric methods let us list other versions of the input-output decoupling problem which have been studied and solved using them.

We start with the input-output decoupling for general control systems. It has been shown [27] that the system $\dot{x} = f(x, u),\ y = h(x)$ is input-output decouplable via an invertible feedback of the form $u = \alpha(x, \tilde{u})$ if and only if the extension $\dot{x} = f(x, u),\ \dot{u} = u^1$ with the original output $y = h(x)$ and with the control u^1 is input-output decouplable via an invertible feedback of the form $u^1 = \alpha^1(x, u) + \beta^1(x, u)\tilde{u}^1$.

Using the concept of controlled invariant distributions and controllability distributions, the block input-output decoupling problem has been solved (see [2,3]). In that problem the output is partioned into components each of them possibly being a vector.

In any practical application of input-output decoupling one would like to achieve not only external stability (which is almost trivial if we have (15)) but also internal stability. This very important and interesting problem has been extensively studied and various interesting results have been obtained (see [28,29,30]).

Finally, we would like to mention input-output decoupling of nonlinear systems $\dot{x} = f(x, u),\ y = h(x)$ via dynamic feedback of the form

$$\begin{aligned} \dot{z} &= g(x, z, v) \\ u &= \alpha(x, z, v) \end{aligned}$$

(see [31,32]).

We listed the above problems to illustrate how wide is the span of various applications of geometric methods to the input-output decoupling problem. Similar remarks apply to other problems and results stated in this survey: they just form starting points for various applications of geometric methods. We do not claim, however, that using geometric methods most of important nonlinear control problems have already been solved. For the input-output decoupling, which we discuss in this Section, the two following important problems still remain open: decoupling via non-invertible feedback and decoupling via output feedback.

8. Conclusions

After recalling geometric concepts of the Lie bracket, distributions, foliations, integrability, we show their usefulness in studying and solving some basic (nonlinear) control theory problems of controllability, observability, systems equivalence, linearization, and decoupling. Besides these, geometric methods have successfully been applied to study such important control problems as stabilizability, constructing observers, output tracking, inversion, use of dynamic feedback. Although many control problems have been well understood and solved using geometric methods some of them still need more investigations. For example, the use of output feedback (which is clearly more natural in applications) or noninvertible feedback (and its role in decoupling and linearization) are still subject to research.

As for the future development of geometric methods we would like to emphasize the feature which has been observed long ago, but whose importance is still growing: the use of combinations of geometry with methods coming from other branches of Mathematics. Examples of that trend are "marriages" of Differential Geometry (also its infinite-dimensional generalizations) with Differential Algebra (study of dynamic feedback, inversion), with Symplectic Geometry (analysis of mechanical control systems), with the singularity theory (constructing input-output representations, study of generic properties, feedback classification), with the center manifold theory (stabilization), with the theory of singular perturbations (high-gain feedback). Also, the use of geometric methods in optimal control (although has always been very significant) is rapidly growing. A classical example of that trend are recent extensive studies focused at the point, where nonlinear control meets sub-riemannian geometry, nonholonomic mechanics, theory of distributions, and optimal control. That "interdisciplinary" character of geometric methods in nonlinear control will be, we believe, one of its most important features in the future.

9. Acknowledgements

I would like to thank Bronisław Jakubczyk for kindly letting me use a part of his notes on the subject and Rafał Żbikowski for convincing me that such a survey should be written and published, and for his help in preparing figures.

10. References

1. R. Hermann, "On the accessibility problem in control theory," in *Differential Equations and Nonlinear Mechanics* (J. P. LaSalle and S. Lefschetz, eds.), pp. 325–332, New York: Academic Press, 1963.

2. A. Isidori, *Nonlinear Control Systems: An Introduction*. New York: Springer-Verlag, Second ed., 1989.

3. H. Nijmeijer and A. J. van der Schaft, *Nonlinear Dynamical Control Systems*. New York: Springer-Verlag, 1990.

4. H. J. Sussmann and V. Jurdjevic, "Controllability of nonlinear systems," *Journal of Differential Equations*, vol. 12, pp. 95–116, 1972.

5. A. J. Krener, "A generalization of Chow's theorem and the bang-bang theorem to nonlinear control systems," *SIAM Journal on Control*, vol. 12, pp. 43–52, 1974.

6. P. E. Crouch and B. Bonnard, "An appraisal of linear analytic system theory with applications to attitude control," Tech. Rep. ESA ESTEC Contract, University of Warwick and University of Bordeaux, 1980.

7. P. E. Crouch, "Spacecraft attitude control and stabilization: Applications of geometric control theory to rigid body models," *IEEE Transactions on Automatic Control*, vol. 29, pp. 321–331, 1984.

8. W. M. Wonham, *Linear Multivariable Control: A Geometric Approach*. New York: Springer-Verlag, Third ed., 1985.

9. R. Hermann and A. J. Krener, "Nonlinear controllability and observability," *IEEE Transactions on Automatic Control*, vol. 22, pp. 728–740, 1977.

10. A. J. Krener, "On the equivalence of control systems and linearization of nonlinear systems," *SIAM Journal on Control*, vol. 11, pp. 670–676, 1973.

11. H. J. Sussmann, "Lie brackets, real analyticity, and geometric control," in *Differential Geometric Control Theory* (R. W. Brockett, R. S. Millman, and H. J. Sussmann, eds.), pp. 1–116, Boston: Birkhäuser, 1983.

12. W. Respondek, "Geometric methods in linearization of control systems," in *Mathematical Control Theory* (C. Olech, B. Jakubczyk, and J. Zabczyk, eds.), pp. 453–467, Warsaw: PWN Polish Scientific Publishers, 1985. (Banach Center Publications, Vol. 14).

13. B. Jakubczyk and W. Respondek, "On linearization of control systems," *Bulletin de l'Académie Polonaise des Sciences Série des Sciences Mathématiques, Astronomiques et Physiques*, vol. 28, pp. 517–522, 1980.

14. L. R. Hunt and R. Su, "Linear equivalents of nonlinear time-varying systems," in *Proceedings of the 4th International Symposium on Mathematical Theory of Networks and Systems, August 5–7, 1981, Santa Monica, CA* (N. Levan, ed.), pp. 119–123, Western Periodicals Co., 1981.

15. P. Brunovsky, "A classification of linear controllable systems," *Kybernetica*, vol. 6, pp. 173–188, 1970.

16. J. J. Craig, *Introduction to Robotics, Mechanics and Control.* Reading: Addison-Wesley, 1986.

17. M. Zribi and J. Chiasson, "Exact linearization control of a PM stepper motor," in *Proceedings of the 8th American Control Conference, Pittsburgh, PA, 21–23 June, 1989*, pp. 320–321, Institute of Electrical and Electronics Engineers, 1989.

18. R. W. Brockett, "Feedback invariants for nonlinear systems," in *Proceedings of the Seventh IFAC World Congress, Helsinki, Finland, 12–16 June 1978* (A. Niemi, B. Wahlström, and J. Virkkunen, eds.), pp. 1115–1120, Pergamon Press, 1978.

19. W. P. Dayawansa, W. M. Boothby, and D. Elliot, "Global state and feedback equivalence of nonlinear systems," *Systems & Control Letters*, vol. 6, pp. 229–234, 1985.

20. W. Respondek, "Global aspects of linearization, equivalence to polynomial forms and decomposition of nonlinear control systems," in *Algebraic and Geometric Methods in Nonlinear Control Theory* (M. Fliess and M. Hazenwinkel, eds.), pp. 257–284, Dordrecht: D. Reidel Publishing Company, 1986.

21. W. Respondek, "Partial linearization, decompositions and fibre linear systems," in *Theory and Applications of Nonlinear Control Systems* (C. I. Byrnes and A. Lindquist, eds.), pp. 137–154, Amsterdam and New York: North-Holland, 1986.

22. A. J. Krener, A. Isidori, and W. Respondek, "Partial and robust linearization by feedback," in *Proceedings of the 22nd Conference on Decision and Control, San Antonio, TX, December 14–16, 1983*, pp. 126–130, Institute of Electrical and Electronics Engineers, 1983.

23. R. Marino, "On the largest feedback linearizable subsystem," *Systems & Control Letters*, vol. 7, pp. 345–351, 1986.

24. R. M. Hirschorn, "(A, B)-invariant distributions and disturbance decoupling of nonlinear systems," *SIAM Journal on Control and Optimization*, vol. 19, pp. 1–19, 1981.

25. A. Isidori, A. J. Krener, C. G. Giorgi, and S. Monaco, "Nonlinear decoupling via feedback: A differential geometric approach," *IEEE Transactions on Automatic Control*, vol. 26, pp. 331–345, 1981.

26. H. Nijmeijer, "Controlled invariance for affine control systems," *International Journal of Control*, vol. 34, pp. 824–833, 1981.

27. A. J. van der Schaft, "Linearization and input-output decoupling for general nonlinear systems," *Systems & Control Letters*, vol. 5, pp. 27–33, 1984.

28. I. J. Ha and E. G. Gilbert, "A complete characterization of decoupling control laws for a general class of nonlinear systems," *IEEE Transactions on Automatic Control*, vol. 31, pp. 823–830, 1986.

29. A. Isidori and J. W. Grizzle, "Fixed modes and nonlinear noninteracting control with stability," *IEEE Transactions on Automatic Control*, vol. 33, pp. 907–914, 1988.

30. S. Battilotti, *Noninteracting Control With Stability for Nonlinear Systems*. London, New York: Springer-Verlag, 1994.

31. J. Descusse and C. H. Moog, "Decoupling with dynamic compensation for strong invertible affine nonlinear systems," *International Journal of Control*, vol. 43, pp. 1387–1398, 1985.

32. H. Nijmeijer and W. Respondek, "Dynamic input-output decoupling of nonlinear control systems," *IEEE Transactions on Automatic Control*, vol. 33, pp. 1065–1070, 1988.

LOCAL REACHABILITY, LOCAL CONTROLLABILITY AND OBSERVABILITY OF A CLASS OF 2-D BILINEAR SYSTEMS

Tadeusz Kaczorek
Institute of Control and Industrial Electronics, Warsaw University of Technology
ul. Koszykowa 75, 00-662 Warszawa, Poland
E-mail: kaczorek@nov.isep.pw.edu.pl

ABSTRACT

New models of Fornasini-Marchesini type and Roesser type of 2-D bilinear systems are introduced and relationships between them are established. It is shown that the models of 2-D bilinear systems are equivalent to models of 2-D linear systems with variable coefficients. The general response formula for the new models are derived. Necessary and sufficient conditions are established for the local reachability, local controllability for given boundary conditions and observability of a class of 2-D bilinear systems. It is shown that the local reachability for given boundary conditions of a 2-D bilinear system implies always its local controllability for given boundary conditions.

1. Introduction

The most popular models of two-dimensional (2-D) linear systems are the state space models introduced by Roesser [1], Fornasini-Marchesini models [2,3] and Kurek [4]. In [5,6] the state space models have been extended for the singular ones. A review of singular 2-D linear discrete systems has been given in [7]. Bilinear 1-D systems and their controllability have been considered in many books and papers [8,9,10,11,12,13,14,15,16]. Recently in [17,18] new models of 2-D bilinear systems have been introduced. In this chapter new more general models of Fornasini-Marchesini type and Roesser type of 2-D bilinear systems will be introduced. The general response formula for the new models will be derived. Necessary and sufficient conditions for the local reachability, local controllability and observability of a class of 2-D bilinear systems will be established.

2. Models of 2-D Bilinear Systems

Consider the following two models of Fornasini-Marchesini type of 2-D bilinear systems

$$\begin{aligned} x_{i+1,j+1} = \; & A_0 x_{ij} + A_1 x_{i+1,j} + A_2 x_{i,j+1} + \sum_{k=1}^{m} u^k_{i+1,j} B_{1k} x_{i+1,j} + \\ & \sum_{k=1}^{m} u^k_{i,j+1} B_{2k} x_{i,j+1} + C_1 u_{i+1,j} + C_2 u_{i,j+1} \end{aligned} \tag{1}$$

$$x_{i+1,j+1} = A_0 x_{ij} + A_1 x_{i+1,j} + A_2 x_{i,j+1} + \sum_{k=1}^{n} x_{i+1,j}^{k} B_{1k}^{'} u_{i+1,j} + \sum_{k=1}^{n} x_{i,j+1}^{k} B_{2k}^{'} u_{i,j+1} + C_1 u_{i+1,j} + C_2 u_{i,j+1} \quad i, j \in \mathbb{Z}_+ \tag{2}$$

where $x_{ij} := \left[x_{ij}^1, x_{ij}^2, \ldots, x_{ij}^n\right]^T$ is the local state vector at the point $(i, j) \in \mathbb{Z}_+ \times \mathbb{Z}_+$, $\mathbb{Z}_+$ is the set of nonnegative inteegers, the upper index T denotes the transposition,

$$u_{ij} := \left[u_{ij}^1, u_{ij}^2, \ldots, u_{ij}^m\right]^T$$

is the input vector, $A_0, A_1, A_2, B_{1k}, B_{2k} \in \mathbb{R}^{n\times n}$, $B_{1k}^{'}, B_{2k}^{'}, C_1, C_2 \in \mathbb{R}^{n\times m}$ and $\mathbb{R}^{p\times q}$ is the set $p \times q$ real matrices.
The output equation of the models has the form

$$y_{ij} = Dx_{ij} + Hu_{ij} \tag{3}$$

where $y_{ij} \in \mathbb{R}^p$ is the output vector at the point (i, j) and $D \in \mathbb{R}^{p\times n}$, $H \in \mathbb{R}^{p\times m}$.

Boundary conditions for 1 and 2 are given by

$$x_{i0} \text{ for } i \in \mathbb{Z}_+ \text{ and } x_{0j} \text{ for } j \in \mathbb{Z}_+ \tag{4}$$

Let B_{1k}^q (B_{2k}^q) be the qth ($q = 1, \ldots, n$) column of the matrix B_{1k} (B_{2k}) and $B_{1k}^{'q}$ ($B_{2k}^{'q}$) be the qth ($q = 1, \ldots, m$) column of the matrix $B_{1k}^{'}$ ($B_{2k}^{'}$).

Note that

$$\sum_{k=1}^{n} x_{i+1,j}^k B_{1k}^{'} u_{i+1,j} = \sum_{k=1}^{n} x_{i+1,j}^k \left(\sum_{q=1}^{m} B_{1k}^{'q} u_{i+1,j}^q\right) = \sum_{q=1}^{m} \left(\sum_{k=1}^{n} x_{i+1,j}^k B_{1k}^{'q}\right) u_{i+1,j}^q \tag{5}$$

and

$$\sum_{k=1}^{m} u_{i+1,j}^k B_{1k} x_{i+1,j} = \sum_{k=1}^{m} u_{i+1,j}^k \left(\sum_{q=1}^{n} B_{1k}^q x_{i+1,j}^q\right) = \sum_{k=1}^{m} \left(\sum_{q=1}^{n} B_{1k}^q x_{i+1,j}^q\right) u_{i+1,j}^k \tag{6}$$

From 5 and 6 it follows that

$$B_{1k}^q = B_{1q}^{'k} \text{ for } q = 1, \ldots, n \text{ and } k = 1, \ldots, m \tag{7}$$

Similarily we may show that

$$B_{2k}^q = B_{2k}^{'k} \text{ for } q = 1, ..., n \text{ and } k = 1, ..., m \tag{8}$$

Using 7, 8 we may write the equation 1 in the form 2 and the equation 2 in the form 1.

Define

$$\bar{u}_{ij} := u_{ij} \otimes I_n = \begin{bmatrix} u_{ij}^1 I_n \\ u_{ij}^2 I_n \\ \vdots \\ u_{ij}^m I_n \end{bmatrix}, \quad \bar{x}_{ij} := x_{ij}^T \otimes I_n = \left[x_{ij}^1 I_n, x_{ij}^2 I_n, \ldots, x_{ij}^n I_n\right]$$

and

$$\bar{B}_1 := [B_{11}, B_{12}, \ldots, B_{1m}], \quad \bar{B}_2 := [B_{21}, B_{22}, \ldots, B_{2m}],$$

$$\bar{B}_1' := \begin{bmatrix} B_{11}' \\ B_{12}' \\ \vdots \\ B_{1n}' \end{bmatrix}, \quad \bar{B}_2' := \begin{bmatrix} B_{21}' \\ B_{22}' \\ \vdots \\ B_{2n}' \end{bmatrix}$$

where I_n is the $n \times n$ identity matrix and $\otimes$ denotes the Kronecker product. Using the above notation we may write 1 and 2 in the form

$$x_{i+1,j+1} = A_0 x_{ij} + \left(A_1 + \bar{B}_1 \bar{u}_{i+1,j}\right) x_{i+1,j} \left(A_2 + \bar{B}_2 \bar{u}_{i,j+1}\right) x_{i,j+1} + C_1 u_{i+1,j} + C_2 u_{i,j+1} \tag{9}$$

$$x_{i+1,j+1} = A_0 x_{ij} + A_1 x_{i+1,j} + A_2 x_{i,j+1} + \left(\bar{x}_{i,j+1} \bar{B}_1' + C_1\right) u_{i+1,j} + \left(\bar{x}_{i,j+1} \bar{B}_2' + C_2\right) u_{i,j+1} \tag{10}$$

respectively.

An 2-D bilinear model of the Roesser type has the form

$$\begin{bmatrix} x_{i+1,j}^h \\ x_{i,j+1}^v \end{bmatrix} = \left(A + \sum_{k=1}^{m} u_{ij}^k B_k\right) \begin{bmatrix} x_{ij}^h \\ x_{ij}^v \end{bmatrix} + C u_{ij} \qquad i, j \in \mathbb{Z}_+ \tag{11}$$

or

$$\begin{bmatrix} x_{i+1,j}^h \\ x_{i,j+1}^v \end{bmatrix} = A \begin{bmatrix} x_{ij}^h \\ x_{ij}^v \end{bmatrix} \left(\sum_{q=1}^{n_1} x_{ij}^{hq} B_{1q} + \sum_{l=1}^{n_2} x_{ij}^{vl} B_{2l} + C\right) u_{ij} \qquad i, j \in \mathbb{Z}_+ \tag{12}$$

where x_{ij}^{hk} is the kth component of the horizontal state vector $x_{ij}^h \in \mathbb{R}^{n_1}$ at the point (i, j), x_{ij}^{vl} is the lth component of the vertical state vector $x_{ij}^v \in \mathbb{R}^{n_1}$ at the point (i, j), u_{ij}^k is the kth component of the input vector $u_{ij} \in \mathbb{R}^m$ and $A, B_k, C, B_{1q}, B_{2l}$ are real matrices of appropriate dimensions.

The output equation of the models has the form

$$y_{ij} = D \begin{bmatrix} x_{ij}^h \\ x_{ij}^v \end{bmatrix} + H u_{ij} \qquad i, j \in \mathbb{Z}_+ \tag{13}$$

where $y_{ij} \in \mathbb{R}^p$ is the output vector at the point (i, j) and $D \in \mathbb{R}^{p\times n}$, $n := n_1 + n_2$, $H \in \mathbb{R}^{p\times m}$.

Boundary conditions for 11 and 12 are given by

$$x_{0j}^h \text{ for } j \in \mathbb{Z}_+ \text{ and } x_{i0}^v \text{ for } i \in \mathbb{Z}_+ \tag{14}$$

Let

$$B_k = \left[B_k^{h1}, \ldots, B_k^{hn_1}, B_k^{v1}, \ldots, B_k^{vn_2}\right], \quad k = 1, \ldots, m$$

$$B_{1q} = \left[B_{1q}^1, \ldots, B_{1q}^m\right], B_{2l} = \left[B_{2l}^1, \ldots, B_{2l}^m\right], \quad q = 1, \ldots, n_1, \quad l = 1, \ldots, n_2$$

From comparison of the relations

$$\sum_{k=1}^{m} u_{ij}^k B_k \begin{bmatrix} x_{ij}^h \\ x_{ij}^v \end{bmatrix} = \sum_{k=1}^{m}\sum_{q=1}^{n_1} u_{ij}^k B_k^{hq} x_{ij}^{hq} + \sum_{k=1}^{m}\sum_{l=1}^{n_2} u_{ij}^k B_k^{vl} x_{ij}^{vl} \tag{15}$$

and

$$\sum_{q=1}^{n_1} x_{ij}^{hq} B_{1q} u_{ij} + \sum_{l=1}^{n_2} x_{ij}^{vl} B_{2l} u_{ij} = \sum_{k=1}^{m}\sum_{q=1}^{n_1} u_{ij}^k B_{1q}^k x_{ij}^{hq} + \sum_{k=1}^{m}\sum_{l=1}^{n_2} u_{ij}^k B_{2l}^k x_{ij}^{vl} \tag{16}$$

it follows that

$$\begin{aligned} B_k^{hq} &= B_{1q}^k \text{ for } k = 1, \ldots, m; \quad q = 1, \ldots, n_1 \\ &\quad \text{and} \\ B_k^{vl} &= B_{2l}^k \text{ for } k = 1, \ldots, m; \quad l = 1, \ldots, n_2 \end{aligned} \tag{17}$$

Therefore, using 17 we may transform the equation 11 into 12 and vice versa the equation 12 into 11.

Note that 11 may be written in the form

$$\begin{bmatrix} x_{i+1,j}^h \\ x_{i,j+1}^v \end{bmatrix} = \left(A + B\bar{u}_{ij}\right) \begin{bmatrix} x_{ij}^h \\ x_{ij}^v \end{bmatrix} + Cu_{ij} \qquad i, j \in \mathbb{Z}_+ \tag{18}$$

where

$$B := [B_1, B_2, \ldots, B_m], \quad \bar{u}_{ij} := u_{ij} \otimes I_n$$

Defining

$$x_{ij} := \begin{bmatrix} x_{ij}^h \\ x_{ij}^v \end{bmatrix}, \quad A := \begin{bmatrix} A_1 \\ A_2 \end{bmatrix}, \quad B := \begin{bmatrix} B_1 \\ B_2 \end{bmatrix} \quad \textit{and} \quad C = \begin{bmatrix} C_1' \\ C_2' \end{bmatrix},$$

$$A_1 \in \mathbb{R}^{n_1\times n}, \quad A_2 \in \mathbb{R}^{n_2\times n}, \quad B_1 \in \mathbb{R}^{n_1\times nm}, \quad B_2 \in \mathbb{R}^{n_2\times nm}, \quad C_1' \in \mathbb{R}^{n_1\times m}, \quad C_2' \in \mathbb{R}^{n_2\times m}$$

from 18 we obtain

$$x_{i+1,j+1} = \begin{bmatrix} 0 \\ A_2 + B_2\bar{u}_{i+1,j} \end{bmatrix} x_{i+1,j} + \begin{bmatrix} A_1 + B_1\bar{u}_{i,j+1} \\ 0 \end{bmatrix} x_{i,j+1} + \begin{bmatrix} 0 \\ C_2' \end{bmatrix} u_{i+1,j} + \begin{bmatrix} C_1' \\ 0 \end{bmatrix} u_{i,j+1} \tag{19}$$

From comparison of 9 and 19 it follows that the 2-D bilinear model of the Roesser type 11 is a particular case of the model 1.
Note that the models 9 and 18 may be written as

$$x_{i+1,j+1} = A_0 x_{ij} + A^1_{i+1} x_{i+1,j} + A^2_{i,j+1} x_{i,j+1} + C_1 u_{i+1,j} + C_2 u_{i,j+1} \tag{20}$$

and

$$\begin{bmatrix} x^h_{i+1,j} \\ x^v_{i,j+1} \end{bmatrix} = \bar{A}_{ij} \begin{bmatrix} x^h_{ij} \\ x^v_{ij} \end{bmatrix} + C u_{ij} \tag{21}$$

where

$$A^1_{i+1,j} = A_1 + \bar{B}_1 \bar{u}_{i+1,j}, \quad A^2_{i,j+1} = A_2 + \bar{B}_2 \bar{u}_{i,j+1}, \quad A_{ij} = A + B\bar{u}_{ij}$$

For given u_{ij} 20 and 21 are models of 2-D linear systems with variable coefficients. Therefore, the following theorem has been proved.

Theorem 1 *The Fornasini-Marchesini type model 1 and the Roesser type model 11 of 2-D bilinear systems are equivalent to the models 20 and 21 of 2-D linear systems with variable coefficients, respectively.*

Similar results can be obtained for the model 12.

3. General Response Formula

We shall only derive the solution to the equation 9 with boundary conditions 4, since the equation 10 may transformed to 9.

Theorem 2 *The solution x_{ij} to the equation 9 with boundary conditions 4 is given by*

$$\begin{aligned} x_{ij} = {} & T_{i-1,j-1} A_0 x_{00} + \sum_{k=l}^{i} T_{i-k-1,j-1} A_0 x_{k0} + \sum_{l=1}^{j} T_{i-1,j-l-1} A_0 x_{0l} + \\ & + \sum_{k=1}^{i} T_{i-k,j-1} \left[\left(A_1 + \bar{B}_1 \bar{u}_{k0} \right) x_{k0} + C_1 u_{k0} \right] + \\ & + \sum_{l=1}^{j} T_{i-1,j-l} \left[\left(A_2 + \bar{B}_2 \bar{u}_{kl} \right) x_{0l} + C_2 u_{0l} \right] + \end{aligned}$$

$$
\begin{aligned}
&+\sum_{k_1=1}^{i}\sum_{l_1=1}^{j}\left[T_{i-k_1,j-l_1-1}\bar{B}_1+T_{i-k_1-1,j-l_1}\bar{B}_2\right]\bar{u}_{k_1l_1}M_{k_1l_1}+\\
&+\left(\sum_{(1,1)\le(k_2,l_2)<(i,j)}\left[T_{i-k_2,j-l_2-1}\bar{B}_1+T_{i-k_2-1,j-l_2}\bar{B}_2\right]\bar{u}_{k_2l_2}\right)\times\\
&\times\left(\sum_{(1,1)\le(k_1,l_1)<(k_2,l_2)}\left[T_{k_2-k_1,l_2-l_1-1}\bar{B}_1+T_{k_2-k_1-1,l_2-l_1}\bar{B}_2\right]\bar{u}_{k_1l_1}M_{k_1l_1}\right)+\ldots+\\
&+\left(\sum_{(r,r)\le(k_r,l_r)<(i,j)}\left[T_{i-k_r,j-l_r-1}\bar{B}_1+T_{i-k_r-1,j-l_r}\bar{B}_2\right]\bar{u}_{k_rl_r}\right)\times\\
&\times\left(\sum_{(r,r)\le(k_{r-1},l_{r-1})<(k_r,l_r)}\left[T_{k_r-k_{r-1},l_r-l_{r-1}-1}\bar{B}_1+T_{k_r-k_{r-1}-1,l_r-l_{r-1}}\bar{B}_2\right]\bar{u}_{k_{r-1}l_{r-1}}\right)\times\ldots\times\\
&\times\left(\sum_{(1,1)\le(k_1,l_1)<(k_2,j_2)}\left[T_{k_2-k_1,l_2-l_1-1}\bar{B}_1+T_{k_2-k_1-1,l_2-l_1}\bar{B}_2\right]\bar{u}_{k_1l_1}M_{k_1l_1}\right)+\\
&+\sum_{k_1=1}^{i}\sum_{l_1=1}^{j}\left[T_{i-k_1,j-l_1-1}C_1+T_{i-k_1-1,j-l_1}C_2\right]\bar{u}_{k_1l_1}+\\
&\times\left(\sum_{(1,1)\le(k_2,l_2)<(i,j)}\left[T_{i-k_2,j-l_2-1}\bar{B}_1+T_{i-k_2-1,j-l_2}\bar{B}_2\right]\bar{u}_{k_2l_2}\right)\times\\
&\times\left(\sum_{(1,1)\le(k_1,l_1)<(k_2,l_2)}\left[T_{k_2-k_1,l_2-l_1-1}C_1+T_{k_2-k_1-1,l_2-l_1}C_2\right]\bar{u}_{k_1l_1}\right)+\ldots+\\
&+\left(\sum_{(r,r)\le(k_r,l_r)<(i,j)}\left[T_{i-k_r,j-l_r-1}\bar{B}_1+T_{i-k_r-1,j-l_r}\bar{B}_2\right]\bar{u}_{k_rl_r}\right)\times\\
&\times\left(\sum_{(r,r)\le(k_{r-1},l_{r-1})<(k_r,l_r)}\left[T_{k_r-k_{r-1},l_r-l_{r-1}-1}\bar{B}_1+T_{k_r-k_{r-1}-1,l_r-l_{r-1}}\bar{B}_2\right]\bar{u}_{k_{r-1}l_{r-1}}\right)\times\ldots\times\\
&+\left(\sum_{(1,1)\le(k_1,l_1)<(k_2,l_2)}\left[T_{k_2-k_1,l_2-l_1-1}C_1+T_{k_2-k_1-1,l_2-l_1}C_2\right]\bar{u}_{k_1l_1}\right)\qquad r\le\max(i,j)
\end{aligned}
\tag{22}
$$

where T_{pq} is the transition matrix defined as follows

$$T_{00}:=I_n\ \text{(the identity matrix)}$$

$$
\begin{gathered}
T_{pq}=T_{p-1,q-1}A_0+T_{p,q-1}A_1+T_{p-1,q}A_2\ \text{ for }\ (i,j)>(0,0)\\
T_{pq}=0\ \text{ (the zero matrix) for }\ i<0\ \text{ and/or }\ j<0
\end{gathered}
\tag{23}
$$

and

$$M_{pq} = T_{p-1,q-1}A_0x_{00} + \sum_{k=1}^{p} T_{p-k-1,q-1}A_0x_{k0} + \sum_{l=1}^{q} T_{p-1,q-l-1}A_0x_{0l} + \\ + \sum_{k=1}^{p} T_{p-k,q-1}\left[\left(A_1 + \bar{B}_1\bar{u}_{k0}\right)x_{k0} + C_1u_{k0}\right] + \\ + \sum_{l=1}^{q} T_{p-1,q-l}\left[\left(A_2 + \bar{B}_2\bar{u}_{0l}\right)x_{0l} + C_2u_{0l}\right]. \quad (24)$$

Proof The proof can be accomplished by induction on the pair (i, j).
From 22 for $i = j = 1,\ i = 2,\ j = 1$ and $i = 1,\ j = 2$ we have

$$\begin{aligned} x_{11} &= A_0x_{00} + \left(A_1 + \bar{B}_1\bar{u}_{10}\right)x_{10} + \left(A_2 + \bar{B}_2u_{01}\right)x_{01} + C_1u_{10} + C_2u_{01} \\ x_{21} &= T_{10}A_0x_{00} + T_{10}\left[\left(A_1 + \bar{B}_1\bar{u}_{10}\right)x_{10} + C_1u_{10}\right] + \left(A_1 + \bar{B}_1\bar{u}_{20}\right)x_{20} + C_1u_{20} + \\ &\quad + T_{10}\left[\left(A_2 + \bar{B}_2\bar{u}_{01}\right)x_{01} + C_2u_{01}\right] + \\ &\quad \bar{B}_2\bar{u}_{11}\left[A_0x_{00} + \left(A_1 + \bar{B}_1\bar{u}_{10}\right)x_{10} + C_1u_{10} + \left(A_2 + \bar{B}_2\bar{u}_{01}\right)x_{01} + C_2u_{01}\right] \\ x_{12} &= T_{01}A_0x_{00} + A_0x_{01} + T_{01}\left[\left(A_1 + \bar{B}_1\bar{u}_{10}\right)x_{10} + C_1u_{10}\right] + \\ &\quad + T_{01}\left[\left(A_2 + \bar{B}_2\bar{u}_{01}\right)x_{01} + C_2u_{01}\right] + \left(A_2 + \bar{B}_2\bar{u}_{02}\right)x_{02} + C_2u_{02} + \\ &\quad + \bar{B}_1\bar{u}_{11}\left[A_0x_{00} + \left(A_1 + \bar{B}_1\bar{u}_{10}\right)x_{10} + C_1u_{10} + \left(A_2 + \bar{B}_2\bar{u}_{01}\right)x_{01} + C_2u_{01}\right] \end{aligned}$$

The same results we obtain from 9 for $i = j = 0,\ \ i = 1, j = 0$ and $i = 0,\ j = 1$ 23 and 24. Therefore, the hypothesis is true for the pairs $(i = 1,\ j = 1)$, $(i = 2,\ j = 1)$ and $(i = 1,\ j = 2)$.
Assuming that the hypothesis holds for the pairs (i, j), $(i+1, j)$ and $(i,\ j+1)$, $i > 0,\ j > 0$ we shall show that it is also valid for the pair $(i + 1, j + 1)$. Using 9, 22, 23 and 24 we may write

$$\begin{aligned} x_{i+1,j+1} &= A_0x_{ij} + \left(A_1 + \bar{B}_1\bar{u}_{i+1,j}\right)x_{i+1,j} + \left(A_2 + \bar{B}_2\bar{u}_{i,j+1}\right)x_{i,j+1} + \\ &+ C_1u_{i+1,j} + C_2u_{i,j+1} = \\ &= A_0\left\{T_{i-1,j-1}A_0x_{00} + \sum_{k=1}^{i} T_{i-k-1,j-1}A_0x_{k0} + \sum_{l=1}^{j} T_{i-1,j-l-1}A_0x_{0l} + \right. \\ &+ \sum_{k=1}^{i} T_{i-k,j-1}\left[\left(A_1 + \bar{B}_1\bar{u}_{k0}\right)x_{k0} + C_1u_{k0}\right] + \\ &+ \sum_{l=1}^{j} T_{i-1,j-l}\left[\left(A_2 + \bar{B}_2\bar{u}_{kl}\right)x_{0l} + C_2u_{0l}\right] + \\ &+ \sum_{k_1=1}^{i}\sum_{l_1=1}^{j}\left[T_{i-k_1,j-l_1-1}\bar{B}_1 + T_{i-k_1-1,j-l_1}\bar{B}_2\right]\bar{u}_{k_1l_1}M_{k_1,l_1} + \end{aligned}$$

$$+ \left(\sum_{(1,1)\le(k_2,l_2)<(i,j)} \left[T_{i-k_2,j-l_2-1}\bar{B}_1 + T_{i-k_2-1,j-l_2}\bar{B}_2 \right] \bar{u}_{k_2 l_2} \right) \times$$

$$\times \left(\sum_{(1,1)\le(k_1,l_1)<(k_2,l_2)} \left[T_{k_2-k_1,l_2-l_1-1}\bar{B}_1 + T_{k_2-k_1-1,l_2-l_1}\bar{B}_2 \right] \bar{u}_{k_1 l_1} M_{k_1 l_1} \right) + \ldots +$$

$$+ \left(\sum_{(r,r)\le(k_r,l_r)<(i,j)} \left[T_{i-k_r,j-l_r-1}\bar{B}_1 + T_{i-k_r-1,j-l_r}\bar{B}_2 \right] \bar{u}_{k_r l_r} \right) \times$$

$$\times \left(\sum_{(r,r)\le(k_{r-1},l_{r-1})<(k_r,l_r)} \left[T_{k_r-k_{r-1},l_r-l_{r-1}-1}\bar{B}_1 + T_{k_r-k_{r-1}-1,l_r-l_{r-1}}\bar{B}_2 \right] \bar{u}_{k_{r-1} l_{r-1}} \right) \times \ldots \times$$

$$\times \left(\sum_{(1,1)\le(k_1,l_1)<(k_2,j_2)} \left[T_{k_2-k_1,l_2-l_1-1}\bar{B}_1 + T_{k_2-k_1-1,l_2-l_1}\bar{B}_2 \right] \bar{u}_{k_1 l_1} M_{k_1 l_1} \right) +$$

$$+ \sum_{k_1=1}^{i} \sum_{l_1=1}^{j} \left[T_{i-k_1,j-l_1-1} C_1 + T_{i-k_1-1,j-l_1} C_2 \right] \bar{u}_{k_1 l_1} +$$

$$+ \left(\sum_{(1,1)\le(k_2,l_2)<(i,j)} \left[T_{i-k_2,j-l_2-1}\bar{B}_1 + T_{i-k_2-1,j-l_2}\bar{B}_2 \right] \bar{u}_{k_2 l_2} \right) \times$$

$$\times \left(\sum_{(1,1)\le(k_1,l_1)<(k_2,l_2)} \left[T_{k_2-k_1,l_2-l_1-1} C_1 + T_{k_2-k_1-1,l_2-l_1} C_2 \right] \bar{u}_{k_1 l_1} \right) + \ldots +$$

$$+ \left(\sum_{(r,r)\le(k_r,l_r)<(i,j)} \left[T_{i-k_r,j-l_r-1}\bar{B}_1 + T_{i-k_r-1,j-l_r}\bar{B}_2 \right] \bar{u}_{k_r l_r} \right) \times$$

$$\times \left(\sum_{(r,r)\le(k_{r-1},l_{r-1})<(k_r,l_r)} \left[T_{k_r-k_{r-1},l_r-l_{r-1}-1}\bar{B}_1 + T_{k_r-k_{r-1}-1,l_r-l_{r-1}}\bar{B}_2 \right] \bar{u}_{k_{r-1} l_{r-1}} \right) \times \ldots \times$$

$$\left. \times \left(\sum_{(1,1)\le(k_1,l_1)<(k_2,j_2)} \left[T_{k_2-k_1,l_2-l_1-1} C_1 + T_{k_2-k_1-1,l_2-l_1} C_2 \right] \bar{u}_{k_1 l_1} \right) \right\} +$$

$$+ \left(A_1 + \bar{B}_1 u_{i+1,j} \right) \left\{ T_{i,j-1} A_0 x_{00} + \sum_{k=1}^{i+1} T_{i-k,j-1} A_0 x_{k0} + \sum_{l=1}^{j} T_{i,j-l-1} A_0 x_{0l} + \right.$$

$$+ \sum_{k=1}^{i+1} T_{i-k+1,j-1} \left[\left(A_1 + \bar{B}_1 \bar{u}_{k0} \right) x_{k0} + C_1 u_{k0} \right] + \sum_{l=1}^{j} T_{i,j-l} \left[\left(A_2 + \bar{B}_2 \bar{u}_{kl} \right) x_{0l} + C_2 u_{0l} \right] +$$

$$+ \sum_{k_1=1}^{i+1} \sum_{l_1=1}^{j} \left[T_{i+1-k_1,j-l_1-1}\bar{B}_1 + T_{i-k_1,j-l_1}\bar{B}_2 \right] \bar{u}_{k_1,l_1} M_{k_1,l_1} +$$

$$+ \left(\sum_{(1,1)\le(k_2,l_2)<(i+1,j)} \left[T_{i-k_2+1,j-l_2-1}\bar{B}_1 + T_{i-k_2,j-l_2}\bar{B}_2 \right] \bar{u}_{k_2 l_2} \right) \times$$

$$\times \left(\sum_{(1,1)\le(k_1,l_1)<(k_2,l_2)} \left[T_{k_2-k_1,l_2-l_1-1}\bar{B}_1 + T_{k_2-k_1-1,l_2-l_1}\bar{B}_2 \right] \bar{u}_{k_1 l_1} M_{k_1 l_1} \right) + \ldots +$$

$$+ \left(\sum_{(r,r)\le(k_r l_r)<(i+1,j)} \left[T_{i-k_r+1,j-l_r-1}\bar{B}_1 + T_{i-k_r,j-l_r}\bar{B}_2 \right] \bar{u}_{k_r l_r} \right) \times$$

$$\times \left(\sum_{(r,r)\le(k_{r-1},l_{r-1})<(k_r,l_r)} \left[T_{k_r-k_{r-1},l_r-l_{r-1}-1}\bar{B}_1 + T_{k_r-k_{r-1}-1,l_r-l_{r-1}}\bar{B}_2 \right] \bar{u}_{k_{r-1} l_{r-1}} \right) \times \ldots \times$$

$$\times \left(\sum_{(1,1)\le(k_1,l_1)<(k_2,j_2)} \left[T_{k_2-k_1,l_2-l_1-1}\bar{B}_1 + T_{k_2-k_1-1,l_2-l_1}\bar{B}_2 \right] \bar{u}_{k_1 l_1} M_{k_1 l_1} \right) +$$

$$+ \sum_{k_1=1}^{i+1} \sum_{l_1=1}^{j} \left[T_{i-k_1+1,j-l_1-1} C_1 + T_{i-k_1+1,j-l_1} C_2 \right] \bar{u}_{k_1 l_1} +$$

$$+ \left(\sum_{(1,1)\le(k_2,l_2)<(i+1,j)} \left[T_{i-k_2+1,j-l_2-1}\bar{B}_1 + T_{i-k_2,j-l_2}\bar{B}_2 \right] \bar{u}_{k_2 l_2} \right) \times$$

$$\times \left(\sum_{(1,1)\le(k_1,l_1)<(k_2,l_2)} \left[T_{k_2-k_1,l_2-l_1-1} C_1 + T_{k_2-k_1-1,l_2-l_1} C_2 \right] \bar{u}_{k_1 l_1} \right) + \ldots +$$

$$+ \left(\sum_{(r,r)\le(k_r,l_r)<(i+1,j)} \left[T_{i-k_r+1,j-l_r}\bar{B}_1 + T_{i-k_r,j-l_r}\bar{B}_2 \right] \bar{u}_{k_r l_r} \right) \times$$

$$\times \left(\sum_{(r,r)\le(k_{r-1},l_{r-1})<(k_r,l_r)} \left[T_{k_r-k_{r-1},l_r-l_{r-1}-1}\bar{B}_1 + T_{k_r-k_{r-1}-1,l_r-l_{r-1}}\bar{B}_2 \right] \bar{u}_{k_{r-1} l_{r-1}} \right) \times \ldots \times$$

$$\times \left. \left(\sum_{(1,1)\le(k_1,l_1)<(k_2,l_2)} \left[T_{k_2-k_1,l_2-l_1-1} C_1 + T_{k_2-k_1-1,l_2-l_1} C_2 \right] \bar{u}_{k_1 l_1} \right) \right\} +$$

$$+ \left(A_2 + \bar{B}_2 \bar{u}_{i,j+1} \right) \left\{ T_{i-1,j} A_0 x_{00} + \sum_{k=1}^{i} T_{i-k-1,j} A_0 x_{k0} + \sum_{l=1}^{j+1} T_{i-1,j-l} A_0 x_{0l} + \right.$$

$$+ \sum_{k=1}^{i} T_{i-k,j} \left[\left(A_1 + \bar{B}_1 \bar{u}_{k0} \right) x_{k0} + C_1 u_{k0} \right] + \sum_{l=1}^{j+1} T_{i-1,j-l+1} \left[\left(A_2 + \bar{B}_2 \bar{u}_{kl} \right) x_{0l} + C_2 u_{0l} \right] +$$

$$+ \sum_{k_1=1}^{i} \sum_{l_1=1}^{j+1} \left[T_{i-k_1,j-l_1}\bar{B}_1 + T_{i-k_1-1,j-l_1+1}\bar{B}_2 \right] \bar{u}_{k_1 l_1} M_{k_1 l_1} +$$

$$
\begin{aligned}
&+ \sum_{(1,1)\le(k_2,l_2)<(i,j+1))} \left[T_{i-k_2,j-l_2}\bar{B}_1 + T_{i-k_2-1,j-l_2+1}\bar{B}_2\right]\bar{u}_{k_2l_2} \times \\
&\times \left(\sum_{(1,1)\le(k_1,l_1)<(k_2l_2)} \left[T_{k_2-k_1,l_2-l_1-1}\bar{B}_1 + T_{k_2-k_1-1,l_2-l_1}\bar{B}_2\right]\bar{u}_{k_1l_1}M_{k_1l_1}\right) + \ldots + \\
&+ \left(\sum_{(r,r)\le(k_r,l_r)<(i,j+1)} \left[T_{i-k_r,j-l_r}\bar{B}_1 + T_{i-k_r-1,j-l_r+1}\bar{B}_2\right]\bar{u}_{k_rl_r}\right) \times \\
&\times \left(\sum_{(r,r)\le(k_r-1,l_r-1)<(k_r,l_r)} \left[T_{k_r-k_{r-1},l_r-l_{r-1}-1}\bar{B}_1 + T_{k_r-k_{r-1}-1,l_r-l_{r-1}}\bar{B}_2\right]\bar{u}_{k_r-1l_r-1}\right) \times \ldots \times \\
&\times \left(\sum_{(1,1)\le(k_1,l_1)<(k_2,j_2+1)} \left[T_{k_2-k_1,l_2-l_1-1}\bar{B}_1 + T_{k_2-k_1-1,l_2-l_1}\bar{B}_2\right]\bar{u}_{k_1l_1}M_{k_1l_1}\right) + \\
&+ \sum_{k_1=1}^{i}\sum_{l_1=1}^{j+1}\left[T_{i-k_1,j-l_1}C_1 + T_{i-k_1-1,j-l_1+1}C_2\right]\bar{u}_{k_1l_1} + \\
&+ \left(\sum_{(1,1)\le(k_2,l_2)<(i,j+1)} \left[T_{i-k_2,j-l_2}\bar{B}_1 + T_{i-k_2-1,j-l_2+1}\bar{B}_2\right]\bar{u}_{k_2l_2}\right) \times \\
&\times \left(\sum_{(1,1)\le(k_1,l_1)<(k_2,l_2)} \left[T_{k_2-k_1,l_2-l_1-1}C_1 + T_{k_2-k_1-1,l_2-l_1}C_2\right]\bar{u}_{k_1l_1}\right) + \ldots + \\
&+ \left(\sum_{(r,r)\le(k_r,l_r)<(i,j+1)} \left[T_{i-k_r,j-l_r}\bar{B}_1 + T_{i-k_r-1,j-l_r+1}\bar{B}_2\right]\bar{u}_{k_rl_r}\right) \times
\end{aligned}
$$

$$
\begin{aligned}
&\times \left(\sum_{(r,r)\le(k_{r-1},l_{r-1})<(k_r,l_r)} \left[T_{k_r-k_{r-1},l_r-l_{r-1}-1}\bar{B}_1 + T_{k_r-k_{r-1}-1,l_r-l_{r-1}}\bar{B}_2\right]\bar{u}_{k_{r-1}l_{r-1}}\right) \times \ldots \times \\
&\times \left.\left(\sum_{(1,1)\le(k_1,l_1)<(k_2,l_2)} \left[T_{k_2-k_1,l_2-l_1-1}C_1 + T_{k_2-k_1-1,l_2-l_1}C_2\right]\bar{u}_{k_1l_1}\right)\right\} + C_1u_{i+1,j} + C_2u_{i,j+1} = \\
&= T_{i,j}A_0x_{00} + \sum_{k=1}^{i+1}T_{i-k,j}A_0x_{k0} + \sum_{l=1}^{j+1}T_{i,j-l}A_0x_{0l} + \\
&+ \sum_{k=1}^{i+1}T_{i-k+1,j}\left[\left(A_1 + \bar{B}_1\bar{u}_{k0}\right)x_{k0} + C_1u_{k0}\right] + \sum_{l=1}^{j+1}T_{i,j-l+1}\left[\left(A_2 + \bar{B}_2\bar{u}_{kl}\right)x_{0l} + C_2u_{0l}\right] + \\
&+ \sum_{k_1=1}^{i+1}\sum_{l_1=1}^{j+1}\left[T_{i-k_1+1,j-l_1}\bar{B}_1 + T_{i-k_1,j-l_1+1}\bar{B}_2\right]\bar{u}_{k_1l_1}M_{k_1l_1} +
\end{aligned}
$$

$$+\left(\sum_{(1,1)\le(k_2,l_2)<(i+1,j+1)}\left[T_{i-k_2+1,j-l_2}\bar{B}_1+T_{i-k_2,j-l_2+1}\bar{B}_2\right]\bar{u}_{k_2l_2}\right)\times$$

$$\times\left(\sum_{(1,1)\le(k_1,l_1)<(k_2,l_2)}\left[T_{k_2-k_1,l_2-l_1-1}\bar{B}_1+T_{k_2-k_1-1,l_2-l_1}\bar{B}_2\right]\bar{u}_{k_1l_1}M_{k_1l_1}\right)+\ldots+$$

$$+\left(\sum_{(r,r)\le(k_r,l_r)<(i+1,j+1)}\left[T_{i-k_r+1,j-l_r}\bar{B}_1+T_{i-k_r,j-l_r+1}\bar{B}_2\right]\bar{u}_{k_rl_r}\right)\times$$

$$\times\left(\sum_{(r,r)\le(k_r-1,l_r-1)<(k_r,l_r)}\left[T_{k_r-k_{r-1},l_r-l_{r-1}-1}\bar{B}_1+T_{k_r-k_{r-1}-1,l_r-l_{r-1}}\bar{B}_2\right]\bar{u}_{k_{r-1}l_{r-1}}\right)\times\ldots\times$$

$$\times\left(\sum_{(1,1)\le(k_1,l_1)<(k_2,j_2+1)}\left[T_{k_2-k_1,l_2-l_1-1}\bar{B}_1+T_{k_2-k_1-1,l_2-l_1}\bar{B}_2\right]\bar{u}_{k_1l_1}M_{k_1l_1}\right)+$$

$$+\sum_{k_1=1}^{i+1}\sum_{l_1=1}^{j+1}\left[T_{i-k_1+1,j-l_1}C_1+T_{i-k_1,j-l_1+1}C_2\right]\bar{u}_{k_1l_1}+$$

$$+\left(\sum_{(1,1)\le(k_2,l_2)<(i+1,j+1)}\left[T_{i-k_2+1,j-l_2}\bar{B}_1+T_{i-k_2,j-l_2+1}\bar{B}_2\right]\bar{u}_{k_2l_2}\right)\times$$

$$\times\left(\sum_{(1,1)\le(k_1,l_1)<(k_2,l_2)}\left[T_{k_2-k_1,l_2-l_1-1}C_1+T_{k_2-k_1-1,l_2-l_1}C_2\right]\bar{u}_{k_1l_1}\right)+\ldots+$$

$$+\left(\sum_{(r,r)\le(k_r,l_r)<(i+1,j+1)}\left[T_{i-k_r+1,j-l_r}\bar{B}_1+T_{i-k_r,j-l_r+1}\bar{B}_2\right]\bar{u}_{k_rl_r}\right)\times$$

$$\times\left(\sum_{(r,r)\le(k_{r-1},l_{r-1})<(k_r,l_r)}\left[T_{k_r-k_{r-1},l_r-l_{r-1}-1}\bar{B}_1+T_{k_r-k_{r-1}-1,l_r-l_{r-1}}\bar{B}_2\right]\bar{u}_{k_{r-1}l_{r-1}}\right)\times\ldots\times$$

$$\times\left(\sum_{(1,1)\le(k_1,l_1)<(k_2,l_2)}\left[T_{k_2-k_1,l_2-l_1-1}C_1+T_{k_2-k_1-1,l_2-l_1}C_2\right]\bar{u}_{k_1l_1}\right)$$

Therefore, the hypothesis is valid for the pair $(i+1,\ j+1)$. ■

To find a solution of 21 we shall consider the equation 21 as a model with variable coefficients [6].

Let

$$\bar{A}_{ij}=A+B\bar{u}_{ij}=\begin{bmatrix}A_{ij}^{11} & A_{ij}^{12}\\ A_{ij}^{21} & A_{ij}^{22}\end{bmatrix},\quad C=\begin{bmatrix}C_1\\ C_2\end{bmatrix},\quad \begin{array}{ll}A_{ij}^{11}\in\mathbb{R}^{n_1\times n_1}, & C_1\in\mathbb{R}^{n_1\times m},\\ A_{ij}^{22}\in\mathbb{R}^{n_2\times n_2}, & C_2\in\mathbb{R}^{n_2\times m}\end{array}$$

The transition matrix T_{ij}^{kl} of 21 is defined as follows
$T_{ij}^{00}=I$ (the identity matrix) for $i,\ j\in\mathbb{Z}_+$

$T_{ij}^{kl} = A_{i-1,j}^{10} T_{i-1,j}^{k-1,l} + A_{ij}^{10} T_{i,j-1}^{k,l-1}$ for $i, j, k, l \in \mathbb{Z}_+$

$$A_{i-1,j}^{10} := \begin{bmatrix} A_{i-1,j}^{11} & A_{i-1,j}^{12} \\ 0 & 0 \end{bmatrix}, \quad A_{i,j-1}^{01} := \begin{bmatrix} 0 & 0 \\ A_{i,j-1}^{21} & A_{i,j-1}^{22} \end{bmatrix} \tag{25}$$

$A_{ij}^{10} = 0, \quad A_{ij}^{01} = 0$ for $i < 0$ or/and $j < 0$
$T_{ij}^{kl} = 0$ (the zero matrix) for $k < 0$ or $l < 0$ or $i < 0$ or $j < 0$
Note that T_{ij}^{kl} depends on u_{ij}.

Theorem 3 *The solution* $\begin{bmatrix} x_{ij}^h \\ x_{ij}^v \end{bmatrix}$ *to the equation 21 with boundary conditions 14 is given by*

$$\begin{bmatrix} x_{ij}^h \\ x_{ij}^v \end{bmatrix} = \sum_{p=0}^{i} T_{ij}^{i-p,j} \begin{bmatrix} 0 \\ x_{p0}^v \end{bmatrix} + \sum_{q=0}^{j} T_{ij}^{i,j-q} \begin{bmatrix} x_{0q}^h \\ 0 \end{bmatrix} +$$

$$+ \sum_{p=0}^{i-1} \sum_{q=0}^{j} T_{ij}^{i-p-1,j-q} \begin{bmatrix} C_1 \\ 0 \end{bmatrix} u_{pq} + \sum_{p=0}^{i} \sum_{q=0}^{j-1} T_{ij}^{i-p,j-q-1} \begin{bmatrix} 0 \\ C_2 \end{bmatrix} u_{pq} \tag{26}$$

The proof can be accomplished by induction on the pair (i, j) in a similar way as for Theorem 2.
The general response formula may be obtained by substitution of 22 into 3 and 26 into 13.

4. Local Reachability and Local Controllability

To simplify our considerations let us consider a homogeneous, single-input 2-D bilinear system described by the equation

$$x_{i+1,j+1} = A_0 x_{ij} + A_1 x_{i+1,j} + A_2 x_{i,j+1} + u_{ij} B x_{ij} \quad i, j \in \mathbb{Z}_+ \tag{27}$$

where $x_{ij} \in \mathbb{R}^n$ is the local state vector, u_{ij} is a scalar input and $A_k, \; B \in \mathbb{R}^{n\times n}, \; k = 0, 1, 2$
Note that 27 my be written in the form

$$x_{i+1,j+1} = A_{ij}^0 x_{ij} + A^1 x_{i+1,j} + A^2 x_{i,j+1} \tag{28}$$

where

$$A_{ij}^0 := A_0 + u_{ij} B, \quad A^1 := A_1, \quad A^2 := A_2$$

28 is a particular case of the model with variable coefficients [6].

Theorem 4 *The solution x_{ij} to 27 or 28 with boundary conditions 4 is given by*

$$\begin{aligned} x_{ij} &= \sum_{p=1}^{i}\left(T_{ij}^{i-p,j-1}A_1 + T_{ij}^{i-p-1,j-1}A_{p0}^0\right)x_{p0} + \\ &+ \sum_{q=1}^{j}\left(T_{ij}^{i-1,j-q}A^2 + T_{ij}^{i-1,j-q-1}A_{p0}^0\right)x_{0q} + T_{ij}^{i-1,j-1}A_{00}^0 x_{00} \quad i,j \in \mathbb{Z}_+ \end{aligned} \tag{29}$$

where T_{ij}^{kl} is the transition matrix of 28 defined as follows

$$\begin{aligned} T_{ij}^{00} &= I \text{ (the identity matrix)} \quad \text{for } i, j \in \mathbb{Z}_+ \\ T_{ij}^{kl} &= A_{i-1,j-1}^0 T_{i-1,j-1}^{k-1,l-1} + A^1 T_{i,j-1}^{k,l-1} + A^2 T_{i-1,j}^{k-1,l} \quad \text{for } i, j, k, l \in \mathbb{Z}_+ \\ T_{ij}^{kl} &= 0 \text{ (the zero matrix)} \quad \text{for } i < 0 \text{ or } j < 0 \text{ or } k < 0 \text{ or } l < 0 \end{aligned} \tag{30}$$

The proof can be accomplished by induction on the pair (i, j) in a similar way as for Theorem 2.

Definition 1 *The 2-D bilinear system 27 is called locally reachable in the rectangle*

$$R_{rt} := [(0,0),(r,t)] := \{(i,j) \in \mathbb{Z}_+ \times \mathbb{Z}_+ : 0 \le i \le r,\ 0 \le j \le t\} \tag{31}$$

if for any boundary conditions

$$x_{i0},\quad i \in [0,r],\quad x_{0j},\quad j \in [0,t] \tag{32}$$

and every nonzero vector $x_f \in \mathbb{R}^n$ there exist a sequence of inputs u_{ij} for $i \in [0, r-1],\ j \in [0, t-1]$ such that $x_{rt} = x_f$.

Definition 2 *The 2-D bilinear system 27 is called controllable in the rectangle 31 if for any given boundary conditions 32 there exists a sequence of inputs u_{ij} for $i \in [0, r-1],\ j \in [0, t-1]$ such that $x_{rt} = 0$.*

The local reachability and local controllability of the system 27 depends on the model matrices A_k, $(k = 0, 1, 2)$, B but also strongly on the boundary conditions 31.

Theorem 5 *If $A_1^r \neq 0$ or/and $A_2^t \neq 0$ then there exist suitable boundary conditions 32 such that the system 27 is not locally controllable (reachable) in the rectangle 31.*

Proof Using 29 and 30 $x_{i0} = 0$, $i \in [0, r-1]$, $x_{r0} \neq 0$ and $x_{0j} = 0$ for $j \in [0, t-1]$, $x_{0t} \neq 0$ we obtain

$$x_{rt} = T_{0,t-1}A_1 x_{r0} + T_{r-1,0}A_2 x_{0t} = A_1^t x_{r0} + A_2^r x_{0t} \tag{33}$$

From 33 it follows that it does not exist a sequence of input vectors u_{ij} for $i \in [0, r-1]$, $j \in [0, t-1]$ such that $x_{rt} = 0$ *(or $x_{rt} = x_f$).* ■

Therefore, new notions of the local reachability and the local controllability for given boundary conditions will be introduced.

Definition 3 *The 2-D bilinear system 27 is called locally reachable for given boundary conditions 32 in the rectangle 31 if for the given boundary conditions and every nonzero vector $x_f \in \mathbb{R}^n$ there exist a sequence of inputs u_{ij} for $i \in [0, r-1]$, $j \in [0, t-1]$ such that $x_{rt} = x_f$.*

Definition 4 *The 2-D bilinear system 27 is called locally controllable for given boundary conditions 31 in the rectangle 30 if for the given boundary conditions there exist a sequence of inputs u_{ij} for $i \in [0, r-1]$, $j \in [0, t-1]$ such that $x_{rt} = 0$.*

It is assumed that

$$\operatorname{rank} B = 1 \tag{34}$$

If 34 holds then B can be written as

$$B = bd \tag{35}$$

where $b \in \mathbb{R}^{n\times 1}$ and $d \in \mathbb{R}^{1\times n}$.
Defining the new input

$$v_{ij} = u_{ij} d x_{ij}, \quad i, j \in \mathbb{Z}_+ \tag{36}$$

and using 35 we may write 27 in the form of first Fornasini-Marchesini model

$$x_{i+1,j+1} = A_0 x_{ij} + A_1 x_{i+1,j} + A_2 x_{i,j+1} + b v_{ij} \tag{37}$$

It is well-known [6,19] that the 2-D linear system 37 is locally reachable in 31 if and only if

$$\operatorname{rank}\left[T_{ij} b : i \in [0, r-1], \quad j \in [0, t-1]\right] = n \tag{38}$$

where T_{ij} is the transmition matrix of 37 defined as follows

$$\begin{cases} T_{00} := I_n \text{ (the identity matrix)} \\ T_{ij} := A_0 T_{i-1,j-1} + A_1 T_{i,j-1} + A_2 T_{i-1,j} \quad \text{for } i, j \in \mathbb{Z}_+ \\ T_{ij} := 0 \text{ (the zero matrix)} \quad \text{if } i < 0 \quad \text{or/and } j < 0 \end{cases} \tag{39}$$

If 38 holds then there exists a sequence of inputs v_{ij} for $i \in [0, r-1]$, $j \in [0, t-1]$ of 37 such that $x_{rt} = x_f$. Knowing the sequence of inputs v_{ij} we may find from 36 a suitable sequence of inputs u_{ij} for $i \in [0, r-1]$, $j \in [0, t-1]$ of 27 if and only if

$$d x_{ij} \neq 0 \text{ for all } i \in [0, r-1], \; j \in [0, t-1] \tag{40}$$

It will be shown that the condition 40 is satisfied if

$$x_{pq} \notin \ker d \quad \text{for all } p \in [0, r-1], \; q \in [0, t-1] \tag{41}$$

where ker denotes the kernel and

$$x_{pq} = x_{bc}(p, q) + \sum_{\alpha=0}^{p-1} \sum_{\beta=0}^{q-1} T_{p-\alpha-1, q-\beta-1} b v_{\alpha\beta} \tag{42}$$

$$x_{bc}(p,q) = \begin{cases} x_{p0} & \text{for } q = 0,\ p \geq 0 \\ x_{0q} & \text{for } p = 0,\ q > 0 \\ \sum_{\alpha=0}^{p-1} T_{p-\alpha-1,q-1} A_0 x_{\alpha 0} + \sum_{\alpha=1}^{p} T_{p-\alpha,q-1} A_1 x_{\alpha 0} + & \\ + \sum_{\beta=1}^{q-1} T_{p-1,q-\beta-1} A_0 x_{0\beta} + \sum_{\beta=1}^{q} T_{p-1,q-\beta} A_2 x_{0\beta} & \text{for } p > 0,\ q > 0 \end{cases} \tag{43}$$

The solution x_{ij} of 37 for $i = p,\ j = q$ is given by 42. Note that 40 holds if $x_{ij} \notin \ker d$ for all $i \in [0, r-1],\ j \in [0, t-1]$. Therefore, if 41 is satisfied then 40 holds.

Therefore, we have proved the following.

Theorem 6 *The 2-D bilinear system 27 with B satisfying 34 is locally reachable for given boundary conditions 32 in the rectangle 31 if and only if the conditions 38 and 41 are satisfied.*

Example 1 Consider the system 27 with

$$A_0 = \begin{bmatrix} 0 & 0 \\ 0 & 0 \end{bmatrix}, \quad A_1 = \begin{bmatrix} 1 & 0 \\ 0 & 1 \end{bmatrix}, \quad A_2 = \begin{bmatrix} 0 & 1 \\ 1 & 1 \end{bmatrix}, \quad B = \begin{bmatrix} 1 & 1 \\ 0 & 0 \end{bmatrix} \tag{44}$$

and the boundary conditions

$$x_{i0} = \begin{bmatrix} 0 \\ 1 \end{bmatrix} \quad \text{for } i = 0, 1, \ldots \quad \text{and} \quad x_{00} = \begin{bmatrix} 1 \\ 0 \end{bmatrix} \quad \text{for } j = 1, 2, \ldots \tag{45}$$

In this case the condition 34 is satisfied and

$$B = bd$$

where

$$b = \begin{bmatrix} 1 \\ 0 \end{bmatrix} \quad \text{and} \quad d = \begin{bmatrix} 1 & 1 \end{bmatrix}$$

Let $r = t = 2$ and $x_f = \begin{bmatrix} 4 \\ 4 \end{bmatrix}$.

The linear model 37 with $A_0,\ A_1,\ A_2$ given by 44 and $b = \begin{bmatrix} 1 \\ 0 \end{bmatrix}$ is reachable in the rectangle $[(0,0),(2,2)]$ since

$$\text{rank } V_{22} = \text{rank}\,[b,\ A_1 b,\ A_2 b,\ (A_0 + A_1 A_2 + A_2 A_1)\, b] = \text{rank} \begin{bmatrix} 1 & 1 & 0 & 0 \\ 0 & 0 & 1 & 2 \end{bmatrix} = 2$$

Taking into account that $\ker d = \ker \begin{bmatrix} 1 & 1 \end{bmatrix} = \left\{ \begin{bmatrix} a \\ -a \end{bmatrix} \text{ for all } a \in R \right\}$ it is easy to check that the condition 41 is also satisfied.

Using 42 for $p = q = 2$ we may compute the sequence of inputs $\{v_{00}, v_{10}, v_{01}, v_{11}\}$ from the equation

$$x_f - x_{bc}(2,2) = V_{22}\begin{bmatrix} v_{11} \\ v_{10} \\ v_{01} \\ v_{00} \end{bmatrix} = \begin{bmatrix} 1 & 1 & 0 & 0 \\ 0 & 0 & 1 & 2 \end{bmatrix}\begin{bmatrix} v_{11} \\ v_{10} \\ v_{01} \\ v_{00} \end{bmatrix} = \begin{bmatrix} -1 \\ -2 \end{bmatrix} \tag{46}$$

Assuming $v_{11} = v_{00} = 1$ and solving 46 we obtain $[v_{11}, v_{10}, v_{01}, v_{00}] = [1, -2, -4, 1]$ Using 37 and 36 we get $x_{11} = A_0 x_{00} + A_1 x_{10} + A_2 x_{01} + b v_{11} = \begin{bmatrix} 1 \\ 2 \end{bmatrix}$ and

$$u_{00} = \frac{v_{00}}{dx_{00}} = v_{00} = 1, \; u_{10} = \frac{v_{10}}{dx_{10}} = v_{10} = -2,$$

$$u_{01} = \frac{v_{01}}{dx_{01}} = v_{01} = -4, \; u_{11} = \frac{v_{11}}{dx_{11}} = \frac{1}{3}$$

Theorem 7 *The 2-D bilinear systems 27 with B satisfying 34 is locally controllable for given boundary conditions 32 in the rectangle 31 if and only if the condition 41 and*

$$\operatorname{rank}\left[T_{ij}b : i \in [0, r-1], \; j \in [0, t-1]\right] =$$
$$\operatorname{rank}\left[T_{ij}b, T_{i,t-1}A_0, T_{i,t-1}A_1, T_{r-1,j}A_0, T_{r-1,j}A_2 : \; i \in [0, r-1], \; j \in [0, t-1]\right] \tag{47}$$

are satisfied.

Proof A sequence of inputs v_{ij} for $i \in [0, r-1], \; j \in [0, t-1]$ of 37 such that $x_{rt} = 0$ may be found if and only if the condition 47 is satisfied [11]. Knowing the sequence of inputs v_{ij} we may find from 36 a suitable sequence of input u_{ij} for $i \in [0, r-1], \; j \in [0, t-1]$ of 27 if 41 holds. ■

Remark 1 *From 47 it follows that the 2-D bilinear system 27 may be locally controllable for given boundary conditions 32 in 31 even if the linear system 37 is not locally reachable.*

Remark 2 *If 38 holds then 47 is satisfied for any* $T_{i,t-1}A_0, T_{i,t-1}A_1, T_{r-1,j}A_0, T_{r-1,j}A_2$. Therefore, the local reachability for given boundary conditions 32 of the system 27 implies always its local controllability for given boundary conditions 32.

The considerations can be easily extended for 2-D bilinear systems described by the equation

$$x_{i+1,j+1} = A_0 x_{ij} + A_1 x_{i+1,j} + A_2 x_{i,j+1} + u_{ij}^0 B x_{ij} + C u_{ij} \tag{48}$$

where u_{ij}^0 is a scalar input,

$$u_{ij}^T := \left[u_{ij}^1, u_{ij}^2, \ldots, u_{ij}^m\right], \qquad C \in \mathbb{R}^{n \times m}$$

and B satisfies the condition 34.
Using 35 and 36 we may write 48 in the form

$$x_{i+1,j+1} = A_0 x_{ij} + A_1 x_{i+1,j} + A_2 x_{i,j+1} + [b, c] \begin{bmatrix} v_{ij}^0 \\ u_{ij} \end{bmatrix} \tag{49}$$

where

$$v_{ij}^0 := u_{ij}^0 d x_{ij} \tag{50}$$

If the linear system 49 is locally reachable in 31 then there exists a sequence of inputs $\begin{bmatrix} v_{ij}^0 \\ u_{ij} \end{bmatrix}$ for $i \in [0, r-1]$, $j \in [0, t-1]$ of 49 such that $x_{rt} = x_f$. Knowing v_{ij}^0 we may find from 50 a suitable sequence of inputs u_{ij}^0 for 40 if the condition 41 is satisfied.
Therefore, we have the following
Theorem 8 *The bilinear system 48 with B satisfying 34 is locally reachable for given boundary conditions 32 in the rectangle 31 if and only if the condition 41 and*

$$\operatorname{rank}\left\{T_{ij}[b, c] : \ i \in [0, r-1], \ j \in [0, t-1]\right\} = n \tag{51}$$

are satisfied, where T_{ij} is defined by 39.
The considerations can be also extended for 2-D bilinear systems described by the equation

$$x_{i+1,j+1} = A_0 x_{ij} + A_1 x_{i+1,j} + A_2 x_{i,j+1} + \sum_{k=1}^{m} u_{ij}^k B_k x_{ij} \tag{52}$$

where $B_k \in \mathbb{R}^{n \times n}$ and $\operatorname{rank} B_k = 1 \quad \text{for} \quad k = 1, \ldots, m.$
Let

$$B_k = b_k d_k \quad k = 1, \ldots, m \tag{53}$$

where

$$b_k \in \mathbb{R}^{n \times 1}, \quad d_k \in \mathbb{R}^{1 \times n}$$

Defining

$$v_{ij}^k := u_{ij}^k d x_{ij} \quad k = 1, \ldots, m \tag{54}$$

and using 53 we may write 52 in the form

$$x_{i+1,j+1} = A_0 x_{ij} + A_1 x_{i+1,j} + A_2 x_{i,j+1} + \bar{B} v_{ij} \tag{55}$$

where

$$\bar{B} := [b_1, b_2, \ldots, b_m], \quad v_{ij}^T := \left[v_{ij}^1, v_{ij}^2, \ldots, v_{ij}^m\right]$$

In a similar way we may prove the following

Theorem 9 *The bilinear system 52 with B_k satisfying the condition* rank $B_k = 1$, $k = 1, \dots, m$ *is locally reachable for given boundary conditions 32 in the rectangle 31 if and only if the condition 41 and*

$$\operatorname{rank}\left[T_{ij}\bar{B} : i \in [0, r-1],\ j \in [0, t-1]\right] = n$$

are satisfied, where T_{ij} is defined by 39.
Corresponding results for the local controllability for given boundary conditions can be obtainded for the both 2-D bilinear systems 48 and 52.
Combining the results of Theorems 8 and 9 we may obtained similar results for 2-D bilinear system described by the equation

$$x_{i+1,j+1} = A_0 x_{ij} + A_1 x_{i+1,j} + A_2 x_{i,j+1} + \sum_{k=1}^{m} u_{ij}^k B_k x_{ij} + C u_{ij} \tag{56}$$

where rank $B_k = 1$ for $k = 1, \dots, m$.

5. Observability of 2-D Bilinear Systems

Consider the 2-D bilinear system described by 27 and 3 with boundary conditions 4.

Definition 5 *The 2-D bilinear system 27, 3 is called observable in the rectangle 31 if knowing the input u_{ij} and the output y_{ij} for $(i, j) \in R_{rt}$ it is possible to compute uniquely the boundary conditions*

$$x_{i0} \quad \text{for } 0 \le i \le r \quad \text{and} \quad x_{0j} \quad \text{for } 0 \le j \le t \tag{57}$$

Substituting 29 into 3 we obtain

$$y'_{ij} = \sum_{p=1}^{i} M_{ij}^1(p) x_{p0} + \sum_{q=1}^{j} M_{ij}^2(q) x_{0q} + M_{ij}^0 x_{00} \tag{58}$$

where

$$\begin{aligned} y'_{ij} &:= y_{ij} - H u_{ij} \quad i, j \in \mathbb{Z}_+ \\ M_{ij}^1(p) &:= D\left(T_{ij}^{i-p,j-1} A^1 + T_{ij}^{i-p-1,j-1} A_{p0}^0\right) \quad p = 1, \dots, i \\ M_{ij}^2(q) &:= D\left(T_{ij}^{i-1,j-q} A^2 + T_{ij}^{i-1,j-q-1} A_{0q}^0\right) \quad q = 1, \dots, j \\ M_{ij}^0(p) &:= D T_{ij}^{i-1,j-1} A_{00}^0 \end{aligned} \tag{59}$$

Using 58 for $i = 0, 1, \dots, r$ and $j = 0, 1, \dots, t$ we may write

$$\bar{y}_{rt} = M_{rt} \begin{bmatrix} x_{00} \\ \bar{x}_{1r} \\ \bar{x}_{2t} \end{bmatrix} \tag{60}$$

where

$$\bar{y}_{rt} := \begin{bmatrix} y'_{00} \\ y'_{10} \\ \vdots \\ y'_{r0} \\ y'_{01} \\ \vdots \\ y'_{rl} \\ y'_{02} \\ \vdots \\ y'_{rt} \end{bmatrix}, \quad \bar{x}_{1r} := \begin{bmatrix} x_{10} \\ x_{20} \\ \vdots \\ x_{r0} \end{bmatrix}, \quad \bar{x}_{2t} := \begin{bmatrix} x_{01} \\ x_{02} \\ \vdots \\ x_{0t} \end{bmatrix}$$

$$M_{rt} := \begin{bmatrix} D & 0 & 0 & \dots & 0 & 0 & 0 & \dots & 0 \\ 0 & D & 0 & \dots & 0 & 0 & 0 & \dots & 0 \\ \vdots & \vdots & \vdots & \ddots & \vdots & \vdots & \vdots & \ddots & \vdots \\ 0 & 0 & 0 & \dots & D & 0 & 0 & \dots & 0 \\ 0 & 0 & 0 & \dots & 0 & D & 0 & \dots & 0 \\ \vdots & \vdots & \vdots & \ddots & \vdots & \vdots & \vdots & \ddots & \vdots \\ M^0_{r1} & M^1_{r1}(1) & M^1_{r1}(2) & \dots & M^1_{r1}(r) & M^2_{r1}(1) & 0 & \dots & 0 \\ 0 & 0 & 0 & \dots & 0 & 0 & D & \dots & 0 \\ M^0_{rt} & M^1_{rt}(1) & M^1_{rt}(2) & \dots & M^1_{rt}(r) & M^2_{rt}(1) & M^2_{rt}(2) & \dots & M^2_{rt}(t) \end{bmatrix}$$

From linear algebra it is well-known that the equation 60 has a solution $\begin{bmatrix} x_{00} \\ \bar{x}_{1r} \\ \bar{x}_{2t} \end{bmatrix}$ if and only if

$$\text{rank } M_{rt} = \text{rank } [M_{rt}, \ \bar{y}_{rt}] \tag{61}$$

The solution is unique if and only if the matrix M_{rt} has full column rank or

$$\ker \ M_{rt} = 0 \tag{62}$$

where ker denotes the kernel.
It is assumed that for given u_{ij} and y_{ij}, $(i, j) \in R_{rt}$ the condition 61 is satisfied.
Therefore, we have the following

Theorem 10 *The 2-D bilinear system 27, 3 is observable in the rectangle 31 if and only if 61 and 62 hold.*

Remark 3 *Note that the matrix M_{rt} depends on u_{ij}, $(i, j) \in R_{rt}$ belonging to some specified set U_{rt}. In this case the 2-D bilinear system is observable in the rectangle 31 for $u_{ij} \in U_{rt}$.*

Example 2 Consider the 2-D bilinear system 27, 3 with

$$A_1 = \begin{bmatrix} 0 & 0 & 1 \\ 0 & 1 & 0 \\ 1 & 0 & 0 \end{bmatrix}, \quad A_2 = \begin{bmatrix} 0 & 1 & 0 \\ 0 & 0 & 1 \\ 0 & 0 & 0 \end{bmatrix}, \quad B = \begin{bmatrix} 1 & 0 & 1 \\ 0 & 1 & 0 \\ -1 & 0 & -1 \end{bmatrix},$$

$$D = \begin{bmatrix} 1 & 0 & 0 \\ 0 & 1 & 0 \end{bmatrix}, \quad H = 0$$

and for $a)$ $A_0 = 0$, $b)$ $A_0 = \begin{bmatrix} 1 & 0 & 0 \\ 0 & 1 & 0 \\ 0 & 0 & 1 \end{bmatrix}$.

Compute the boundary conditions 57 for $r = 2,\ t = 1$. For the inputs $u_{00} = 1,\ u_{10} = -1$ the outputs of the system are equal

$$y_{00} = \begin{bmatrix} 1 \\ 0 \end{bmatrix}, \quad y_{10} = \begin{bmatrix} 0 \\ 1 \end{bmatrix}, \quad y_{20} = \begin{bmatrix} -1 \\ 1 \end{bmatrix},$$

$$y_{01} = \begin{bmatrix} 1 \\ -1 \end{bmatrix}, \quad y_{11} = \begin{bmatrix} -2 \\ 2 \end{bmatrix}, y_{21} = \begin{bmatrix} 2 \\ 0 \end{bmatrix}$$

Using 59 and 30 we obtain:
case a)

$$M_{21}^a = \begin{bmatrix} D & 0 & 0 & 0 \\ 0 & D & 0 & 0 \\ 0 & 0 & D & 0 \\ 0 & 0 & 0 & D \\ DBu_{00} & DA_1 & 0 & DA_2 \\ DA_2Bu_{00} & D\left(A_2A_1 + Bu_{10}\right) & DA_1 & DA_2^2 \end{bmatrix} =$$

$$= \begin{bmatrix} 1 & 0 & 0 & 0 & 0 & 0 & 0 & 0 & 0 & 0 & 0 & 0 \\ 0 & 1 & 0 & 0 & 0 & 0 & 0 & 0 & 0 & 0 & 0 & 0 \\ 0 & 0 & 0 & 1 & 0 & 0 & 0 & 0 & 0 & 0 & 0 & 0 \\ 0 & 0 & 0 & 0 & 1 & 0 & 0 & 0 & 0 & 0 & 0 & 0 \\ 0 & 0 & 0 & 0 & 0 & 0 & 1 & 0 & 0 & 0 & 0 & 0 \\ 0 & 0 & 0 & 0 & 0 & 0 & 0 & 1 & 0 & 0 & 0 & 0 \\ 0 & 0 & 0 & 0 & 0 & 0 & 0 & 0 & 0 & 1 & 0 & 0 \\ 0 & 0 & 0 & 0 & 0 & 0 & 0 & 0 & 0 & 0 & 1 & 0 \\ u_{00} & 0 & u_{00} & 0 & 0 & 1 & 0 & 0 & 0 & 0 & 1 & 0 \\ 0 & u_{00} & 0 & 0 & 1 & 0 & 0 & 0 & 0 & 0 & 0 & 1 \\ 0 & u_{00} & 0 & u_{10} & 1 & u_{10} & 0 & 0 & 0 & 0 & 0 & 1 \\ -u_{00} & 0 & -u_{00} & 1 & u_{10} & 0 & 0 & 1 & 0 & 0 & 0 & 0 \end{bmatrix} \qquad (63)$$

case b)

$$M_{21}^b = \begin{bmatrix} D & 0 & 0 & 0 \\ 0 & D & 0 & 0 \\ 0 & 0 & D & 0 \\ 0 & 0 & 0 & D \\ D\,(A_0 B u_{00}) & DA_1 & 0 & DA_2 \\ DA_2\,(A_0 B u_{00}) & D\,(A_0 + A_2 A_1 + B u_{10}) & DA_1 & DA_2^2 \end{bmatrix} =$$

$$= \begin{bmatrix} 1 & 0 & 0 & 0 & 0 & 0 & 0 & 0 & 0 & 0 & 0 & 0 \\ 0 & 1 & 0 & 0 & 0 & 0 & 0 & 0 & 0 & 0 & 0 & 0 \\ 0 & 0 & 0 & 1 & 0 & 0 & 0 & 0 & 0 & 0 & 0 & 0 \\ 0 & 0 & 0 & 0 & 1 & 0 & 0 & 0 & 0 & 0 & 0 & 0 \\ 0 & 0 & 0 & 0 & 0 & 0 & 1 & 0 & 0 & 0 & 0 & 0 \\ 0 & 0 & 0 & 0 & 0 & 0 & 0 & 1 & 0 & 0 & 0 & 0 \\ 0 & 0 & 0 & 0 & 0 & 0 & 0 & 0 & 0 & 1 & 0 & 0 \\ 0 & 0 & 0 & 0 & 0 & 0 & 0 & 0 & 0 & 0 & 1 & 0 \\ 1+u_{00} & 0 & u_{00} & 0 & 0 & 1 & 0 & 0 & 0 & 0 & 1 & 0 \\ 0 & 1+u_{00} & 0 & 0 & 1 & 0 & 0 & 0 & 0 & 0 & 0 & 1 \\ 0 & 1+u_{00} & 0 & 1+u_{10} & 1 & u_{10} & 0 & 0 & 0 & 0 & 0 & 1 \\ -u_{00} & 0 & 1-u_{00} & 1 & 1-u_{10} & 0 & 0 & 1 & 0 & 0 & 0 & 0 \end{bmatrix} \tag{64}$$

It is easy to check that

$$\det\ M_{21}^a = u_{00} \quad \text{and} \quad M_{21}^b = 1 - u_{00}$$

Therefore, the matrix 63 is nonsingular if $u_{00} \neq 0$ and the matrix 64 is nonsingular if $u_{00} \neq 1$. Note that in the case b) the system is observable in the rectangle R_{21} even for the zero inputs.

In the case a) the equation 60 takes the form

$$y_{21} = \bar{M}_{21}^a \begin{bmatrix} x_{00} \\ \bar{x}_{12} \\ \bar{x}_{21} \end{bmatrix}, \tag{65}$$

where

$$y_{21} = \left[y_{00}^T, y_{10}^T, y_{20}^T, y_{01}^T, y_{11}^T, y_{21}^T\right]^T, \quad \bar{x}_{12} = \left[x_{10}^T, x_{20}^T\right]^T, \quad \bar{x}_{21} = x_{01}$$

and $\bar{M}_{21}^a$ is equal to the matrix 63 for $u_{00} = 1,\ u_{10} = -1$.

From 65 we have

$$\left[x_{00}^T, x_{10}^T, x_{20}^T, x_{01}^T\right]^T = \left[\bar{M}_{21}^a\right]^{-1} y_{21} =$$

$$\begin{bmatrix} 1 & 0 & -1 & 0 & 1 & -1 & -1 & 1 & -1 & 1 & -1 & 1 \end{bmatrix}^T$$

The above considerations can be extended for more general models of 2-D bilinear systems.

6. Concluding Remarks

New models of Fornasini-Marchesini type 1, 2 and Roesser type 11 of 2-D bilinear systems have been introduced. Relationships between the models have been established. It has been shown that the models of 2-D bilinear systems are equivalent to suitable models of 2-D systems with variable coefficients. Solutions 22 and 26 to the new models have been derived. Necessary and sufficient conditions have been established for the local reachability and local controllability for given boundary conditions of a class of 2-D bilinear systems satisfying the restrictive assumption 34. It has been shown that the local reachability for given boundary conditions of a 2-D bilinear system implies always its local controllability for given boundary conditions. The observability of 2-D bilinear systems have been defined and its necessary and sufficient conditions have been established. Some of the considerations can be extended for singular [6] 2-D bilinear systems; for example it is easy to show that singular models of 2-D bilinear systems are equivalent to suitable singular models of 2-D systems with variable coefficients. An extension of the necessary and sufficient conditions for the models 1, 2 which do not satisfy the condition 34 and for the model 11 is an open problem. An other open problem is the minimum energy control problem [19,6] for the models 1, 2 and 11 of the 2-D bilinear systems.

7. References

1. R. P. Roesser, "A discrete state-space model for linear image processing," *IEEE Transactions on Automatic Control*, vol. 20, no. 1, pp. 1–10, 1975.

2. E. Fornasini and G. Marchesini, "State space realization theory of two-dimensional filters," *IEEE Transactions on Automatic Control*, vol. 21, no. 4, pp. 484–491, 1976.

3. E. Fornasini and G. Marchesini, "Doubly indexed dynamical systems: State space models and structural properties," *Mathematical Systems Theory*, vol. 12, pp. 59–72, 1978.

4. J. Kurek, "The general state-space model for two-dimensional linear digital system," *IEEE Transactions on Automatic Control*, vol. 30, pp. 600–602, 1985.

5. T. Kaczorek, "Singular general model of 2-D systems and its solution," *IEEE Transactions on Automatic Control*, vol. 33, no. 11, pp. 1060–1061, 1988.

6. T. Kaczorek, *Linear Control System*, vol. 1,2. Research Studies Press, 1993.

7. F. L. Lewis, "A review of 2-D implicit systems," *Automatica*, vol. 28, no. 2, pp. 345–354, 1992.

8. D. L. Elliot, T. J. Tarn, and T. Gorka, "Controllability of bilinear systems with bounded controls," *IEEE Transactions on Automatic Control*, vol. 18, no. 2, pp. 298–301, 1973.

9. M. Espana and I. D. Landau, "Reduced order bilinear models for distillation columns," *Automatica*, vol. 14, pp. 345–355, 1978.

10. M. E. Evans and D. N. P. Murthy, "Controllability of a class of discrete-time bilinear systems," *IEEE Transactions on Automatic Control*, vol. 22, no. 1, pp. 78–83, 1977.

11. M. E. Evans and D. N. P. Murthy, "Controllability of the discrete-time inhomogeneous bilinear systems," *Automatica*, vol. 14, pp. 147–151, 1978.

12. Y. Funahashi, "Comments on controllability of a class of discrete-time bilinear systems," *IEEE Transactions on Automatic Control*, vol. 24, no. 4, 1979.

13. T. Gorka, T. J. Tarn, and I. Zaborszky, "On controllability of a class of discrete-time bilinear systems," *Automatica*, vol. 9, no. 4, pp. 615–622, 1979.

14. O. M. Grasselli, A. Isidori, and F. Nicolo, "Dead-beat control of discrete-time bilinear systems," *International Journal of Control*, vol. 32, no. 1, pp. 31–39, 1980.

15. P. Hollis and D. N. P. Murthy, "Study of uncontrollable discrete bilinear systems," *IEEE Transactions on Automatic Control*, vol. 27, no. 1, 1982.

16. R. Mohler, *Bilinear Control Processes: With Applications to Engineering, Ecology, and Medicine*. New York: Academic Press, 1973.

17. T. Kaczorek, "2-D bilinear systems," *Bulletin of the Polish Academy of Sciences, Technical Sciences*, vol. 41, no. 3, pp. 207–214, 1993.

18. T. Kaczorek, "General response formula for 2-D bilinear systems," *Applied Mathematics and Computer Science*, vol. 4, no. 1, pp. 79–86, 1994.

19. J. Klamka, *Controllability of Dynamical Systems*. Warszawa-London: PWN and Kluwer Academic Publisher, 1991.

STABLE ADAPTIVE CONTROL OF A GENERAL CLASS OF NON-LINEAR SYSTEMS

TOR A. JOHANSEN
SINTEF Automatic Control, N-7034 Trondheim, Norway.
E-mail: Tor.Arne.Johansen@regtek.sintef.no

and

MARIOS M. POLYCARPOU
Department of Electrical and Computer Engineering,
University of Cincinnati, Cincinnati, Ohio 45221-0030, USA.
E-mail: polycarpou@uc.edu.

ABSTRACT

The properties of an adaptive control system based on a general class of non-linear models are analyzed. The class of models contains NARX models represented by feedforward neural networks with sigmoidal non-linearity, some radial basis-function expansions, local model networks, and some fuzzy systems. We apply a simple adaptive feedback linearizing controller. The analysis takes into account modeling error and slowly time-varying nominal model parameters. We discuss stability of the closed loop, as well as robustness, and derive performance bounds on the tracking error. Due to the general setup, the results are mainly of a qualitative nature. Finally, we discuss the relevance of the results, and outline some practical modifications and possible extensions.

1. Introduction

The analytical study of adaptive control loops involving complicated non-linear model structures and controllers such as neural networks and fuzzy systems has evolved considerably over the past five years. Some stability results for linearly parameterized model structures using feedback linearizing adaptive control structures include [1,2,3,4], while non-linearly parameterized control structures using feedback linearizing adaptive controllers are analyzed in [5,6,7,8,9,10].

Here we consider discrete-time, nonlinear systems whose input/output behavior can be adequately described by a predictor of the form

$$y(t+d) \quad = \quad f(y(t), ..., y(t-m), u(t), ..., u(t-r), \theta^{\star}(t)) + \nu(t+d) \qquad (1)$$

where f is a non-linear function. The input and output sequences u and y are assumed to be scalar, and m and r are non-negative integers. The integer $d-1 \geq 0$ is the system's time-delay. Alternatively, d can be interpreted as the system's relative degree [11]. The output of the function f is essentially the output predicted using the nominal model, so the sequence ν can be interpreted as unstructured uncertainty, which contains unmodeled dynamics (modeling error and reduced order modeling effects), disturbances, noise, effects due to sampling of continuous-time signals etc. The nominal parameter sequence θ^* may be slowly time-varying.

The NARX-like model (nonlinear autoregressive with exogenous input) represented by (1) is quite general, although there may exist continuous- or discrete-time state-space models that do not have a global NARX representation of this form [12]. On the other hand, we introduce very weak assumptions on the structure of the function f. In particular, it can be represented by a wide range of series expansions, neural networks, fuzzy systems, wavelet expansions etc.

Compared to the other papers on control using non-linear model structures mentioned initially, the results here (see also [13]) rely on significantly weaker assumptions, which in particular allow slowly time-varying nominal parameters and a quite general form of unstructured model uncertainty. Weighted l_2-norms are applied in the analysis [14], which is the key to the strong qualitative results. Here we show that the global stability results in [13] quite naturally reduce to local stability results when the predictor is non-linearly parameterized. It is interesting to observe that the analysis based on weighted l_2-norms is quite analogous to the analysis of adaptive control systems based on discrete-time linear models [14]. There are also close links to the use of weighted L_2-norms when the model is continuous-time [15,16,17].

The continuation of this paper is as follows. First we review some basic results on weighted l_2-norms and state some general assumptions. The adaptive feedback linearizing controller is analyzed for the case of linear and non-linear parameterization of the predictor. Furthermore, the relevance to NARX model structures based on various neural and fuzzy model representations is studied. The assumptions are also closely examined. The paper ends with some practical considerations regarding some of the important design issues and concluding remarks.

2. Preliminaries

The Euclidean norm of a vector $x \in R^n$ is $||x|| = \sqrt{x^T x}$. The difference sequence Δs of a sequence $s = (s(t))$ is defined by $\Delta s(t) = s(t) - s(t-1)$ for all $t \geq 0$. The truncation of a sequence s at time t is denoted s_t. The exponentially weighted l_2-norm of a sequence is defined by

$$||s_t||_2^\delta = \sqrt{\sum_{\tau=0}^{t} \delta^{t-\tau} ||s(\tau)||^2}$$

for any $\delta \in [0, 1]$ and $t \geq 0$. The norm of a discrete-time transfer function $H(q^{-1})$, where q^{-1} is the one-step delay operator and $H(q^{-1})$ has all poles strictly inside the unit disc, is defined by

$$||H(q^{-1})||_\infty = \sup_{\omega \in [0,2\pi]} |H\left(e^{-j\omega}\right)|$$

If $H(q^{-1})$ is proper and has all poles on or inside the disc $|c| \leq \sqrt{\delta}$, then the exponentially weighted norm is defined by $||H(q^{-1})||_\infty^\delta = ||H(\sqrt{\delta}q^{-1})||_\infty$. The mixed notation $s_2(t) = H(q^{-1})s_1(t)$, where $H(q^{-1})$ is an exponentially stable transfer function, means the convolution $s_2(t) = h(t) \star s_1(t)$, where $h(t)$ is the inverse z-transform of the transfer function $H(q^{-1})$. The exponentially decaying term, which may appear as a consequence of non-zero initial conditions, is consistently neglected in this work, as it does not affect the stability analysis. The following lemmas will be used frequently:

Lemma 1 *Let* $s_2(t) = H(q^{-1})s_1(t)$, *where* $H(q^{-1})$ *is a proper causal transfer function that has all poles in the disc* $|c| \leq \sqrt{\delta}$, *where* $\delta \in (0, 1]$, *and suppose* $\sum_{\tau=0}^{t} s_1^2(\tau) < \infty$. *Then* $||(s_2)_t||_2^\delta \leq ||H(q^{-1})||_\infty^\delta ||(s_1)_t||_2^\delta$ *for all* $t \geq 0$

Proof: Given in [14].

■

Lemma 2 *Let* s_1 *be a positive sequence, i.e.* $s_1(t) \geq 0$ *for all* t. *Let the transfer function* $H(q^{-1})$ *be defined by* $H(q^{-1}) = K(1 - \sigma q^{-1})^{-1}$ *for some* $K \geq 0$ *and* $\sigma \in (0, 1)$. *Suppose the sequence* s_2 *satisfies*

$$s_2(t) \leq H(q^{-1})\left(\gamma_0 s_1(t) + \gamma_1 s_1(t-1) + ... + \gamma_n s_1(t-n)\right)$$

for some non-negative integer n *and constants* $\gamma_0, ..., \gamma_n \geq 0$. *Then,*

$$s_2(t) \leq H(q^{-1})\left(\gamma_0 + \gamma_1\sigma^{-1} + ... + \gamma_n\sigma^{-n}\right) s_1(t)$$

Proof:

$$s_2(t) \leq \sigma^t s_2(0) + K\sum_{\tau=0}^{t} \sigma^{t-\tau}\left(\gamma_0 s_1(\tau) + ... + \gamma_n s_1(\tau - n)\right) \tag{2}$$

Clearly, for any $k \in \{0, ..., n\}$

$$\begin{aligned}
\sum_{\tau=0}^{t} \sigma^{t-\tau}\gamma_k s_1(\tau - k) &= \sum_{\tau=-k}^{t-k} \sigma^{t-\tau-k}\gamma_k s_1(\tau) \\
&= \sigma^{-k}\sum_{\tau=0}^{t-k} \sigma^{t-\tau}\gamma_k s_1(\tau) + \text{exp. decaying term} \\
&\leq \sigma^{-k}\sum_{\tau=0}^{t} \sigma^{t-\tau}\gamma_k s_1(\tau) + \text{exp. decaying term}
\end{aligned}$$

where the last inequality follows because s_1 and γ_k are positive. Substituting into (2) gives the desired result.

■

3. Problem Formulation and Basic Assumptions

Assumption AR. *Let $y^\star$ be a bounded reference sequence that is known d steps ahead in time. The bound is denoted $K^\star = \sup_{t\geq 0} |y^\star(t)|$.* ■

The problem we address here is the one of asymptotically tracking this reference, while rejecting noise and disturbances. In other words, if the tracking error is defined as $\tilde{y} = y^\star - y$, then the control objective is to make $|\tilde{y}(t)|$ as small as possible as $t \to \infty$.

Next, we will introduce some assumptions on the model and system, in terms of the model structure f, its sequence of nominal parameters $\theta^\star$, and the unstructured uncertainty ν. Thereafter, we will describe an on-line estimator for (4), as well as an adaptive control structure. The main result in this section concerns the stability, robustness, and performance of this adaptive controller.

Assumption AP. *There exists a constant $\underline{\theta} \geq 0$, such that $||\Delta\theta^\star(t)|| \leq \underline{\theta}$ for all $t \geq 0$. In addition, there exists a known convex and compact set Θ such that $\theta^\star(t) \in \Theta$ for all $t \geq 0$.* ■

This means that the parametric variation from time $t-1$ to t is bounded in norm by the small constant $\underline{\theta}$. As a consequence, there exists a $\overline{\theta} \geq 0$ such that $\theta_1, \theta_2 \in \Theta$ implies $||\theta_1 - \theta_2|| \leq \overline{\theta}$. This assumption is quite a standard one in the analysis of adaptive control loops, see e.g. [18,19].

Assumption AB. *The sign of the high-frequency gain is uniformly constant and known. Without loss of generality, we assume here that it is positive:*

$$\inf_{\theta\in\Theta} \inf_{y,u} \frac{\partial f(y(t), ..., y(t-m), u(t), ..., u(t-r), \theta)}{\partial u(t)} \geq \beta_0 > 0 \tag{3}$$

It is assumed that f is a continuously differentiable function of $u(t)$. Moreover, f is bounded by an affine function, i.e. there exists a constant $C \geq 0$ such that

$$\begin{aligned} &\sup_{\theta\in\Theta} ||f(y(t), ..., y(t-m), u(t), ..., u(t-r), \theta)|| \\ &\qquad \leq C(1 + |y(t)| + ... + |y(t-m)| + |u(t)| + ... + |u(t-r)|) \end{aligned}$$

■

The first part of AB is essentially a strong global controllability assumption. In the usual global controllability definition, it is required that there exists an admissible control input such that any state can be reached from any initial state within an unspecified, but finite time [20]. Our definition requires this time to be d, which is the shortest possible. On the other hand, we are only concerned about the output, not the full state.

The second part of AB is made necessary by the use of input/output stability theory, and is essentially the reason why the same tools that are used to study adaptive control on the basis of linear models can be used to study adaptive control using quite general nonlinear models of the form (1). This growth condition is not too restrictive, since all systems and models where the inputs and outputs do not grow or decay at a rate that is faster than exponential will satisfy this assumption.

4. Linearly Parameterized Model Representation

To begin with, we assume the predictor (1) is linearly parameterized:

$$y(t+d) \quad = \quad \varphi^T(y(t), ..., y(t-m), u(t), ..., u(t-r))\theta^\star(t) + \nu(t+d) \qquad (4)$$

where the definition of φ follows directly from the identity $f(\cdot, \theta) = \varphi^T(\cdot)\theta$, cf. (1).

4.1. Adaptive Control Structure

The fundamentals of discrete-time feedback linearization are studied in [11]. The idea is to render the non-linear nominal system linear with the aid of a static feedback. In general, this is not always possible, as one should expect, but with the above model representation, the adaptive certainty equivalence feedback linearizing controller is well defined by the implicit equation

$$v(t) \quad = \quad \varphi^T(y(t), ..., y(t-m), u(t), ..., u(t-r))\hat{\theta}(t) \qquad (5)$$

where $\hat{\theta}(t)$ is an estimate of $\theta^\star(t)$. From (3) it follows that for each $v(t)$ there always exists a unique control input $u(t)$ that satisfies (5). The closed loop behavior is now described by

$$y(t) \quad = \quad q^{-d}v(t) + \varphi^T(t-d)(\theta^\star(t-d) - \hat{\theta}(t-d)) + \nu(t) \qquad (6)$$

The nominal behavior from v to y is simply the linear system with transfer function q^{-d}, because of the feedback linearization. The second term is the parametric error due to the fact that the controller only has estimates of the nominal parameters. Finally, the third term is the unstructured uncertainty. The following linear control algorithm is applied to control the above nominally linear system

$$v(t) \quad = \quad y^\star(t+d) + G(q^{-1})(y^\star(t) - y(t)) \qquad (7)$$

where $G(q^{-1}) = P(q^{-1})/Q(q^{-1})$. Combining (6) and (7), we see that the closed loop system satisfies

$$\tilde{y}(t) \quad = \quad M(q^{-1})\left(\varphi^T(t-d)(\hat{\theta}(t-d) - \theta^\star(t-d)) - \nu(t)\right) \qquad (8)$$

where

$$M(q^{-1}) = \frac{1}{1+G(q^{-1})q^{-d}} = \frac{Q(q^{-1})}{Q(q^{-1})+q^{-d}P(q^{-1})}$$

is by design a stable transfer function with a bandwidth that equals the desired bandwidth of the closed loop. The effect of the parametric and unstructured uncertainty in (8) will be analyzed in the remaining of this section. Notice that this is a non-trivial task because boundedness of φ and ν are not established a priori.

4.2. Parameter Estimation

The estimate $\hat{\theta}(t)$ of $\theta^\star(t)$ is based on the predictor $\hat{y}(t|t-1) = \varphi^T(t-d)\hat{\theta}(t-1)$ which gives the prediction error

$$\begin{aligned} e(t) &= \hat{y}(t|t-1) - y(t) \\ &= \varphi^T(t-d)\hat{\theta}(t-1) - \varphi^T(t-d)\theta^\star(t-d) - \nu(t) \end{aligned} \tag{9}$$

and normalized prediction error $\varepsilon(t) = e(t)/n(t)$, where the scalar normalizing sequence n is defined by

$$n^2(t) = \delta n^2(t-1) + ||\varphi(t-d)||^2 + 1$$

with arbitrary $n(0) > 0$ and $\delta \in (0, 1)$ given. The normalization is introduced to justify that the normalized unstructured uncertainty is bounded [21]:

Assumption AU. *The unstructured uncertainty satisfies* $|\nu(t)| \leq \mathcal{V}n(t)$ *for some known constant* $\mathcal{V} \geq 0$. ■

The normalized error signal is a standard tool for dealing with unstructured uncertainty. This formulation is widely applied in the field of adaptive control, both in a continuous and discrete time framework, e.g. [16]. As we shall see later, it is straightforward to show that this assumption allows a quite general class of additive and multiplicative modeling error due to, for instance, neglected higher-order dynamics or structural mismatch of the model non-linearities, see also [22].

To estimate the unknown nominal parameters, we apply a standard recursive parameter estimation algorithm with normalization and relative dead-zone [22]:

$$\hat{\theta}'(t) = \hat{\theta}(t-1) - \lambda \frac{\varphi(t-d)}{1+\varphi^T(t-d)\varphi(t-d)} n(t)\mathcal{D}(\varepsilon(t)) \tag{10}$$

$$\hat{\theta}(t) = \mathcal{P}_\Theta\left(\hat{\theta}'(t)\right) \tag{11}$$

$$\mathcal{D}(\varepsilon(t)) = \begin{cases} \varepsilon(t) + d_0 & \text{if } \varepsilon(t) < -d_0 \\ 0 & \text{if } |\varepsilon(t)| \leq d_0 \\ \varepsilon(t) - d_0 & \text{if } \varepsilon(t) > d_0 \end{cases} \tag{12}$$

where $\mathcal{P}_\Theta$ is a continuous parameter projection that projects its argument to the closest point (using Euclidean norm) in Θ. The continuous function $\mathcal{D}$ is referred to as a dead-zone function, the constant $d_0 \geq 0$ is the magnitude of the dead-zone, and $\lambda > 0$ is the estimator gain. For convenience, the parameter estimation error sequence is defined by $\tilde{\theta}(t) = \hat{\theta}(t) - \theta^\star(t)$. Next, we examine the properties of this algorithm:

Theorem 1 *Suppose the outputs are generated by*

$$y(t+d) \;=\; \varphi^T(t)\theta^\star(t) + \nu(t+d)$$

and consider the parameter estimation algorithm (10)-(12) with initial estimate $\hat{\theta}(0) \in \Theta$*. If AP and AU hold, the estimator gain satisfies* $\lambda \in (0,2)$*, and the dead-zone is chosen such that* $d_0 = \mathcal{V}$*, then the algorithm has the properties*

$$\sum_{\tau=0}^{t} \mathcal{D}^2(\varepsilon(\tau)) \;\leq\; \left(\frac{2}{\lambda(2-\lambda)}\left((d\underline{\theta})^2 + 2(1+\lambda)d\underline{\theta}\overline{\theta}\right) + 2(d-1)^2\underline{\theta}^2\right)t + \frac{2}{\lambda(2-\lambda)}\overline{\theta}^2 \tag{13}$$

$$|\varepsilon(t)| \;\leq\; \mathcal{V} + |\mathcal{D}(\varepsilon(t))| \tag{14}$$

$$\sum_{\tau=0}^{t} ||\Delta\hat{\theta}(\tau)||^2 \;\leq\; \left(\frac{\lambda}{2-\lambda}\left((d\underline{\theta})^2 + 2(1+\lambda)d\underline{\theta}\overline{\theta}\right) + (d-1)^2\underline{\theta}^2 + 2\lambda(d-1)\underline{\theta}\overline{\theta}\right)t + \frac{\lambda}{2-\lambda}\overline{\theta}^2 \tag{15}$$

for all $t \geq 0$.

Proof: Rewriting the equation for the prediction error (9) gives

$$\varepsilon(t) \;=\; \frac{\varphi^T(t-d)\tilde{\theta}(t-1)}{n(t)} - \frac{\varphi^T(t-d)}{n(t)}\left(\theta^\star(t-d) - \theta^\star(t-1)\right) - \frac{\nu(t)}{n(t)}$$

Since $d_0 = \mathcal{V}$ there exists a sequence β with the properties $0 \leq \beta(t) \leq 1$ and

$$\mathcal{D}(\varepsilon(t)) \;=\; \beta(t)\frac{\varphi^T(t-d)\tilde{\theta}(t-1)}{n(t)} - \beta(t)\frac{\varphi^T(t-d)}{n(t)}\left(\theta^\star(t-d) - \theta^\star(t-1)\right)$$

From (10) we find

$$\tilde{\theta}'(t) \;=\; \left(I - \lambda\beta(t)\frac{\varphi(t-d)\varphi^T(t-d)}{1+\varphi^T(t-d)\varphi(t-d)}\right)\tilde{\theta}(t-1) + \gamma(t) \tag{16}$$

where

$$\gamma(t) \;=\; -\Delta\theta^\star(t) + \lambda\beta(t)\frac{\varphi(t-d)\varphi^T(t-d)}{1+\varphi^T(t-d)\varphi(t-d)}\left(\theta^\star(t-d) - \theta^\star(t-1)\right)$$

Defining the function $V(t) = ||\tilde{\theta}(t)||^2$, we find $V(t) \leq ||\tilde{\theta}'(t)||^2$ from the convexity of Θ. Hence,

$$\begin{aligned} V(t) - V(t-1) \leq & \left(\gamma^T(t) + \tilde{\theta}^T(t-1)\left(I - \lambda\beta(t)\frac{\varphi(t-d)\varphi^T(t-d)}{1+\varphi^T(t-d)\varphi(t-d)}\right)\right) \\ & \cdot \left(\left(I - \lambda\beta(t)\frac{\varphi(t-d)\varphi^T(t-d)}{1+\varphi^T(t-d)\varphi(t-d)}\right)\tilde{\theta}(t-1) + \gamma(t)\right) \\ & -\tilde{\theta}^T(t-1)\tilde{\theta}(t-1) \end{aligned}$$

Assume $\beta(t) \neq 0$, then

$$\begin{aligned} V(t) - V(t-1) \leq & \ \gamma^T(t)\gamma(t) \\ & +2\gamma^T(t)\left(I - \lambda\beta(t)\frac{\varphi(t-d)\varphi^T(t-d)}{1+\varphi^T(t-d)\varphi(t-d)}\right)\tilde{\theta}(t-1) \\ & -\left(\frac{2\lambda}{\beta(t)} - \lambda^2\right)\left(\frac{n^2(t)}{1+\varphi^T(t-d)\varphi(t-d)}\right) \\ & \cdot \left(\mathcal{D}(\varepsilon(t)) + \beta(t)\frac{\varphi^T(t-d)}{n(t)}(\theta^\star(t-d) - \theta^\star(t-1))\right)^2 \end{aligned}$$

or

$$\begin{aligned} & \left(\mathcal{D}(\varepsilon(t)) + \beta(t)\frac{\varphi^T(t-d)}{n(t)}(\theta^\star(t-d) - \theta^\star(t-1))\right)^2 \leq \\ & \quad \frac{\beta(t)}{\lambda(2-\beta(t)\lambda)}\left(||\gamma(t)||^2 + 2(1+\lambda)||\gamma(t)|| \cdot ||\tilde{\theta}(t-1)|| - \Delta V(t)\right) \end{aligned} \tag{17}$$

If $\beta(t) = 0$, then $V(t) = V(t-1)$ and $\mathcal{D}(\varepsilon(t)) = 0$. Hence, from the properties of β it follows that

$$\begin{aligned} \mathcal{D}^2(\varepsilon(t)) \leq & \ \frac{2}{\lambda(2-\lambda)}\left(||\gamma(t)||^2 + 2(1+\lambda)||\gamma(t)|| \cdot ||\tilde{\theta}(t-1)|| - \Delta V(t)\right) \\ & +2||\theta^\star(t-d) - \theta^\star(t-1)||^2 \end{aligned}$$

Summing, and using the fact that $||\tilde{\theta}(t)|| \leq \overline{\theta}$, which implies $V(t) \leq \overline{\theta}^2$ for all $t \geq 0$, together with the fact that AP implies $||\gamma(t)|| \leq d\underline{\theta}$, we get (13). By the definition of the dead-zone function, (14) follows. Finally, we get from (16) that

$$\begin{aligned} \sum_{\tau=0}^{t} ||\Delta\hat{\theta}(\tau)||^2 & \leq \sum_{\tau=0}^{t} ||\hat{\theta}'(\tau) - \hat{\theta}(\tau-1)||^2 \\ & \leq \sum_{\tau=0}^{t} \left((\gamma(\tau) + \Delta\theta^\star(\tau))^T(\gamma(\tau) + \Delta\theta^\star(\tau))\right) \end{aligned}$$

$$-2(\gamma(\tau) + \Delta\theta^{\star}(\tau))^T \lambda\beta(\tau)\frac{\varphi(\tau-d)\varphi^T(\tau-d)}{1+\varphi^T(\tau-d)\varphi(\tau-d)}\tilde{\theta}(\tau-1)$$

$$+\lambda^2\beta^2(\tau)\tilde{\theta}^T(\tau-1)\frac{\varphi(\tau-d)\varphi^T(\tau-d)\varphi(\tau-d)\varphi^T(\tau-d)}{(1+\varphi^T(\tau-d)\varphi(\tau-d))^2}\tilde{\theta}(\tau-1)\Bigg)$$

$$\leq \sum_{\tau=0}^{t}\Bigg(||\gamma(\tau)+\Delta\theta^{\star}(\tau)||^2 + 2\lambda||\gamma(\tau)+\Delta\theta^{\star}(\tau)||\cdot||\tilde{\theta}(\tau-1)||$$

$$+\lambda^2\frac{n^2(\tau)}{1+\varphi^T(\tau-d)\varphi(\tau-d)}$$

$$\left(\mathcal{D}(\varepsilon(\tau)) + \beta(\tau)\frac{\varphi^T(\tau-d)}{n(\tau)}(\theta^{\star}(\tau-d)-\theta^{\star}(\tau-1))\right)^2\Bigg)$$

Now, (15) follows from (17). ■

The relative dead-zone is a modification that turns off the adaptation of the parameter estimate when the normalized prediction error becomes "small", therefore preventing drift phenomena that otherwise might be excited by the unstructured uncertainty. There are, however, several other modifications available that lead to parameter estimation algorithms with similar properties, including different variations of the σ-modification, ϵ-modification, and parameter projection [23,16,14,24]. The design of adaptive schemes that retain their closed loop stability properties in the presence of not only large parametric uncertainty but also modeling errors (such as additive disturbances and unmodelled dynamics) is referred to as *robust adaptive control*; a unified treatment of this area can be found in [16].

4.3. Closed Loop Stability

It is well known that input-output feedback linearizing control structures may render some of the system states unobservable [11,25]. In order to ensure boundedness of these states, we must study the behavior of the inverse system. By the Implicit Function Theorem [20], and assumption AB, it is evident that the inverse system is globally defined by a function g:

$$u(t) = g(u(t-1), ..., u(t-r), y(t+d), y(t), ..., y(t-m), \nu(t+d), \theta^{\star}(t)) \quad (18)$$

where y and ν are viewed as inputs, and u as the output.

Assumption AI. *The inverse nominal model (18) is globally uniformly exponentially stable, in the sense that for arbitrary initial conditions, there exists constants K_σ and $\sigma \in [0, 1)$ such that the impulse response coefficients of (18) are bounded by an exponentially decaying sequence $(K_\sigma\sqrt{\sigma}^t)$.* ■

Stability of the inverse nominal model is often referred to as a minimum phase property of the model [26]. It should be stressed that with the model representation and control structure outlined above, the system can actually be non-minimum phase, for the following two reasons:

1. The particularly simple form of the model has allowed us to factor out the time delay $q^{-(d-1)}$ from the nominal model, which is clearly a non-minimum phase phenomenon. Notice that this is in contrast to the continuous-time case where the representation of time-delays is less straightforward. Also notice that the factorization of non-linear systems is in general a complicated and very much unsolved problem [27,28,29]. This is in sharp contrast to the linear case, where non-minimum phase phenomena can easily be factored out and identified as zeros outside the unit circle in the complex plane.

2. There may be non-minimum phase phenomena hidden in the unstructured uncertainty. As we shall see, stability of the adaptive control system requires the unstructured uncertainty to be small, so these non-minimum phase phenomena should not be too large. In practise, we expect that non-minimum phase phenomena due to fast sampling of a continuous-time system [30,31] are small enough to be regarded as unmodeled dynamics.

Under Assumption AI, we can prove the following lemma, which essentially shows that the input is bounded by the output and the unstructured uncertainty:

Lemma 3 *Suppose Assumptions AB and AI hold, then for any $\kappa \in (\sigma, 1)$*

$$||(q^{-d}u)_t||_2^\kappa \leq \frac{K_\sigma}{1-\sqrt{\sigma/\kappa}}\left(||y_t||_2^\kappa + ||\nu_t||_2^\kappa\right)$$

Proof. It follows directly from Assumption AI that

$$|u(t)| \leq K_\sigma \sum_{\tau=0}^{t} \sqrt{\sigma}^{t-\tau}\left(|y(\tau+d)| + |\nu(\tau+d)|\right)$$

The result follows from Lemma 1 together with

$$||(1-\sqrt{\sigma}q^{-1})^{-1}||_\infty^\kappa = (1-\sqrt{\sigma/\kappa})^{-1}$$

■

We are now in position to prove the main result:

Theorem 2 *Suppose the system (4) is controlled by (5) and (7), and the algorithm (10)-(12) is applied to estimate the parameter sequence $\theta^\star$. Suppose Assumptions AB, AR, AP, AU and AI hold, and let $\rho \in (0, 1)$ and K_ρ be constants such that the impulse response coefficients of $M(q^{-1})$ are bounded by $K_\rho\sqrt{\rho}^t$. Suppose the magnitude of the dead-zone is chosen as $d_0 = \mathcal{V}$. Let κ be arbitrary in the open* *interval* $\max(\delta, \sigma, \rho) < \kappa < 1$. *If* $\underline{\theta}$ *and* $\mathcal{V}$ *are both sufficiently small, then for arbitrary initial conditions and* $\hat{\theta}(0) \in \Theta$, *the input and output sequences* u *and* y *are bounded, and the tracking error satisfies*

$$\frac{1}{t}\sum_{\tau=0}^{t} \tilde{y}^2(\tau) \;\leq\; c_1 + \frac{1}{t}c_2 \tag{19}$$

for all $t > 0$, *where* c_2 *is proportional to* $\overline{\theta}^2$, *and* $c_1 \to 0$ *when* $\underline{\theta} \to 0$ *and* $\mathcal{V} \to 0$.

Proof: Let us define

$$\tilde{Y}(t) \;=\; \varphi^T(t-d)\tilde{\theta}(t-d) - \nu(t) \tag{20}$$

Combining (9) and (20), we get

$$\frac{\tilde{Y}(t)}{n(t)} \;=\; \varepsilon(t) + \frac{\varphi^T(t-d)}{n(t)}\left(\hat{\theta}(t-d) - \hat{\theta}(t-1)\right) \tag{21}$$

From the definition of the normalizing signal, assumption AB and Lemma 2, it is evident that there exists a constant C such that

$$n^2(t) \;\leq\; \frac{1}{1-\delta} + C^2 \sum_{\tau=0}^{t} \kappa^{t-\tau}\left(1 + |y(\tau-d)| + |u(\tau-d)|\right)^2$$

Using Lemma 3 and the triangle inequality, we find

$$n^2(t) \;\leq\; \frac{1+3C^2}{1-\kappa} + 3C^2\left(||(q^{-d}y)_t||_2^{\kappa}\right)^2 + \frac{3C^2K_\sigma^2}{(1-\sqrt{\sigma/\kappa})^2}\left(||(|y|+|\nu|)_t||_2^{\kappa}\right)^2$$

because $\kappa > \sigma$. Using Lemma 2, we get

$$n^2(t) \;\leq\; \frac{1+3C^2}{1-\kappa} + \left(K_y\,||y_t||_2^{\kappa}\right)^2 + \left(K_\nu\,||\nu_t||_2^{\kappa}\right)^2$$

where $K_y = \sqrt{3}C(\kappa^{-d/2} + \sqrt{2}K_\sigma(1-\sqrt{\sigma/\kappa})^{-1})$ and $K_\nu = \sqrt{6}CK_\sigma(1-\sqrt{\sigma/\kappa})^{-1}$. By the triangle inequality, the bound on $y^\star$, and (8), we get

$$n^2(t) \;\leq\; K_0 + \left(K_y(1-\sqrt{\rho/\kappa})^{-1}||\tilde{Y}_t||_2^{\kappa}\right)^2 + \left(K_\nu||\nu_t||_2^{\kappa}\right)^2$$

where $K_0 = (1 + 3C^2 + 3C^2K_y^2(K^\star)^2)/(1-\kappa)$. This can be written

$$n^2(t) \;\leq\; K_0 + K_y^2\frac{\kappa}{(\sqrt{\kappa} \quad \sqrt{\rho})^2}\sum_{\tau=0}^{t}\kappa^{t-\tau}\tilde{Y}^2(\tau) + K_\nu^2\sum_{\tau=0}^{t}\kappa^{t-\tau}\nu^2(\tau)$$

After some straightforward manipulations and using (21)

$$n^2(t) \;\leq\; K_0 + K_y^2 \frac{\kappa}{(\sqrt{\kappa}-\sqrt{\rho})^2} \sum_{\tau=0}^{t} \kappa^{t-\tau} s^2(\tau) n^2(\tau) + K_\nu^2 \sum_{\tau=0}^{t} \kappa^{t-\tau} \nu^2(\tau)$$

where we have defined

$$s^2(t) \;=\; \left(\varepsilon(t) + \frac{\varphi^T(t-d)}{n(t)} \left(\hat{\theta}(t-d) - \hat{\theta}(t-1) \right) \right)^2 \tag{22}$$

It is straightforward to show that there exists a finite constant $B \geq 1$ such that

$$\frac{n^2(t+1)}{n^2(t)} \;\leq\; B$$

for all $t \geq 0$. Hence

$$n^2(t) \;\leq\; K_0 + \sum_{\tau=0}^{t-1} \kappa^{t-\tau} w^2(\tau) n^2(\tau) \tag{23}$$

where $w^2(t) = B\kappa^{-1}K_\nu^2\mathcal{V}^2 + BK_y^2(\sqrt{\kappa}-\sqrt{\rho})^{-2}s^2(t)$. By the Gronwall lemma, e.g. [32,14], and the relation between arithmetic and geometric means, e.g. [33], we get from (23)

$$n^2(t) \;\leq\; K_0 + K_0 \sum_{\tau=0}^{t-1} \kappa^{t-\tau} w^2(\tau) \left(1 + \frac{\kappa^{-1}}{t-\tau-1} \sum_{i=\tau+1}^{t-1} w^2(i) \right)^{t-\tau-1}$$

From Lemma 1, it follows after some manipulation that

$$\sum_{\tau=0}^{t} w^2(\tau) \;\leq\; \mu_0 t + \mu_1 \tag{24}$$

where

$$\begin{aligned} \mu_0 \;&=\; \left(B\kappa^{-1}K_\nu^2 + \frac{2BK_y^2}{(\sqrt{\kappa}-\sqrt{\rho})^2} \right) \mathcal{V}^2 + \frac{4Bd^2K_y^2}{(\sqrt{\kappa}-\sqrt{\rho})^2} \left(\frac{3+\lambda}{\lambda(2-\lambda)} \right) \underline{\theta}^2 \\ &\quad + \frac{2BdK_y^2}{(\sqrt{\kappa}-\sqrt{\rho})^2} \left(\frac{\lambda^3+4\lambda^2+4\lambda+4}{\lambda(2-\lambda)} \right) \overline{\theta}\underline{\theta} \\ \mu_1 \;&=\; \frac{2BK_y^2}{(\sqrt{\kappa}-\sqrt{\rho})^2} \left(\frac{\lambda^2+2}{\lambda(2-\lambda)} \right) \overline{\theta}^2 \end{aligned} \tag{25}$$

Now,

$$n^2(t) \leq K_0 + K_0 \sum_{\tau=0}^{t-1} \kappa^{t-\tau} w^2(\tau) (1+\kappa^{-1}\mu_0)^{t-\tau-1} \left(1 + \frac{\mu_1}{(\kappa+\mu_0)(t-\tau-1)} \right)^{t-\tau-1}$$

Since $(1+x^{-1})^x \le e$ for all $x > 0$, it follows that

$$n^2(t) \le K_0 + K_0(1+\kappa^{-1}\mu_0)^{-1}\exp\left(\frac{\mu_1}{\kappa+\mu_0}\right)\sum_{\tau=0}^{t-1}(\kappa+\mu_0)^{t-\tau}w^2(\tau)$$

Taking $\underline{\theta}$ and $\mathcal{V}$ sufficiently small, it follows from (25) that $\mu_0+\kappa < 1$. Hence, (cf. Lemma 2.4 in [14])

$$n^2(t) \le K_0 + \frac{K_0(\mu_0+\mu_1)\exp\left(\frac{\mu_1}{\kappa+\mu_0}\right)}{(1-\kappa-\mu_0)(1+\kappa^{-1}\mu_0)} = \overline{n}^2$$

and it is proved that n is a bounded sequence. From (8) and Lemma 1 it follows that

$$\begin{aligned}\sum_{\tau=0}^{t}\tilde{y}^2(\tau) &\le \frac{1}{(1-\sqrt{\rho})^2}\sum_{\tau=0}^{t}\tilde{Y}^2(\tau) \le \frac{1}{(1-\sqrt{\rho})^2}\sum_{\tau=0}^{t}s^2(\tau)n^2(\tau)\\ &\le \frac{2\overline{n}^2}{(1-\sqrt{\rho})^2}\sum_{\tau=0}^{t}\left(\varepsilon^2(\tau)+(d-1)^2\left\|\Delta\hat{\theta}(\tau)\right\|^2\right)\end{aligned}$$

From (22) and Lemma 1, it is evident that (19) holds, and c_1 and c_2 have the stated properties. It is easy to see that $\tilde{y}$ must be a bounded sequence. Hence, y is bounded since $y^\star$ is bounded. Since ν is bounded by n, it follows from Lemma 3 that u is bounded as well. ■

Theorem 2 provides sufficient conditions for robust stability of the adaptive control loop. Stability requires the bounds $\mathcal{V}$ and $\underline{\theta}$ on the unstructured uncertainty and nominal parameter variations both to be sufficiently small. The degree of robustness can the quantified in terms of the inequality $\mu_0+\kappa < 1$ since $\kappa < 1$ is a given constant and μ_0 depends on $\mathcal{V}$ and $\underline{\theta}$. Hence, μ_0 can be interpreted as a sort of stability margin. However, both the stability margin and the performance result (19) can potentially be very conservative, and are best interpreted in a qualitative way.

We see from (19) that the average squared tracking error is bounded by a sum of two terms. The first term is the asymptotic performance bound that scales with the uncertainty and parameter variations. The second term is a bound on the transient performance that scales with the bound $\overline{\theta}^2$ on the parameter set, and vanishes asymptotically at the rate $1/t$. Hence, the average squared tracking error will tend towards a neighborhood of the origin at the rate $1/t$. Unfortunately, we cannot provide any strong results about the instantaneous value of the tracking error, except that it is bounded. Various phenomena like bursting may therefore be present [24].

5. Non-linearly Parameterized Model Structure

In this section we extend the results of the previous section to models of the form (1). In other words, the predictor is allowed to be nonlinearly parameterized.

Suppose the certainty equivalence adaptive feedback linearizing controller

$$v(t) = f(y(t), ..., y(t-m), u(t), ..., u(t-r), \hat{\theta}(t)) \tag{26}$$

is applied. Under smoothness assumptions on f, Taylor's theorem immediately gives the following error equation [34]

$$\tilde{y}(t) = M(q^{-1})\left(\frac{\partial f}{\partial \theta}(\cdot, \hat{\theta}(t-d))\tilde{\theta}(t-d) + \frac{1}{2}\tilde{\theta}(t-d)\frac{\partial^2 f}{\partial \theta^2}(\cdot, \overline{\theta}(t-d))\tilde{\theta}(t-d) - \nu(t)\right)$$

where $\overline{\theta}(t) = \alpha(t)\theta^\star(t) + (1-\alpha(t))\hat{\theta}(t)$ for some $\alpha(t) \in [0, 1]$. Defining

$$\varphi(t) = \frac{\partial f}{\partial \theta}(\cdot, \hat{\theta}(t))$$

$$\eta(t) = \nu(t) - \frac{1}{2}\tilde{\theta}(t-d)\frac{\partial^2 f}{\partial \theta^2}(\cdot, \overline{\theta}(t-d))\tilde{\theta}(t-d)$$

we get

$$\tilde{y}(t) = M(q^{-1})(\varphi^T(t-d)\tilde{\theta}(t-d) - \eta(t))$$

By substituting the unstructured uncertainty $\nu(t)$ by $\eta(t)$, which also includes higher-order terms (with respect to the parameter error $\tilde{\theta}$), it is now straightforward to extend the results of Theorem 2 to nonlinearly parameterized systems.

Theorem 3 *Suppose the system (1) is controlled by (26) and (7), and the algorithm (10)-(12) is applied to estimate the parameter sequence $\theta^\star$. Suppose all the assumptions of Theorem 2 hold, and in addition*

1. *The function f is twice continuously differentiable with respect to θ, there exists a constant $K \geq 0$ such that its Hessian satisfies*

$$\sup_{y,u}\sup_{\theta \in \Theta}\left|\left|\frac{\partial f^2}{\partial \theta^2}(y(t), ..., y(t-m), u(t), ..., u(t-r), \theta)\right|\right|_2 \leq K$$

where $||\cdot||_2$ is the induced matrix norm.

2. *The bound on the parameter set $\overline{\theta}$ is sufficiently small.*

3. *The magnitude of the dead-zone is chosen as $d_0 = \mathcal{V} + K\overline{\theta}^2/2$.*

Then the conclusion of Theorem 2 still holds.

Proof: It is clear that $\eta(t) \leq \mathcal{V}' n(t)$ where

$$\mathcal{V}' = \mathcal{V} + \frac{K}{2}\overline{\theta}^2 = d_0$$

Hence, since $\underline{\theta}$, $\overline{\theta}$ and $\mathcal{V}$ are all sufficiently small, then Lemma 1 and Theorem 2 hold. ■

As one should expect, the global stability result in Theorem 2 reduces to a local stability result when the predictor is not linearly parameterized. Hence, the main price to pay for the relaxed assumption is that the bound on the parametric uncertainty $\overline{\theta}$ must be sufficiently small. In addition, it was necessary to introduce a regularity assumption on f and to increase the size of the dead-zone to account for the higher-order effects. Since Assumption AB requires f to be uniformly bounded by an affine function, it is clear that the uniform boundedness of the Hessian is a quite reasonable assumption. For instance, if f is Lipschitz, it will satisfy the regularity condition. Unfortunately, the increased dead-zone will potentially give even more conservative performance bounds than in the linear parameterization case.

6. Nominal Model Uncertainty, Uniform Approximation, Neural Networks and other Model Structures

The choice of model structure is of major importance in model-based control. In this section we investigate to what extent certain generic non-linear model structures satisfy the assumptions made in the analysis. We concentrate on Assumptions AB, AP and AU.

6.1. *Model Uncertainty*

A major difference between Assumption AB on one side, and Assumptions AP and AU on the other side, is that assumption AB concerns the model structure and parameter set, which are subject to design, while assumptions AP and AU are concerned with the relationship between the nominal model and the underlying system, and are not only difficult to verify, but are also outside the influence of the control law designer. In general, assumptions AP and AU require the nominal model to be capable of closely approximating the input/output behavior of the system. It may therefore be desirable to choose the function f such that the resulting model structure is a close approximation to a large class of non-linear systems.

Next, we proceed with an example of unstructured uncertainty that satisfies assumption AU. Assume the nominal model defined in (1) can be described as a possibly nonlinear operator H_n, and we have additive unmodeled dynamics H_u and multiplicative unmodeled dynamics H_y in addition to external disturbances and noise w as

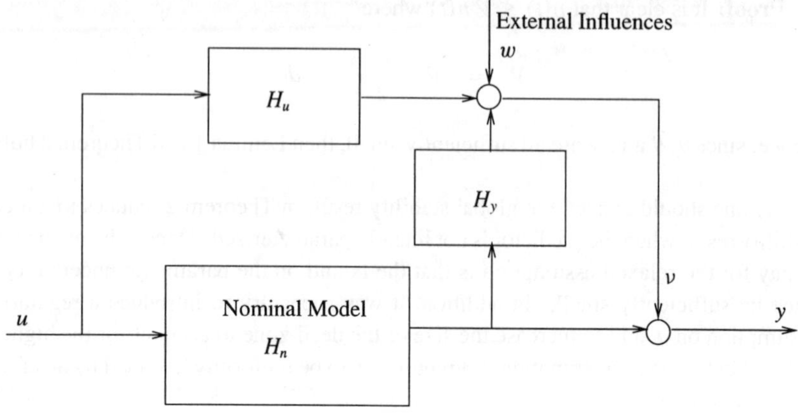

Figure 1: Nominal model and uncertainty.

shown in Figure 1:

$$y = H_n u + \nu \tag{27}$$

$$\nu = H_u u + H_y H_n u + w \tag{28}$$

which gives

$$\begin{aligned} y &= H_n u + H_u u + H_y H_n u + w \\ &= (I + H_y) H_n u + H_u u + w \end{aligned}$$

where the three terms clearly contain multiplicative uncertainty, additive uncertainty, and external influences, respectively.

Lemma 4 *Suppose Assumption AB holds, and in addition*

$$\begin{aligned} H_y(q^{-1}) &= \varepsilon_y \sum_{i=0}^{\infty} \alpha_i q^{-i}, \quad \sum_{i=0}^{\infty} |\alpha_i| \sqrt{\delta}^{-i} < \infty \\ H_u(q^{-1}) &= \varepsilon_u \sum_{i=0}^{\infty} \beta_i q^{-i}, \quad \sum_{i=0}^{\infty} |\beta_i| \sqrt{\delta}^{-i} < \infty \\ |w(t)| &\leq \varepsilon_w n(t) \end{aligned}$$

Then $\nu(t) \leq \mathcal{V} n(t)$ *where* $\mathcal{V} = K \max(\varepsilon_y, \varepsilon_u, \varepsilon_w)$ *and* K *is a constant.*

Proof. It is straightforward to see that there exist a constant K_y such that

$$\begin{aligned} |(H_n u)(t)| &= |f(y(t-d), ..., y(t-d-m), u(t-d), ..., u(t-d-r), \theta^\star(t))| \\ &\leq C\sqrt{1 + ||\varphi(t-d)||^2} \\ &\leq K_y n(t) \end{aligned} \tag{29}$$

Likewise, from Taylor's theorem there exists a sequence $\overline{u}$ such that

$$\begin{aligned} f(y(t), ..., y(t-m), u(t), u(t-1), ..., u(t-r)) \;=&\\ & f(y(t), ..., y(t-m), 0, u(t-1), ..., u(t-r)) \\ & + \frac{\partial f}{\partial u(t)}(y(t), ..., y(t-m), \overline{u}(t), u(t-1), ..., u(t-r))u(t) \end{aligned}$$

This gives

$$\begin{aligned} u(t) \;=\; & \left(\frac{\partial f}{\partial u(t)}(y(t), ..., y(t-m), \overline{u}(t), u(t-1), ..., u(t-r))\right)^{-1} \\ & \cdot(f(y(t), ..., y(t-m), u(t), u(t-1), ..., u(t-r)) \\ & \;-f(y(t), ..., y(t-m), 0, u(t-1), ..., u(t-r))) \end{aligned}$$

and

$$|u(t)| \;\leq\; \beta_0^{-1} 2K_y n(t) \tag{30}$$

Hence, from (28), (29) and (30)

$$\begin{aligned} \nu(t) \;&\leq\; \sum_{i=0}^{\infty} \left(\varepsilon_y K_y |\alpha_i| + \varepsilon_u K_u |\beta_i|\right) n(t-i) + \varepsilon_w n(t) \\ &\leq\; \sum_{i=0}^{\infty} \left(\varepsilon_y K_y |\alpha_i| + \varepsilon_u K_u |\beta_i|\right) \sqrt{\delta}^{-i} n(t) + \varepsilon_w n(t) \end{aligned}$$

and the stated result follows. ■

In the above lemma, the unmodeled dynamics are modeled as impulse responses that are bounded by exponentially decaying sequences, which is a quite general class of unmodeled dynamics that ensures that assumption AU is satisfied.

6.2. *Uniform Approximation*

Some classes of functions that can approximate arbitrary continuous function to an arbitrary uniform accuracy on compact subsets include radial basis-functions [35], operating regime based models [36], piecewise constant or linear models and splines, fuzzy models [37], and neural networks [38,39,40,41]. Unfortunately, these approximation properties are of very limited practical value. One reason is that the approximation results only hold on compact sets, which means that the system inputs and outputs must be known to be bounded a priori, which significantly reduces the relevance of these approximation results in the analysis of stability. One way that at least in principle resolves this problem is to apply a supervisory control system, like for instance a sliding mode controller [3,2] to ensure that the system state remains within a compact subset of the state-space. Another reason for the inadequacy of uniform approximation results to address the issue of stability in closed loop control systems is related to the number of adjustable parameters required to approximate a given function. Without significant prior knowledge, the number of parameters in the generic model structures mentioned above may be prohibitively large in order to guarantee a uniformly accurate model in a sufficiently large state-space region. It is easy to see that the performance bounds in Theorems 2 and 3 scales with the number of parameters, in the sense that an over-parameterized model may lead to poor robustness and performance bounds. It is therefore desirable to employ a low complexity model structure that matches the structure of the system well. This seems difficult to satisfy, unless significant prior knowledge is applied to choose the function f and the parameter set Θ. The fact that a given class of function approximators have good approximation properties is certainly not sufficient, in general.

Before we proceed with a further discussion of these assumptions, let us first briefly review function approximation using the abovementioned representations. With radial basis-functions, the function f is represented as

$$f(\psi, \theta) = \sum_{i=1}^{N} \theta_i \rho(||\psi - \psi_i||)$$

where the information vector is

$$\psi(t) = (y(t), ..., y(t-m), u(t), ..., u(t-r))^T$$

and $\rho(\cdot)$ is a given scalar function, e.g. an Gaussian, spline, or multi-quadric function [42,43]. Closely related are normalized radial basis-functions [44] of the form

$$f(\psi, \theta) = \sum_{i=1}^{N} \theta_i w_i(\psi)$$

where the basis-function

$$w_i(\psi) = \frac{\rho(||\psi - \psi_i||)}{\sum_{j=1}^{N} \rho(||\psi - \psi_j||)}$$

form a convex combination of the parameter values, since $\sum_{i=1}^{N} w_i(\psi) = 1$ for all ψ. It is assumed that

$$\sum_{i=1}^{N} \rho(||\psi - \psi_i||) > 0$$

for all ψ. Operating regime based models [45] (also closely related to adaptive connectionist spline networks [46], some fuzzy models [47], and mixtures of local experts [48]) can also be viewed as generalizations of normalized radial basis-function expansions, since they typically take the form

$$f(\psi, \theta) = \sum_{i=1}^{N} (\theta_{0,i} + \theta_{1,i}^T \psi) w_i(\psi)$$

which leads to a convex combination of linear ARX models. In other words, the constant coefficients in the normalized radial basis-function expansion have been replaced by local linear models.

Finally, consider sigmoidal basis-function models (usually called neural networks) on the form

$$f(\psi, \theta) = \theta_0 + \sum_{i=1}^{N} \theta_{1,i} \sigma \left(\theta_{2,i} + \theta_{3,i}^T \psi\right)$$

where the function $\sigma(\cdot)$ is sigmoidal (soft saturation function). Contrary to the approximation methods above, the sigmoidal basis-function is non-linearly parameterized.

Models based on all the abovementioned generic function approximation methods satisfy the exponential growth condition in assumption AB. On the other hand, it is in general difficult to design a convex set Θ such that (3) holds, using only knowledge of β_0. Of course, significant prior knowledge is also used to restrict Θ such that one can guarantee that $\overline{\theta}$ is sufficiently small in the case of non-linear parameterization.

7. Extensions and Modifications

In this section we discuss some modifications and extensions of the adaptive control scheme analyzed in this paper. The aim is to develop procedures that are more suitable for practical applications.

7.1. Extensions to the MIMO case

Although the theory has been developed for the SISO case, the extension to the MIMO case is conceptually straightforward, provided the MIMO predictor is written in the form

$$\begin{aligned} y_1(t+d_1) &= f_1(y(t), ..., y(t-m), u(t), ..., u(t-r), \theta^\star(t)) + \nu_1(t+d_1) \\ &\vdots \\ y_{m^\star}(t+d_{m^\star}) &= f_{m^\star}(y(t), ..., y(t-m), u(t), ..., u(t-r), \theta^\star(t)) + \nu_{m^\star}(t+d_{m^\star}) \end{aligned}$$

where the dimension of the input and output vectors $u(t)$ and $y(t)$ are $r^\star$ and $m^\star$, respectively. The vector relative degree is $d = (d_1, ..., d_{m^\star})$.

The certainty equivalence feedback linearizing control equation is now

$$\begin{aligned} v_1(t) &= f_1(y(t), ..., y(t-m), u(t), ..., u(t-r), \hat{\theta}(t)) \\ &\vdots \\ v_{m^\star}(t) &= f_{m^\star}(y(t), ..., y(t-m), u(t), ..., u(t-r), \hat{\theta}(t)) \end{aligned} \tag{31}$$

With straightforward modifications of the assumptions, the analysis in this paper can easily be extended to also cover the MIMO case, using similar techniques as in [4]. Of course, in the MIMO case, the results will typically be even more conservative and qualitative. This will in particular concern systems that are difficult to decouple robustly, like for instance ill-conditioned systems.

7.2. *Reformulating the Feedback Linearization Control Equation*

Although Assumption AB ensures existence of a solution of the certainty equivalence feedback linearizing control law (26), the actual computation of the control input $u(t)$ is a non-trivial problem, unless this equation is affine in the input $u(t)$. Moreover, Assumption AB is quite restrictive since it requires the sign of the high frequency gain to be constant and bounded from below. Hence, it excludes input saturation and systems where the gain changes sign with the operating point, for instance. Finally, the design of a parameter set Θ that simultaneously ensures that Assumption AB is satisfied and also contains a sequence of slowly time-varying nominal parameters $\theta^\star(t)$ that provide a sufficiently accurate nominal model is a very difficult problem.

The abovementioned problems suggests an alternative approach, namely approximate feedback linearization. The idea is to develop an algorithm for computing a control input $u(t)$ with the property that it approximates the exact adaptive feedback linearizing control law, simplifies the design of the parameter set Θ, and avoids the restrictive assumptions made above.

The design of approximate feedback linearization controllers has been extensively addressed in the framework of geometric nonlinear control theory [49] for the case of continuous-time systems. Although geometric nonlinear control theory provides powerful methods for controller design, an important class of nonlinear systems do not satisfy the restrictive conditions for feedback linearization. Several approaches have been proposed for designing stable control systems that are based on approximate feedback linearization techniques. These approaches include the pseudo-linearization [50,51], extended linearization [52], linearization families [53], and high-order approximate linearization [54].

On the basis of the MIMO adaptive feedback linearizing control law (31), we suggest the following optimization formulation of the approximate adaptive feedback linearizing control problem:

Minimize

$$
\begin{aligned}
J(u(t)) &= \left\| v(t) - f(y(t), ..., y(t-m), u(t), ..., u(t-r), \hat{\theta}(t)) \right\|^2 \\
&\quad + \lambda_1 ||u(t)||^2 + \lambda_2 ||u(t) - u(t-1)||^2
\end{aligned}
$$

subject to the constraints

$$
\begin{aligned}
\underline{u} &\leq u(t) \leq \overline{u} \\
\underline{\Delta u} &\leq u(t) - u(t-1) \leq \overline{\Delta u}
\end{aligned}
$$

where λ_1 and λ_2 are non-negative constants. We see that when $\lambda_1 = \lambda_2 = 0$ and the constraints are not violated, the solution of the problem of minimizing $J(u(t))$ coincides with the solution of (31).

The above optimization formulation has the following advantages over the exact formulation (31):

- Saturation and constraints on the rate of change of the control input are incorporated in the formulation. More general constraints on inputs and outputs can, of course, be included as well. Hence, the formulation of the control objective is similar to the model predictive control formulation, which has proven very engineering friendly [55].

- By analogy to the linear control case [56], we expect that the use of penalty terms prevents a direct cancellation of the zero dynamics. Hence, it is not always necessary that the inverse nominal model be stable to ensure internal stability of the control system.

- It is no longer necessary to restrict Θ to avoid singularity of the implicit certainty equivalence feedback linearizing control law. The reason is that the quadratic penalty terms in $J(u(t))$ regularizes the inversion of the implicit control law, in the sense that there always exists a $u(t)$ that minimizes $J(u(t))$. In other words, close to singular operating points where the sensitivity of the output $y(t+d)$ with respect to the input $u(t)$ is very small, a more or less arbitrary small input value $u(t)$ is chosen. Because of the low sensitivity at such operating points, this introduces only a small difference on the output compared to the exact feedback linearizing approach.

- One can handle over- or under-determined control problems, i.e. $m^\star \neq r^\star$.

Hence, many of the difficult design problems are resolved with this approach, and new features are added. Unfortunately, the theoretical properties developed in previous sections remains to be extended. Moreover, the problem of actually computing the control input $u(t)$ is not simplified. In general, a non-linear programming algorithm must be applied, and convergence to a global minimum cannot be guaranteed.

7.3. Robust Parameter Estimation

For some problems, a least squares type parameter estimator algorithm may be better suited than the one suggested above, for instance in order to improve the convergence rate. In order to reduce phenomena such as bursting and parameter drift, some further modifications of the parameter estimator may be needed. This can be achieved in different ways using for instance directional forgetting [57,58,59].

8. Concluding Remarks

We have derived theoretical results for an adaptive feedback linearizing control structure for a general class of non-linear systems. The results address some stability, robustness, and performance issues.

The analysis is based on standard input/output stability theory, and in particular the use of weighted l_2-norms. The analysis is completely analogous to the case with a linear model [14]. It is interesting to observe that the complexity of the analysis is only slightly higher compared to the linear case.

The assumptions are in general weak, but difficult to verify. Hence, the results are best interpreted in a qualitative way. Consequently, their immediate practical implications are not very significant, since one in practise is more concerned with quantitative analysis of performance and robustness, rather than a qualitative one. However, the qualitative analysis is still useful, since it illustrates the fundamental limitations and general design trade-offs in a transparent way.

Practical modifications, which are hard to analyze, must often be introduced. These are in particular related to solvability of the certainty equivalence feedback linearizing control law, as well as the need to explicitly take into account fundamental limitations like saturation and non-minimum phase phenomena. Moreover, the on-line parameter estimation algorithm should be modified to improve its robustness.

In practise, significant empirical or mechanistic a priori knowledge must be available for modeling and adaptive controller design. This does not exclude the use of neural networks or fuzzy models in adaptive controllers, but it means that a priori knowledge should be incorporated directly in the neural or fuzzy model like in [47,60,36]. Alternatively, a non-linear black-box model can be combined with for instance a known model, where the non-linear black-box model accounts for the dynamics that are not captured by the known model [61,62,63,64,65].

9. References

1. S. S. Sastry and A. Isidori, "Adaptive control of linearizable systems," *IEEE Trans. Automatic Control*, vol. 34, pp. 1123–1131, 1989.

2. R. M. Sanner and J.-J. Slotine, "Gaussian networks for direct adaptive control," *IEEE Trans. Neural Networks*, vol. 3, pp. 837–863, 1992.

3. M. M. Polycarpou and P. A. Ioannou, "Identification and control of nonlinear systems using neural network models: Design and stability analysis," Tech. Rep. 91-09-01, Dept. Electrical Enginering – Systems, University of Southern California, Los Angeles, September 1991.

4. T. A. Johansen, "Fuzzy model based control: Stability, robustness, and performance issues," *IEEE Trans. Fuzzy Systems*, vol. 2, pp. 221–234, 1994.

5. F.-C. Chen and H. K. Khalil, "Adaptive control of nonlinear systems using neural networks - a dead-zone approach," in *Proc. American Control Conference, Boston*, pp. 667–672, 1991.

6. F.-C. Chen and H. K. Khalil, "Adaptive control of nonlinear systems using neural networks," *Int. J. Control*, vol. 55, pp. 1299–1317, 1992.

7. M. M. Polycarpou and P. A. Ioannou, "Modeling, identification and stable adaptive control of continuous-time nonlinear dynamical sytems using neural networks," in *Proc. American Control Conference, Chicago*, pp. 36–40, 1992.

8. R. Marino and P. Tomei, "Global adaptive output-feedback control of nonlinear systems, Part II: Nonlinear parametrization," *IEEE Trans. Automatic Control*, vol. 38, pp. 33–49, 1993.

9. F. L. Lewis, K. Liu, and A. Yesildirek, "Multilayer neural net robot controller with guaranteed tracking performance," in *Proc. American Control Conference, San Francisco*, pp. 2785–2791, 1993.

10. F.-C. Chen and C.-C. Liu, "Adaptively controlling nonlinear continuous-time systems using multilayer neural networks," *IEEE Trans. Automatic Control*, vol. 39, pp. 1306–1310, 1994.

11. S. Monaco and D. Normand-Cyrot, "Minimum-phase nonlinear discrete-time systems and feedback stabilization," in *Proc. IEEE Conf. Decision and Control, Los Angeles, Ca.*, pp. 979–986, 1987.

12. I. J. Leontaritis and S. A. Billings, "Input-output parametric models for non-linear systems," *Int. J. Control*, vol. 41, pp. 303–344, 1985.

13. T. A. Johansen, "Weighted l_2-norms for analysis of an adaptive control loop based on a non-linear model." Accepted for IEEE Conf. Decision and Control, New Orleans, LA, 1995.

14. A. Datta, "Robustness of discrete-time adaptive controllers: An input-output approach," *IEEE Trans. Automatic Control*, vol. 38, pp. 1852–1857, 1993.

15. K. S. Tsakalis, "Robustness of model reference adaptive controllers: Input-Output properties," *IEEE Trans. Automatic Control*, vol. 37, pp. 556–565, 1992.

16. P. Ioannou and A. Datta, "Robust adaptive control: A unified approach," *Proceedings of the IEEE*, vol. 79, pp. 1736–1768, 1991. Also in *Foundations of Adaptive Control*, P. V. Kokotović, Ed., Springer-Verlag, Berlin, 1991.

17. T. A. Johansen and P. A. Ioannou, "Robust adaptive control of minimum phase nonlinear systems." Accepted for publication in Int. J. Adaptive Control and Signal Processing, 1995.

18. R. H. Middleton and G. C. Goodwin, "Adaptive control of time-varying linear systems," *IEEE Trans. Automatic Control*, vol. 33, pp. 150–155, 1988.

19. G. Kreisselmeier, "Adaptive control of a class of slowly time-varying plants," *Systems and Control Letters*, vol. 8, pp. 97–103, 1986.

20. H. Nijmeijer and A. J. van der Schaft, *Nonlinear Dynamical Control Systems*. Springer-Verlag, New York, 1990.

21. L. Praly, "Robust model reference adaptive controllers, Part I: Stability analysis," in *Proc. IEEE Conf. Decision and Control, Las Vegas, Nevada*, pp. 1009–1014, 1984.

22. G. Kreisselmeier and B. D. O. Anderson, "Robust model reference adaptive control," *IEEE Trans. Automatic Control*, vol. 31, pp. 127–133, 1986.

23. K. S. Narendra and A. M. Annaswamy, *Stable Adaptive Systems*. Prentice-Hall, Englewood Cliffs, NJ, 1989.

24. B. E. Ydstie, "Transient performance and robustness of direct adaptive control," *IEEE Trans. Automatic Control*, vol. 37, pp. 1091–1105, 1992.

25. C. I. Byrnes and A. Isidori, "Local stabilization of minimum phase nonlinear systems," *Systems and Control Letters*, vol. 11, pp. 9–17, 1988.

26. C. I. Byrnes and A. Isidori, "A frequency domain philosophy for nonlinear systems," in *Proceedings of the 1984 IEEE Conf. Decision and Control, Las Vegas, Nevada*, pp. 1569–1573, 1984.

27. J. Hammer, "Non-linear systems, stabilization and coprimeness," *Int. J. Control*, vol. 42, pp. 1–20, 1985.

28. C. A. Desoer and M. G. Kabuli, "Stability and robustness of non-linear unity-feedback system: Factorization approach," *Int. J. Control*, vol. 47, pp. 1133–1148, 1988.

29. A. Baños, F. N. Bailey, and M. A. Armada, "Factorization theory and feedback linearization of a class of non-linear systems," in *Proc. IEEE Conf. Decision and Control, Tucson, Az.*, pp. 2023–2026, 1992.

30. K. J. Åström, P. Hagander, and J. Sternby, "Zeros of sampled systems," *Automatica*, vol. 20, pp. 31–38, 1984.

31. D. W. Clarke, "Self-tuning control of non-minimum phase systems," *Automatica*, vol. 20, pp. 501–517, 1984.

32. C. A. Desoer and M. Vidyasagar, *Feedback Systems: Input-Output Properies*. Academic Press, New York, 1975.

33. W. Rudin, *Real and Complex Analysis*. McGraw-Hill, New York, 1986.

34. M. M. Polycarpou and P. A. Ioannou, "Neural networks as on-line approximators of nonlinear system," in *Proc. 31st IEEE Conf. Decision and Control, Tucson, Az.*, pp. 7–12, 1992.

35. J. Park and I. W. Sandberg, "Approximation and radial-basis-function networks," *Neural Computation*, vol. 5, pp. 305–316, 1993.

36. T. A. Johansen, "Operating regime based process modeling and identification," Tech. Rep. 94-109-W. Dr. Ing. thesis, Department of Engineering Cybernetics, Norwegian Institute of Technology, Trondheim, Norway, 1994. http://www.itk.unit.no/ansatte/Johansen,Tor.Arne/dring_taj.ps.gz.

37. B. Kosko, "Fuzzy systems as universal approximators," *IEEE Trans. Computers*, vol. 43, pp. 1329–1333, 1994.

38. G. Cybenko, "Approximations by superpositions of a sigmoidal function," *Mathematics of Control, Signals and Systems*, vol. 2, pp. 303–314, 1989.

39. A. R. Barron, "Universal approximation bounds for superpositions of a sigmoidal function," *IEEE Trans. Information Theory*, vol. 39, pp. 930–945, 1993.

40. T. Chen and H. Chen, "Approximation of continuous functionals by neural networks with application to dynamical systems," *IEEE Trans. Neural Networks*, vol. 4, pp. 910–918, 1993.

41. R. Żbikowski and A. Dzieliński, "Neural approximation: A control perspective," in *Neural Network Engineering in Dynamic Control Systems* (K. Kunt, G. Irwin, and K. Warwick, eds.), pp. 1–25, Springer-Verlag, London, 1995.

42. N. Dyn, "Interpolation and approximation by radial and related functions," in *Topics in multivariate approximation* (C. K. Chui, L. L. Schumaker, and D. J. Ward, eds.), Academic Press, New York, NY, 1987.

43. F. Girosi, M. Jones, and T. Poggio, "Priors, stabilizers and basis functions: From regularization to radial, tensor and additive splines," Tech. Rep. AI Memo 1430, MIT, Cambridge, 1994.

44. J. Moody and C. J. Darken, "Fast learning in networks of locally-tuned processing units," *Neural Computation*, vol. 1, pp. 281–294, 1989.

45. T. A. Johansen and B. A. Foss, "Constructing NARMAX models using ARMAX models," *Int. J. Control*, vol. 58, pp. 1125–1153, 1993.

46. R. D. Jones and co-workers, "Nonlinear adaptive networks: A little theory, a few applications," Tech. Rep. 91-273, Los Alamos National Lab., NM, 1991.

47. T. Takagi and M. Sugeno, "Fuzzy identification of systems and its application to modeling and control," *IEEE Trans. Systems, Man, and Cybernetics*, vol. 15, pp. 116–132, 1985.

48. R. A. Jacobs, M. I. Jordan, S. J. Nowlan, and G. E. Hinton, "Adaptive mixtures of local experts," *Neural Computation*, vol. 3, pp. 79–87, 1991.

49. A. Isidori, *Nonlinear Control Systems, 2nd Ed.* Springer Verlag, Berlin, 1989.

50. C. Reboulet and C. Champetier, "A new method for linerizing non-linear systems: The pseudolinearization," *Int. J. Control*, vol. 40, pp. 631–638, 1984.

51. J. Wang and W. J. Rugh, "On the pseudo-linearization problem for nonlinear systems," *Systems and Control Letters*, vol. 12, pp. 161–167, 1989.

52. W. T. Baumann and W. J. Rugh, "Feedback control of nonlinear systems by extended linearization," *IEEE Trans. Automatic Control*, vol. 31, pp. 40–46, 1986.

53. J. Wang and W. J. Rugh, "Feedback linearization families for nonlinear systems," *IEEE Trans. Automatic Control*, vol. 32, pp. 935–940, 1987.

54. Z. Xu and J. Hauser, "Higher order approximate feedback linearization about a manifold," *J. Math Syst., Estimation and Control*, vol. 4, pp. 451–465, 1994.

55. J. B. Rawlings, E. S. Meadows, and K. R. Muske, "Nonlinear model predictive control: A tutorial and survey," in *Preprints IFAC Symposium ADCHEM, Kyoto, Japan*, pp. 203–214, 1994.

56. D. W. Clarke and P. J. Gawthrop, "A self-tuning controller," *IEE Proc. D*, vol. 122, pp. 927–934, 1975.

57. T. R. Fortescue, L. S. Kershenbaum, and B. E. Ydstie, "Implementation of self-tuning regulators with variable forgetting factors," *Automatica*, vol. 17, pp. 831–835, 1981.

58. S. Sælid, O. Egeland, and B. Foss, "A solution to the blow-up problem in adaptive controllers," *Modeling, Identification and Control*, vol. 6, no. 1, pp. 39–56, 1985.

59. J. E. Parkum, N. K. Poulsen, and J. Holst, "Selective forgetting in adaptive procedures," in *Proc. 11th IFAC World Congress, Tallin, Estonia*, 1990.

60. R. A. Jacobs and M. I. Jordan, "Learning piecewise control strategies in a modular neural network architecture," *IEEE Trans. Systems, Man, and Cybernetics*, vol. 23, pp. 337–345, 1993.

61. M. A. Kramer, M. L. Thompson, and P. M. Phagat, "Embedding theoretical models in neural networks," in *Proceedings American Control Conference, Chicago, Il.*, pp. 475–479, 1992.

62. T. A. Johansen and B. A. Foss, "Representing and learning unmodeled dynamics with neural network memories," in *Proceedings of the American Control Conference, Chicago, Il.*, pp. 3037–3043, June 1992.

63. S. Tan, "A combined PID and neural control scheme for nonlinear dynamical systems," in *Proceeding of the IEEE Int. Conf. Intelligent Control and Instrumentation, Singapore*, vol. 1, pp. 377–383, 1992.

64. D. C. Psichogios and L. H. Ungar, "A hybrid neural network - first principles approach to process modeling," *AIChE J.*, vol. 38, pp. 1499–1511, 1992.

65. M. L. Thompson and M. A. Kramer, "Modeling chemical processes using prior knowledge and neural networks," *AIChE J.*, vol. 40, pp. 1328–1340, 1994.

Part III

Neural Techniques and Applications

Part III

Special Techniques and Applications

ROBUST ADAPTIVE NEUROCONTROL OF MIMO CONTINUOUS-TIME PROCESSES BASED ON THE e_1-MODIFICATION SCHEME

Jean-Michel RENDERS and Marco SAERENS
Automatic Control Laboratory and IRIDIA Laboratory, Université Libre de Bruxelles
Av. F. Roosevelt 50, 1050 Bruxelles, BELGIUM

ABSTRACT

In this manuscript, we introduce two different weight adaptation laws, based on Lyapunov stability theory, for the adaptive control of a certain class of MIMO continuous-time processes. The adaptation strategy is indirect in that it uses a sensitivity model of the plant (in our case, the sensitivity derivatives are estimated with a neural network that identifies the plant) in order to implement the adaptation law. The resulting adjustment rules are surprisingly simple in comparison with the gradient-based algorithm (also similar to the dynamic back-propagation, Williams & Zipser 1989), introduced by Narendra & Parthasarathy (1990, 1992). However, the stability statements, based on a first-order expansion, are only valid if the initial weight values are not too far from their optimal values that allow perfect model matching (local stability). We therefore propose to initialize the weights with values that solve the linear problem. After the treatment of the ideal case, we derive a robust adjustment law that takes noise and modelling errors into account. Indeed, due to noise, disturbances and modelling errors, the perfect model matching condition cannot be verified, and it is not possible to ensure that the error converges to zero, but only that it remains bounded. Moreover, it has been shown that in certain cases, the simple adaptive scheme designed for the ideal case can be driven unstable in the presence of bounded disturbances and unmodelled dynamics. Finally, in practical situations, the sensitivity model needed for the adaptation strategy is always imperfect. We therefore use a modified adjustment law known as the "e1-modification" scheme, originally introduced by Narendra & Annaswamy (1987) for the adaptive control of linear systems, that does not require information about the bounds of the perturbation effects. This adaptive control scheme is proved to be robust in that all the signals in the adaptive loop are bounded for bounded disturbances. Finally, we show the effectiveness of the robust adaptation law by applying it to the control of a simulated two inputs - two outputs flow-mixer.

1. Introduction

There has been a recent resurgence of interest for the old cybernetic project to exploit neural networks for control applications (Miller et al. 1990; Narendra and Parthasarathy 1990; Hunt et al. 1992; White & Sofge 1992; Rao Vemuri 1992; Gupta & Sinha 1994). Several approaches are possible when using neural nets for the control of plants (Psaltis, Sideris & Yamamura 1987; Barto 1990; Hunt et al. 1992). One possible strategy that fits in perfectly with the classical adaptive control attitude consists in using the difference between the actual output of the plant and the desired output in order to adapt the weights of connections. However, in order to implement this adjustment law, some prior knowledge on the way the plant reacts to slight input modifications, i.e. the Jacobian matrix of the plant, is needed. One possible, while inelegant, solution consists in approximating the partial derivatives by plotting the plant reactions to slight control modifications at the operating points (Psaltis, Sideris & Yamamura 1987, 1988).

Another, more sophisticated, technique has been proposed (Jordan 1989; Jordan & Rumelhart 1991; Nguyen & Widrow 1989, 1990), which incorporates the default prior knowledge directly in a network and links the neural controller to a neural model of the plant – a neural network that identifies the process. As pointed out by Narendra & Parthasarathy (Narendra & Parthasarathy 1990), it corresponds to the classical indirect control approach: the parameters of the plant are estimated at any instant, and the parameters of the controller (in this case, the weights of the neural controller) are adjusted assuming that the estimated parameters of the plant represent the true values. However it needs, in compensation, either a preceding learning stage (the identification of the plant) or a "neurally expressed" prior knowledge of the dynamics of the plant. Narendra & Parthasarathy (1990, 1992) use a similar indirect technique to show that identification and adaptive control schemes are practically feasible.

Indeed, Narendra & Parthasarathy (1990) showed by simulations that a neural network can be used effectively for the identification and control of nonlinear dynamical processes. The results are based on the universal approximation properties of three-layer neural networks. As a matter of fact, it has been shown (Hornik, Stinchombe & White 1989 and 1990; Stinchombe & White 1989; for a review, see White 1992) that a three-layer network (one hidden layer) with an arbitrarily large number of nodes in the hidden layer can approximate any continuous function $\Re^n \rightarrow \Re^m$ over a compact subset of $\Re^n$. Other recent results indicate that, in some cases, two hidden layers are required for the approximation of inverses of continuous functions (Sontag 1992). These are remarkable properties that make neural nets suitable for modelling nonlinear processes and designing nonlinear controllers. However, this is not enough. Indeed, as mentioned before, in the indirect approach, most of the neurocontrol algorithms use a differentiable model of the process in order to compute adaptation laws. For instance, gradient descent algorithms involve the Jacobian matrix of the process that must be deduced from the model. In some cases, the model is itself a neural net which, for this reason, must also approximate the derivative. So, authors developed conditions in which a net will also approximate the derivative of the mapping (Cardaliaguet & Euvrard 1992; Gallant & White 1992).

Another important property relies on the use of the least mean square criterion. It is well known from the mathematical statistics literature (and estimation theory) that a neural net trained by minimizing a least mean square will approximate the conditional expectation of the output, given the input. The output of the feedforward neural nets trained with a least mean square criterion will therefore be an estimate of this conditional expectation, which has important properties (for instance, for classification, it minimizes the risk of error) (White 1989; Ruck et al. 1990; Wan 1990; for a review, see Richard & Lippmann 1991).

These facts provide some motivations to study neural networks in the framework of nonlinear systems theory (see Narendra & Parthasarathy 1990, for more details). However, the choice of identification and controller models for nonlinear plants is a formidable problem and successful identification and control has to depend upon several strong assumptions regarding the input-output behavior of the plant. For instance, complete controllability and observability must be assumed (see Narendra & Parthasarathy 1990, for a discussion).

Recently, researchers tried to design adaptation laws that ensure the stability of the closed-loop process (see Kawato 1990; Renders 1991; Gomi & Kawato 1993 in the context of robotics applications; see Parthasarathy & Narendra 1991; Sanner & Slotine 1992; Narendra 1992; Tzirkel-Hancock & Fallside 1992; Renders 1993; Jin, Pipe & Winfield 1993; Chen & Liu 1993; Jin, Nikiforuk & Gupta 1993; Saerens, Renders & Bersini 1995; Renders, Saerens & Bersini 1995; Rovithakis & Christodoulou 1994 in the context of adaptive control with neural nets), or to ensure the viability of the regulation law (Seube 1990; Seube & Macias 1991). Stability of the overall system is certainly a prime requirement that must be met before thinking of real-world applications. Our aim in this paper is *first* (section 2) to introduce a weight adjustment law, based on Lyapunov stability theory, for the adaptive control of a certain class of MIMO continuous-time processes. This extends our previous work (Renders 1993; Saerens, Renders & Bersini 1995), where only single-input single-output (SISO) systems were considered; moreover, in the present work we introduce an integral term in order to avoid the steady-state errors. As previously mentioned, the adjustment strategy is indirect in that it uses a sensitivity model of the plant (some sensitivity derivatives have to be known; in our case, we estimate them with a neural network that identifies the plant) in order to implement the adaptation law. The resulting adaptation law is surprisingly simple in comparison with the gradient-based algorithm (also known as the dynamic backpropagation algorithm used to train recurrent neural nets, Williams & Zipser 1989) introduced by Narendra & Parthasarathy (1990; 1992). The stability statement, based on a first-order expansion, is, however, only valid if the initial weight values are not too far from their optimal values that allow perfect model matching (local stability). Therefore, following Kawato (1990), we propose to initialize the weights with values that solve the linear problem. Notice that the proposed adaptation law can be applied to more general processes than the algorithms of Parthasarathy & Narendra (1991), Narendra (1992), Sanner & Slotine (1992), Tzirkel-Hancock & Fallside (1992), Chen & Liu (1993), Jin, Nikiforuk & Gupta (1993), Johansen (1994), and Rovithakis & Christodoulou (1994), where affine processes were considered.

In the *second part* of the paper (section 3), we derive a robust adjustment law that takes account of noise, disturbances, and modelling errors. Indeed, due to noise, disturbances, and modelling errors, the perfect model matching condition cannot be verified, and it is no longer possible to ensure that the error converges to zero, but only that it remains bounded. Moreover, it has been shown that in certain cases, the simple adaptive scheme proposed in section 2, designed for the ideal case, can be driven unstable in the presence of bounded disturbances and unmodelled dynamics (see for instance Rohrs, Valavani, Athans & Stein 1985). The underlying instability mechanisms are basically the parameter drift phenomenon or the excitation of unmodelled high-frequency dynamics by the reference signal or by the high gain feedback generated by the adaptive control law (Ioannou & Datta 1991). Finally, in practical situations, a perfect sensitivity model of the plant is not available, which results in disturbances in the adaptation law.

In order to counteract these instability phenomena, one possible choice is to adopt a "dead-zone" strategy (see Ortega & Tang for a review). However, the dead-zone approach requires some information about the bounds of the perturbations effects, which in general are not available. We therefore use a modified adjustment law known as the "e_1-modification" scheme, originally introduced by Narendra & Annaswamy (1987) for the adaptive control of linear systems, that does not require this additional information. This adaptive control scheme is proved to be robust in that all the signals in the adaptive loop are bounded for bounded disturbances. Moreover, it has been shown that in the ideal case, that is when the disturbances are removed, the origin of the error equations is exponentially stable when the reference input is persistently exciting with a sufficiently large amplitude (Narendra & Annaswamy 1987).

In the *third part* (section 4) of the paper, we propose an architecture that allows us to initialize the weights of the net with a linearized controller. This is important (and this will be confirmed in the simulations section), since the local stability results are only valid if we are in the neighborhood of the optimal weight values.

In the *last part*, we apply the robust adaptation law to the control of a simulated two-inputs, two-outputs flow-mixer. We show that the control law is effective for a tracking task, with and without measurement noise. The simulation results also illustrate the advantage of initializing the weights with a rough linear controller. The benefit of the robust adaptation law on the simple one is also shown on a non-persistently exciting situation.

2. Adaptive neurocontrol in the ideal case

2.1. Introduction

If $y_i^{(d_i)}$ is the d_i order time derivative of the signal $y_i(t)$, we assume the the process can be described by

$$y_i^{(d_i)} = f_i(\mathbf{x}(t), \mathbf{u}(t)) \qquad i = 1, ..., n \tag{1a}$$

where the state $\mathbf{x}(t)$ is supposed to be measurable and can be obtained from the n outputs $y_i(t)$ and their time derivatives up to order (d_i-1). $\mathbf{u}(t) = [u_1, u_2, ..., u_n]^T$ and $\mathbf{y}(t) = [y_1, y_2, ..., y_n]^T$ are respectively the inputs and the outputs of the plant at time t. d_i is the relative degree and, by definition of d_i, $\partial f_i/\partial u_j \neq 0$ for at least one $u_j \in \{u_1, u_2, ..., u_n\}$. We also require that the relationship between the n inputs $u_j(t)$ and the n differentiated outputs $y_i^{(d_i)}$ given by equation (1a) is a global diffeomorphism on the operating region. This means that, whatever the value of $\mathbf{x}(t)$ in the operating region, the set of equations $y_i^{(d_i)}=f_i(\mathbf{x}(t), \mathbf{u}(t))$ ($i = 1, ..., n$) can be solved uniquely in terms of $\mathbf{u}(t)$, for every set of values $y_i^{(d_i)}$ ($i = 1, ..., n$) in the operating region. This implies that the Jacobian matrix is non-singular (of rank n) in the operating region: $\det[\partial f_i/\partial u_j(t)] \neq 0$. If we define the interactor matrix as $[\boldsymbol{\xi}(s)]_{ij} = s^{d_i}\delta_{ij}$, equations (1a) can be rewritten as

$$\boldsymbol{\xi}(s)\mathbf{y}(t) = \mathbf{F}[\mathbf{x}(t); \mathbf{u}(t)] \tag{1b}$$

where $\mathbf{F} = [f_1, f_2, ..., f_n]^T$. Note that (1b) involves a mixture of time-domain and frequency-domain notations (s is the Laplace variable); such hybrid notations are common in the adaptive control literature and their interpretation is obvious.

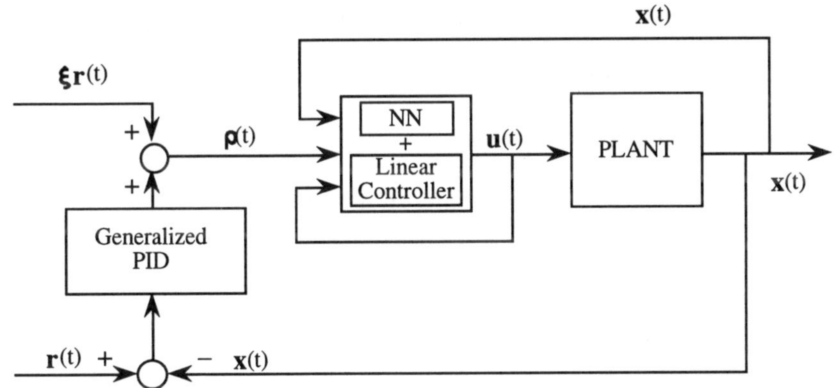

Figure 1. Overall architecture used for adaptive neurocontrol of a certain class of unknown continuous-time nonlinear processes.

We now consider that the output of a reference filter $\tilde{y}_i^{(d_i)}$ (which is supposed to be both computable and achievable) is obtained by filtering a reference signal $\mathbf{r}(t)$, taking account of the error signal (see Figure 1):

$$\begin{aligned}\tilde{y}_i^{(d_i)} - r_i^{(d_i)} &= \lambda_i^{d_i-1}\left(r_i^{(d_i-1)} - y_i^{(d_i-1)}\right) + ... + \lambda_i^0 (r_i - y_i) + \lambda_i^{\mathrm{I}} \int (r_i - y_i)\, dt \\ &= \lambda_l^{d_i-1} \varepsilon_i^{(d_i-1)} + ... + \lambda_i^0 \varepsilon_i + \lambda_i^{\mathrm{I}} \int \varepsilon_i\, dt \qquad i = 1, ..., n \end{aligned} \tag{2a}$$

where $\mathbf{r}(t) = [r_1, r_2, ..., r_n]^T$ is the reference signal, $r_i^{(j)}$ the time derivative of order j, and

$$\varepsilon_i^{(j)} = (r_i^{(j)} - y_i^{(j)}) \qquad i = 1, ..., n \tag{3}$$

In a more compact way, we have

$$\boldsymbol{\xi}(s)\,(\tilde{\mathbf{y}}(t) - \mathbf{r}(t)) = \mathbf{H}(s)\,\boldsymbol{\varepsilon} \tag{2b}$$

where $\mathbf{H}(s)$ is a diagonal polynomial matrix and $\tilde{\mathbf{y}}(t) = [\tilde{y}_1, \tilde{y}_2, ..., \tilde{y}_n]^T$.

We observe that when perfect tracking is achieved, that is, by taking $y_i^{(j)}$ instead of $\tilde{y}_i^{(j)}$ in (2), the equation allows us to design the transient of the initial error. Indeed, we have

$$0 = \varepsilon_i^{(d_i)} + \lambda_i^{d_i-1}\,\varepsilon_i^{(d_i-1)} + ... + \lambda_i^0\,\varepsilon_i + \lambda_i^I \int \varepsilon_i\,dt \qquad i = 1, ..., n \tag{4}$$

and the coefficients λ_i^j will be chosen in order to obtain a desired transient behaviour.

If we make the assumption that the system is asymptotically minimum phase, then, there exists a stable control law, which allows perfect tracking of the signal $\boldsymbol{\xi}(s)\tilde{\mathbf{y}}(t)$:

$$\mathbf{u} = \mathscr{U}[\boldsymbol{\xi}(s)\tilde{\mathbf{y}}(t), \mathbf{x}(t)] = \mathscr{U}[\boldsymbol{\rho}(t), \mathbf{x}(t)] \tag{5}$$

where we defined $\boldsymbol{\rho}(t)=\boldsymbol{\xi}(s)\tilde{\mathbf{y}}(t)$. We have

$$\boldsymbol{\xi}(s)\mathbf{y}(t) = \boldsymbol{\xi}(s)\tilde{\mathbf{y}}(t) \qquad \text{if } \mathbf{u} = \mathscr{U}[\boldsymbol{\rho}(t), \mathbf{x}(t)] \tag{6}$$

We adopt a control law generated by a neural network:

$$\mathbf{u}(t) = \mathbf{N}[\boldsymbol{\rho}(t), \mathbf{x}(t); \mathbf{w}(t)] \tag{7}$$

where $\mathbf{w}$ is the connection weight vector of the network. The network has inputs $\boldsymbol{\rho}(t)$, $\mathbf{x}(t)$ and output $\mathbf{u}(t)$. We assume that the neural controller $\mathbf{u}(t)=\mathbf{N}[\boldsymbol{\rho}(t), \mathbf{x}(t); \mathbf{w}(t)]$ can approximate the control law (5) to any degree of accuracy in the operating region. This is the standard perfect model matching condition.

2.2. *Preliminary lemma (error equations)*

In this paragraph, we establish a relationship between the error at the output of the process and the error on the weight values, based on a first-order expansion around the perfectly tuned values.

Let us suppose that at a given time we have $\mathbf{u}(t) = \mathbf{N}[\boldsymbol{\rho}(t), \mathbf{x}(t); \mathbf{w}(t)]$. Now, we define $\tilde{\mathbf{w}} = (\mathbf{w}(t) - \mathbf{w}^*)$ where $\mathbf{w}^*$ is an optimal weight vector that allows perfect model matching with $\mathbf{u}(t) = \mathbf{N}[\boldsymbol{\rho}(t), \mathbf{x}(t); \mathbf{w}^*]$. We suppose that $\mathbf{w}^*$ is sufficiently close to $\mathbf{w}(t)$ to permit a first order expansion around $\mathbf{w}(t)$:

$$\mathbf{N}[\boldsymbol{\rho}(t), \mathbf{x}(t); \mathbf{w}^*] \cong \mathbf{N}[\boldsymbol{\rho}(t), \mathbf{x}(t); \mathbf{w}(t)] - [\frac{\partial \mathbf{N}}{\partial \mathbf{w}}[\boldsymbol{\rho}(t), \mathbf{x}(t); \mathbf{w}(t)]]\, \tilde{\mathbf{w}}(t) \quad (8)$$

$\cong$ indicates that the relationship is true at the first order. Notice that if the output of the net is linear in its parameters (for instance, a radial basis function network), this development is strictly valid. We now define a modified signal $\boldsymbol{\rho}'(t)$ such that

$$\begin{aligned}\mathbf{N}[\boldsymbol{\rho}'(t), \mathbf{x}(t); \mathbf{w}^*] &= \mathbf{N}[\boldsymbol{\rho}(t), \mathbf{x}(t); \mathbf{w}^*] + [\frac{\partial \mathbf{N}}{\partial \mathbf{w}}[\boldsymbol{\rho}(t), \mathbf{x}(t); \mathbf{w}(t)]]\, \tilde{\mathbf{w}}(t) \quad (9)\\ &\cong \mathbf{N}[\boldsymbol{\rho}(t), \mathbf{x}(t); \mathbf{w}(t)]\end{aligned}$$

Since $\mathbf{w}(t)$ is in the neighborhood of $\mathbf{w}^*$, $\boldsymbol{\rho}'(t)$ will be sufficiently close to $\boldsymbol{\rho}(t)$ in order to make a first-order expansion of $\mathbf{N}[\boldsymbol{\rho}'(t), \mathbf{x}(t); \mathbf{w}^*]$ around $\mathbf{N}[\boldsymbol{\rho}(t), \mathbf{x}(t); \mathbf{w}^*]$. We have

$$\begin{aligned}\mathbf{N}[\boldsymbol{\rho}'(t), \mathbf{x}(t); \mathbf{w}^*] &\cong \mathbf{N}[\boldsymbol{\rho}(t), \mathbf{x}(t); \mathbf{w}^*]\\ &+ \frac{\partial \mathbf{N}}{\partial \boldsymbol{\rho}(t)}[\boldsymbol{\rho}(t), \mathbf{x}(t); \mathbf{w}^*]\, (\boldsymbol{\rho}'(t) - \boldsymbol{\rho}(t)) \quad (10)\end{aligned}$$

According to the definition of $\boldsymbol{\rho}'(t)$, we find

$$\boldsymbol{\rho}'(t) \cong \boldsymbol{\rho}(t) + [\frac{\partial \mathbf{N}}{\partial \boldsymbol{\rho}(t)}[\boldsymbol{\rho}(t), \mathbf{x}(t); \mathbf{w}^*]]^{-1}\, [\frac{\partial \mathbf{N}}{\partial \mathbf{w}}[\boldsymbol{\rho}(t), \mathbf{x}(t); \mathbf{w}(t)]]\, \tilde{\mathbf{w}}(t) \quad (11)$$

Now, since we have $\tilde{\mathbf{y}}(t)=\boldsymbol{\xi}^{-1}(s)\boldsymbol{\rho}(t)$, from the definition of $\boldsymbol{\rho}'(t)$, we also have $\mathbf{y}(t)\cong\boldsymbol{\xi}^{-1}(s)\, \boldsymbol{\rho}'(t)$ so that

$$\mathbf{e}(t) = (\mathbf{y}(t) - \tilde{\mathbf{y}}(t)) \cong \boldsymbol{\xi}^{-1}(s)\, (\boldsymbol{\rho}'(t) - \boldsymbol{\rho}(t))$$

or

$$\boldsymbol{\xi}(s)\, \mathbf{e}(t) \cong (\boldsymbol{\rho}'(t) - \boldsymbol{\rho}(t)) \quad (12)$$

Therefore

$$\begin{aligned}\boldsymbol{\xi}(s)\, \mathbf{e}(t) &\cong [\frac{\partial \mathbf{N}}{\partial \boldsymbol{\rho}(t)}[\boldsymbol{\rho}(t), \mathbf{x}(t); \mathbf{w}^*]]^{-1}\, [\frac{\partial \mathbf{N}}{\partial \mathbf{w}}[\boldsymbol{\rho}(t), \mathbf{x}(t); \mathbf{w}(t)]]\, \tilde{\mathbf{w}}(t)\\ &\cong \mathbf{v}(t)\, \tilde{\mathbf{w}}(t) \quad (13)\end{aligned}$$

where we defined

$$\mathbf{v}(t) = [\frac{\partial \mathbf{N}}{\partial \boldsymbol{\rho}(t)}[\boldsymbol{\rho}(t), \mathbf{x}(t); \mathbf{w}^*]]^{-1}\, [\frac{\partial \mathbf{N}}{\partial \mathbf{w}}[\boldsymbol{\rho}(t), \mathbf{x}(t); \mathbf{w}(t)]]$$

Now, let us show that

$$[\frac{\partial \mathbf{N}}{\partial \boldsymbol{\rho}(t)}[\boldsymbol{\rho}(t), \mathbf{x}(t); \mathbf{w}^*]]^{-1} = \frac{\partial(\boldsymbol{\xi}(s)\mathbf{y}(t))}{\partial \mathbf{u}(t)} , \tag{14}$$

the derivative on the left side of (14) being computed at $\mathbf{w}^*$.

Indeed, for $\mathbf{w}=\mathbf{w}^*$ (perfect tuning), we perform perfect tracking, so that

$$\frac{\partial(\boldsymbol{\xi}(s)\mathbf{y}(t))}{\partial \boldsymbol{\rho}(t)} = \mathbf{I} = \frac{\partial(\boldsymbol{\xi}(s)\mathbf{y}(t))}{\partial \mathbf{u}(t)} \frac{\partial \mathbf{N}}{\partial \boldsymbol{\rho}(t)}[\boldsymbol{\rho}(t), \mathbf{x}(t); \mathbf{w}^*]$$

which proves (14).

We therefore have

$$\begin{aligned}\mathbf{v}(t) &= \frac{\partial(\boldsymbol{\xi}(s)\mathbf{y}(t))}{\partial \mathbf{u}(t)} \frac{\partial \mathbf{N}}{\partial \mathbf{w}}[\boldsymbol{\rho}(t), \mathbf{x}(t); \mathbf{w}(t)] \\ &= \frac{\partial \mathbf{F}[\mathbf{x}(t); \mathbf{u}(t)]}{\partial \mathbf{u}(t)} \frac{\partial \mathbf{N}}{\partial \mathbf{w}}[\boldsymbol{\rho}(t), \mathbf{x}(t); \mathbf{w}(t)] \end{aligned} \tag{15}$$

The time-varying vector $\mathbf{v}(t)$ is supposed to be measurable and computable at any time. The control is therefore indirect: the term $\partial\mathbf{F}[\mathbf{x}(t); \mathbf{u}(t)]/\partial\mathbf{u}(t)$ (matrix of sensitivity derivatives) can be computed thanks to a differentiable model of the process.

However, in (13), while the state of the process is supposed to be measurable, the determination of $\boldsymbol{\xi}(s)\mathbf{y}(t)$ is not always feasible. In order to avoid this potential problem, let us decompose $\tilde{y}_i^{(d_i)}$ given by (2a) as

$$\begin{aligned}\tilde{y}_i^{(d_i)} &= r_i^{(d_i)} + (\alpha_i^2 + k_i)(r_i^{(d_i-1)} - y_i^{(d_i-1)}) + (\alpha_i^3 + k_i\,\alpha_i^2)(r_i^{(d_i-2)} - y_i^{(d_i-2)}) + \ldots \\ &+ (\alpha_i^{d_i} + k_i\,\alpha_i^{d_i-1})(r_i^{(1)} - y_i^{(1)}) + (\alpha_i^{d_i+1} + k_i\,\alpha_i^{d_i})(r_i - y_i) + (k_i\,\alpha_i^{d_i+1}) \int (r_i - y_i)\,\mathrm{d}t \\ &= r_i^{(d_i)} + (\alpha_i^2 + k_i)\,\varepsilon_i^{(d_i-1)} + (\alpha_i^3 + k_i\,\alpha_i^2)\,\varepsilon_i^{(d_i-2)} + \ldots \\ &+ (\alpha_i^{d_i} + k_i\,\alpha_i^{d_i-1})\,\varepsilon_i^{(1)} + (\alpha_i^{d_i+1} + k_i\,\alpha_i^{d_i})\,\varepsilon_i + (k_i\,\alpha_i^{d_i+1}) \int \varepsilon_i\,\mathrm{d}t \\ &= y_i^{(d_i)} + \varepsilon_i^{(d_i)} + \alpha_i^2\,\varepsilon_i^{(d_i-1)} + \alpha_i^3\,\varepsilon_i^{(d_i-2)} + \ldots + \alpha_i^{d_i}\,\varepsilon_i^{(1)} + \alpha_i^{d_i+1}\,\varepsilon_i \\ &+ k_i\,[\varepsilon_i^{(d_i-1)} + \alpha_i^2\,\varepsilon_i^{(d_i-2)} + \ldots + \alpha_i^{d_i-1}\,\varepsilon_i^{(1)} + \alpha_i^{d_i}\,\varepsilon_i + \alpha_i^{d_i+1} \int \varepsilon_i\,\mathrm{d}t\,]\end{aligned}$$

$$= y_i^{(d_i)} + \dot{\delta}_i + k_i\, \delta_i \tag{16}$$

where we introduced the auxiliary error

$$\delta_i = \varepsilon_i^{(d_i-1)} + \alpha_i^2\, \varepsilon_i^{(d_i-2)} + \ldots + \alpha_i^{d_i-1}\, \varepsilon_i^{(1)} + \alpha_i^{d_i}\, \varepsilon_i + \alpha_i^{d_i+1} \int \varepsilon_i\, dt \tag{17}$$

which is computable because directly related to the state. The coefficients α_i^j must be chosen in such a way that the corresponding transfer function is Hurwitz. In particular, this implies that if $\delta_i \to 0$, we have $\varepsilon_i \to 0$.

Equation (16) can be rewritten as

$$e_i^{(d_i)} = y_i^{(d_i)} - \tilde{y}_i^{(d_i)} = -(s + k_i)\, \delta_i(t) \tag{18}$$

and by defining $\mathbf{K} = \mathrm{diag}[k_i]$ and $\boldsymbol{\delta}(t) = [\delta_1, \delta_2, \ldots, \delta_n]^T$, we find

$$\boldsymbol{\xi}(s)\, \mathbf{e}(t) = \boldsymbol{\xi}(s)\, (\mathbf{y}(t) - \tilde{\mathbf{y}}(t)) = -(s\mathbf{I} + \mathbf{K})\, \boldsymbol{\delta}(t) \tag{19}$$

From equation (13):

$$\boldsymbol{\delta}(t) = -(s\mathbf{I} + \mathbf{K})^{-1}\, \mathbf{v}(t)\, \tilde{\mathbf{w}}(t) \tag{20}$$

where $(s\mathbf{I} + \mathbf{K})^{-1}$ is SPR if all the $k_i > 0$.

2.3. Local stability proof

Let us first state the stability result:

Convergence statement*. If* $\mathbf{v}(t) = \dfrac{\partial \mathbf{F}[\mathbf{x}(t); \mathbf{u}(t)]}{\partial \mathbf{u}(t)}\, \dfrac{\partial \mathbf{N}}{\partial \mathbf{w}}[\boldsymbol{\rho}(t), \mathbf{x}(t); \mathbf{w}(t)]$ *is globally bounded, the adaptation rule*

$$\dot{\mathbf{w}} = \eta\, \mathbf{v}^T \boldsymbol{\delta}(t) \qquad (\eta > 0) \tag{21}$$

will drive the auxiliary error $\boldsymbol{\delta}(t)$ *asymptotically to zero:* $\boldsymbol{\delta}(t) \to 0$; *that is,* $(\varepsilon_i^{(d_i-1)} + \alpha_i^2\, \varepsilon_i^{(d_i-2)} + \ldots + \alpha_i^{d_i-1}\, \varepsilon_i^{(1)} + \alpha_i^{d_i}\, \varepsilon_i + \alpha_i^{d_i+1} \int \varepsilon_i\, dt\,) \to 0$, *and therefore* $\varepsilon_i \to 0$.

This convergence statement can be proved using the stability lemma given in Appendix.

Proof of the convergence statement*.* We can easily verify that we satisfy the conditions of the stability lemma, with

$$\mathbf{v}(t) = \frac{\partial \mathbf{F}[\mathbf{x}(t);\, \mathbf{u}(t)]}{\partial \mathbf{u}(t)} \frac{\partial \mathbf{N}}{\partial \mathbf{w}}[\boldsymbol{\rho}(t), \mathbf{x}(t);\, \mathbf{w}(t)]$$

$$\mathbf{y}(t) = \boldsymbol{\delta}(t)$$

Notice that the term $\partial\mathbf{N}/\partial\mathbf{w}[\boldsymbol{\rho}(t), \mathbf{x}(t);\, \mathbf{w}(t)]$ appearing in (21) can be computed thanks to backpropagation algorithm, while $\partial\mathbf{F}[\mathbf{x}(t);\, \mathbf{u}(t)]/\partial\mathbf{u}(t)$ is a matrix of sensitivity derivatives (the Jacobian matrix of the process defined by (1)). As proposed by Jordan (1989) and Narendra & Parthasarathy (1990), this Jacobian matrix can be deduced from a model of the process; possibly in the form of a neural identifier. As mentioned above, the adjustment strategy is indirect in that it uses a sensitivity model of the plant (in our case, we use a neural network that identifies the plant) in order to implement the adaptation law.

For the training of the neural identifier, in order to avoid the computation of the derivatives $y_i^{(d_i)}$, we can apply the same trick as for the training of the neural controller, or adapt the linear adaptive identification algorithms (Narendra & Annaswamy 1989; Rimon & Narendra 1992).

Various simulations of a similar adaptation law have been carried out for the SISO case (Renders 1993; Renders, Bersini & Saerens 1993; Saerens, Renders & Bersini 1995; Saerens et al. 1992).

2.4. Link with the external linearization approach

The existence of a unique stable control law which allows perfect tracking as expressed in (5) and (6) implies that the process is linearizable/decouplable by static state feedback and that the zero-dynamics is asymptotically stable (asymptotically minimum phase system). In certain cases, when these conditions are not directly satisfied, it is possible to design a dynamic pre-compensator, usually consisting of time-delays assigned to the different process inputs, in such a way that the new sytem (pre-compensator + process) fulfills the conditions of linearization by static state feedback. This design necessitates some prior knowledge about the process to be controlled, namely a generalized "Interactor Matrix". Further developments should focus on relaxing this need.

Now the neural controller can be seen as a linearizing/decoupling block and, when adopting a reference model as given in (2), a generalized PID (D^2, D^3, etc) controller plays the role of correcting adequately the residual error (see Figure 1). Mathematically, when the controller law is expressed by (7), the closed-loop behaviour is described by the following equations:

$$\boldsymbol{\xi}(s)\mathbf{y}(t) = \mathbf{F}[\mathbf{x}(t);\, \mathbf{N}[\boldsymbol{\rho}(t), \mathbf{x}(t);\, \mathbf{w}(t)]]$$

$$= \boldsymbol{\rho}(t) + \boldsymbol{\Delta}(t)$$

where $\mathbf{\Delta}(t) = [\Delta_1, \Delta_2, ..., \Delta_n]^{\mathrm{T}}$ represents the effect of modelling errors. Ideally (without modelling error), we should have:

$$\boldsymbol{\rho}(t) = \mathbf{F}[\mathbf{x}(t); \mathcal{U}[\boldsymbol{\rho}(t), \mathbf{x}(t)]]$$

so that $\mathbf{\Delta}(t)$ is:

$$\mathbf{\Delta}(t) = \mathbf{F}[\mathbf{x}(t); \mathbf{N}[\boldsymbol{\rho}(t), \mathbf{x}(t); \mathbf{w}(t)]] - \mathbf{F}[\mathbf{x}(t); \mathcal{U}[\boldsymbol{\rho}(t), \mathbf{x}(t)]]$$

$$= \frac{\partial \mathbf{F}[\mathbf{x}(t); \mathbf{u}(t)]}{\partial \mathbf{u}(t)}\Big|_{\mathbf{u}=\mathbf{N}[\boldsymbol{\rho}(t), \mathbf{x}(t); \mathbf{w}(t)]} (\mathbf{N}[\boldsymbol{\rho}(t), \mathbf{x}(t); \mathbf{w}(t)] - \mathcal{U}[\boldsymbol{\rho}(t), \mathbf{x}(t)])$$

$$+ \mathrm{O}(\|\mathbf{N}[\boldsymbol{\rho}(t), \mathbf{x}(t); \mathbf{w}(t)] - \mathcal{U}[\boldsymbol{\rho}(t), \mathbf{x}(t)]\|^2)$$

Therefore, from equation (2a), the error terms $\varepsilon_i = (r_i - y_i)$ satisfy the following equations:

$$\Delta_i(t) = \varepsilon_i^{(d_i)} + \lambda_i^{d_i-1} \varepsilon_i^{(d_i-1)} + ... + \lambda_i^0 \varepsilon_i + \lambda_i^{\mathrm{I}} \int \varepsilon_i \, \mathrm{d}t \qquad i = 1, ..., n$$

This indicates that if $\mathbf{\Delta}(t)$ tends towards a constant value $\mathbf{\Delta}_0$, then $(\mathbf{r} - \mathbf{y})$ tends towards zero: this is the reason why we introduced an integral term in the reference model.

3. Adaptive neurocontrol in the presence of disturbances

3.1. Introduction

Due to noise, disturbances, and modelling errors, the perfect model matching condition cannot be verified, and it will not be possible to ensure that $\varepsilon_i \to 0$ asymptotically, but only that it will remain bounded. Moreover, in practice, a perfect model of the process will certainly not be available, so that $\mathbf{v}(t)$ will also be subject to disturbances.

In certain cases, the simple adaptive scheme designed for the ideal case in section 2 can be driven unstable in the presence of bounded disturbances and unmodelled dynamics. The underlying instability mechanisms are basically the parameter drift phenomenon or the excitation of unmodelled high-frequency dynamics by the reference signal or by the high gain feedback generated by the adaptive control law (Ioannou & Datta 1991).

In order to counteract these instability phenomena, one possible choice is to adopt a "dead-zone" strategy, so that the adaptive law (21) is modified as follows:

$$\dot{\mathbf{w}} = 0, \qquad \text{if } \boldsymbol{\zeta}^{\mathrm{T}}\mathbf{P}\boldsymbol{\zeta} \leq \mathrm{E}_0 \qquad (22a)$$

$$\dot{\mathbf{w}} = \eta\, \mathbf{v}^{\mathrm{T}}\, \boldsymbol{\delta}(t), \qquad \text{if } \boldsymbol{\zeta}^{\mathrm{T}}\mathbf{P}\boldsymbol{\zeta} > \mathrm{E}_0 \tag{22b}$$

where $\boldsymbol{\zeta}(t)$ is the state vector of the state representation of the SPR transfer function matrix of the system defined by (20). If this transfer function is chosen as $(s\mathbf{I} + \mathbf{K})^{-1}$, with $\mathbf{K}$ diagonal, then the conditions (22ab) are $\boldsymbol{\varepsilon}^{\mathrm{T}}\boldsymbol{\varepsilon} \le \mathrm{E}_0$ and $\boldsymbol{\varepsilon}^{\mathrm{T}}\boldsymbol{\varepsilon} > \mathrm{E}_0$. The corresponding Lyapunov function is (for details, see Liu & Chen 1993)

$$V = \mathrm{E}_0 + \frac{|\lambda|}{\eta}\, \boldsymbol{\psi}^{\mathrm{T}}\, \boldsymbol{\psi}, \qquad \text{if } \boldsymbol{\zeta}^{\mathrm{T}}\mathbf{P}\boldsymbol{\zeta} \le \mathrm{E}_0 \tag{23a}$$

$$V = \boldsymbol{\zeta}^{\mathrm{T}}\, \mathbf{P}\, \boldsymbol{\zeta} + \frac{|\lambda|}{\eta}\, \boldsymbol{\psi}^{\mathrm{T}}\, \boldsymbol{\psi}, \qquad \text{if } \boldsymbol{\zeta}^{\mathrm{T}}\mathbf{P}\boldsymbol{\zeta} > \mathrm{E}_0 \tag{23b}$$

However, the dead-zone approach requires some information about the bounds of the perturbation effects, which is not available in general. We therefore propose to use a modified adjustment law known as the "e_1-modification" scheme, originally introduced by Narendra & Annaswamy (1987) for the adaptive control of linear systems, that does not require any additional information.

3.2. A modified adjustment law

Let us first give the stability statement:

Stability statement. *If* $\mathbf{v}(t) = \dfrac{\partial\mathbf{F}[\mathbf{x}(t);\, \mathbf{u}(t)]}{\partial\mathbf{u}(t)}\, \dfrac{\partial\mathbf{N}}{\partial\mathbf{w}}[\boldsymbol{\rho}(t), \mathbf{x}(t); \mathbf{w}(t)]$ *is globally bounded, the adaptation rule*

$$\dot{\mathbf{w}} = \eta\, \mathbf{v}^{\mathrm{T}} \boldsymbol{\delta}(t) - \eta\, \gamma\, \|\boldsymbol{\delta}\|\, \mathbf{w}, \qquad \text{with } \eta, \gamma > 0 \tag{24}$$

will ensure that the weight vector $\mathbf{w}$ *as well as the auxiliary error* $\boldsymbol{\delta}(t)$ *remain bounded, and therefore each* ε_i *remains bounded.*

By defining robustness as boundedness of the signals in the adaptive loop for bounded disturbances, we will show below that the adaptive system with adustment law (24) is robust. We will consider two different cases, depending on the kind of disturbance considered: **(i)** bounded disturbances due to noise and neural network modelling errors for the neural controller, and **(ii)** bounded disturbances due to imperfect modelling of the process (which results in errors in the estimation of the sensitivity derivatives).

3.2.1. Bounded disturbances due to noise and neural controller modelling errors

Let us assume that the uncertainty effects can be expressed as (Narendra & Annaswamy 1989)

$$\boldsymbol{\delta}(t) = -(s\mathbf{I} + \mathbf{K})^{-1} (\mathbf{v}(t)\,\tilde{\mathbf{w}}(t) + \boldsymbol{\phi}(t)) \tag{25}$$

where the vector $\boldsymbol{\phi}(t)$ represents the effect of the different sources of perturbations:

$$\boldsymbol{\phi}(t) = \mathbf{f}[\tilde{\mathbf{w}}(t), \text{neural controller modelling errors, noise}] \tag{26}$$

The neural controller modelling errors come from the fact that the neural controller has only a finite number of parameters; therefore, even in the absence of noise and in the case of perfect parameter tuning, it cannot approximate the control law to any degree of accuracy.

As for the ideal case, since $(s\mathbf{I} + \mathbf{K})^{-1}$ is SPR, the error equation (25) admits a state-space representation

$$\dot{\boldsymbol{\zeta}} = \mathbf{A}\,\boldsymbol{\zeta} + \mathbf{B}\,(\mathbf{v}(t)\,\tilde{\mathbf{w}}(t) + \boldsymbol{\phi}(t)) \tag{27a}$$

$$\boldsymbol{\delta}(t) = \mathbf{C}^{\mathrm{T}}\,\boldsymbol{\zeta} \tag{27b}$$

and the relationships (A5ab – in appendix) hold. This representation is assumed to be in normal form, so that we have $\|\boldsymbol{\zeta}\| \geq \|\boldsymbol{\delta}\|$.

Proof of the stability statement. Let us define a positive definite function V of the form

$$V = \boldsymbol{\zeta}^{\mathrm{T}}\,\mathbf{P}\,\boldsymbol{\zeta} + \frac{1}{\eta}\,\tilde{\mathbf{w}}^{\mathrm{T}}\,\tilde{\mathbf{w}} \tag{28}$$

Now, we compute

$$\begin{aligned}
\dot{V} &= \dot{\boldsymbol{\zeta}}^{\mathrm{T}}\,\mathbf{P}\,\boldsymbol{\zeta} + \boldsymbol{\zeta}^{\mathrm{T}}\,\mathbf{P}\,\dot{\boldsymbol{\zeta}} + 2\,\frac{1}{\eta}\,\tilde{\mathbf{w}}^{\mathrm{T}}\,\dot{\tilde{\mathbf{w}}} \\
&= -\boldsymbol{\zeta}^{\mathrm{T}}\,\mathbf{Q}\,\boldsymbol{\zeta} + 2\,\boldsymbol{\phi}^{\mathrm{T}}\,\boldsymbol{\delta}(t) - 2\,\tilde{\mathbf{w}}^{\mathrm{T}}\,\gamma\,\|\boldsymbol{\delta}\|\,\mathbf{w} \\
&= -\boldsymbol{\zeta}^{\mathrm{T}}\,\mathbf{Q}\,\boldsymbol{\zeta} + 2\,\boldsymbol{\phi}^{\mathrm{T}}\,\boldsymbol{\delta}(t) - 2\,\gamma\,\|\boldsymbol{\delta}\|\,\tilde{\mathbf{w}}^{\mathrm{T}}\,\tilde{\mathbf{w}} - 2\,\gamma\,\|\boldsymbol{\delta}\|\,\tilde{\mathbf{w}}^{\mathrm{T}}\,\mathbf{w}^{*}
\end{aligned}$$

and let us introduce $\lambda_{\mathrm{Q}}^{\max}$ and $\lambda_{\mathrm{Q}}^{\min}$, the largest and smallest eigenvalue of the positive definite matrix $\mathbf{Q}$. We have

$$0 \leq \lambda_{\mathrm{Q}}^{\min}\,\boldsymbol{\zeta}^{\mathrm{T}}\,\boldsymbol{\zeta} \leq \boldsymbol{\zeta}^{\mathrm{T}}\,\mathbf{Q}\,\boldsymbol{\zeta} \leq \lambda_{\mathrm{Q}}^{\max}\,\boldsymbol{\zeta}^{\mathrm{T}}\,\boldsymbol{\zeta}$$

Now, let us consider two different conditions **(a)** and **(b)**:

(a) $\|\tilde{\mathbf{w}}\| \geq \|\mathbf{w}^{*}\| + [\frac{\phi_0}{\gamma}]^{1/2}$, with ϕ_0 being an upper bound of $\|\boldsymbol{\phi}\|$ (29)

Since $\boldsymbol{\phi}(t)$ is a function of $\tilde{\mathbf{w}}(t)$, and we are trying to demonstrate that $\tilde{\mathbf{w}}(t)$ is bounded, it is in fact not obvious that $\|\boldsymbol{\phi}(t)\|$ admits an upper bound ϕ_0. However, we will show later that this is the case if the initial values $\tilde{\mathbf{w}}_0$ and $\boldsymbol{\zeta}_0$ satisfy

$$\eta\, \boldsymbol{\zeta}_0^{\mathrm{T}}\, \mathbf{P}\, \boldsymbol{\zeta}_0 + \tilde{\mathbf{w}}_0^{\mathrm{T}}\, \tilde{\mathbf{w}}_0 \leq \delta^2$$

where δ is a positive number chosen such that the first order expansion is still valid for $\|\tilde{\mathbf{w}}\| \leq \delta$.

Let us show that $\dot{V} < 0$ in condition **(a)**.

Preliminary lemma. For $a, b, c \in \Re^+$ $(a, b, c > 0)$, we have:
if $a > b + c$, then $a^2 > ab + c^2$.

Proof. Indeed, we have

$$a > b + c$$

so that

$$a^2 > ab + ac > ab + c^2$$

In particular, from (29), it implies that

$$\|\tilde{\mathbf{w}}\|^2 \geq \|\tilde{\mathbf{w}}\|\, \|\mathbf{w}^*\| + \frac{\phi_0}{\gamma} \tag{30}$$

Proof of boundedness for condition (a). We compute the time derivative of the positive definite function (28), which gives

$$\begin{aligned}
\dot{V} &= -\boldsymbol{\zeta}^{\mathrm{T}}\mathbf{Q}\,\boldsymbol{\zeta} + 2\,\boldsymbol{\phi}^{\mathrm{T}}\,\boldsymbol{\delta}(t) - 2\gamma\,\|\boldsymbol{\delta}\|\;\tilde{\mathbf{w}}^{\mathrm{T}}\,\tilde{\mathbf{w}} - 2\gamma\,\|\boldsymbol{\delta}\|\;\tilde{\mathbf{w}}^{\mathrm{T}}\,\mathbf{w}^* \\
&= -\boldsymbol{\zeta}^{\mathrm{T}}\mathbf{Q}\,\boldsymbol{\zeta} + 2\gamma\|\boldsymbol{\delta}\|\left[\frac{\boldsymbol{\phi}^{\mathrm{T}}\,\boldsymbol{\delta}(t)}{\gamma\,\|\boldsymbol{\delta}\|} - \tilde{\mathbf{w}}^{\mathrm{T}}\,\mathbf{w}^*\right] - 2\gamma\,\|\boldsymbol{\delta}\|\;\tilde{\mathbf{w}}^{\mathrm{T}}\,\tilde{\mathbf{w}} \\
&= -\boldsymbol{\zeta}^{\mathrm{T}}\mathbf{Q}\,\boldsymbol{\zeta} + 2\gamma\|\boldsymbol{\delta}\|\,\left\|\frac{\boldsymbol{\phi}^{\mathrm{T}}\,\boldsymbol{\delta}(t)}{\gamma\,\|\boldsymbol{\delta}\|} - \tilde{\mathbf{w}}^{\mathrm{T}}\,\mathbf{w}^*\right\| - 2\gamma\,\|\boldsymbol{\delta}\|\;\|\tilde{\mathbf{w}}\|^2
\end{aligned}$$

But, from (30), we have

$$\begin{aligned}
\left\|\frac{\boldsymbol{\phi}^{\mathrm{T}}\,\boldsymbol{\delta}(t)}{\gamma\,\|\boldsymbol{\delta}\|} - \tilde{\mathbf{w}}^{\mathrm{T}}\,\mathbf{w}^*\right\| &\leq \left\|\frac{\boldsymbol{\phi}^{\mathrm{T}}\,\boldsymbol{\delta}(t)}{\gamma\,\|\boldsymbol{\delta}\|}\right\| + \|\tilde{\mathbf{w}}^{\mathrm{T}}\,\mathbf{w}^*\| \\
&\leq \frac{\|\boldsymbol{\phi}\|}{\gamma} + \|\tilde{\mathbf{w}}\|\,\|\mathbf{w}^*\| \\
&\leq \frac{\phi_0}{\gamma} + \|\tilde{\mathbf{w}}\|\,\|\mathbf{w}^*\| \leq \|\tilde{\mathbf{w}}\|^2
\end{aligned}$$

which implies that $\dot{V} \le 0$.

(b) $\|\boldsymbol{\zeta}\| \ge \dfrac{\gamma \|\mathbf{w}^*\|}{\lambda_Q^{min}} + \dfrac{2\phi_0}{\lambda min;Q}$ (31)

Preliminary lemma. Let us show that for $\mathbf{x}, \mathbf{a} \in \Re^n$, the expression $(\mathbf{a}^T \mathbf{x} + \mathbf{x}^T \mathbf{x})$ is a minimum for $\mathbf{x} = -\mathbf{a}/2$.

Proof. By direct computation. This also implies that $(\mathbf{a}^T \mathbf{x} + \mathbf{x}^T \mathbf{x}) \ge -\dfrac{\|\mathbf{a}\|^2}{4}$.

Proof of boundedness for condition (b). We have

$$\begin{aligned}\dot{V} &= -\boldsymbol{\zeta}^T \mathbf{Q} \boldsymbol{\zeta} + 2\, \boldsymbol{\phi}^T \boldsymbol{\delta}(t) - 2\gamma \|\boldsymbol{\delta}\|\, \tilde{\mathbf{w}}^T \tilde{\mathbf{w}} - 2\gamma \|\boldsymbol{\delta}\|\, \tilde{\mathbf{w}}^T \mathbf{w}^* \\ &\le -\lambda_Q^{min} \boldsymbol{\zeta}^T \boldsymbol{\zeta} + 2\, \boldsymbol{\phi}^T \boldsymbol{\delta}(t) - 2\gamma \|\boldsymbol{\delta}\| [\tilde{\mathbf{w}}^T \tilde{\mathbf{w}} + \tilde{\mathbf{w}}^T \mathbf{w}^*] \\ &\le -\lambda_Q^{min} \boldsymbol{\zeta}^T \boldsymbol{\zeta} + 2\, \boldsymbol{\phi}^T \boldsymbol{\delta}(t) + \frac{\gamma}{2} \|\boldsymbol{\delta}\|\, \mathbf{w}^{*T}\mathbf{w}^*\end{aligned}$$

The last inequality come from the direct application of the preliminary lemma.

Moreover, we have

$$\begin{aligned}\left\| \frac{2\,\boldsymbol{\phi}^T \boldsymbol{\delta}}{\lambda min;Q} + \frac{\gamma \|\boldsymbol{\delta}\|\, \|\mathbf{w}^*\|^2}{2\lambda min;Q} \right\| &\le \frac{2\,\|\boldsymbol{\phi}\|\, \|\boldsymbol{\delta}\|}{\lambda min;Q} + \frac{\gamma \|\boldsymbol{\delta}\|\, \|\mathbf{w}^*\|^2}{2\lambda min;Q} \\ &\le \frac{2\,\phi_0 \|\boldsymbol{\delta}\|}{\lambda min;Q} + \frac{\gamma \|\boldsymbol{\delta}\|\, \|\mathbf{w}^*\|^2}{2\lambda min;Q} \\ &= \|\boldsymbol{\delta}\| \left[\frac{2\phi_0}{\lambda min;Q} + \frac{\gamma \|\mathbf{w}^*\|^2}{2\lambda min;Q}\right] \\ &\le \|\boldsymbol{\delta}\|\, \|\boldsymbol{\zeta}\| \\ &\le \|\boldsymbol{\zeta}\|^2\end{aligned}$$

which implies that $\dot{V} \le 0$.

Now, let us consider again the assumption that $\|\boldsymbol{\phi}(t)\|$ admits an upper bound ϕ_0. We decompose the signal $\boldsymbol{\phi}(t)$ as

$$\boldsymbol{\phi}(t) = O(\|\tilde{\mathbf{w}}\|^2) + O(\|\boldsymbol{\varepsilon}\|) + \boldsymbol{\nu}(t)$$

where $\boldsymbol{\varepsilon}$ represents the intrinsic modelling error of the network, i.e. the approximation error when the parameters are tuned ($\tilde{\mathbf{w}}=0$), and $\boldsymbol{\nu}(t)$ represents the effect of noise. Let us define the set $\mathcal{M}$ such that

$$\mathcal{M} = \{[\boldsymbol{\zeta}, \tilde{\mathbf{w}}]^T \mid \eta\, \boldsymbol{\zeta}^T \mathbf{P}\, \boldsymbol{\zeta} + \tilde{\mathbf{w}}^T \tilde{\mathbf{w}} \leq \delta^2\}$$

where δ is a positive number chosen such that the first order expansion is still valid for $\|\tilde{\mathbf{w}}\| \leq \delta$ and such that

$$\mathcal{M} \supseteq \mathcal{Q} = \{[\boldsymbol{\zeta}, \tilde{\mathbf{w}}]^T \mid (\|\tilde{\mathbf{w}}\| \geq \|\mathbf{w}^*\| + [\frac{\Phi_0}{\gamma}]^{1/2}) \text{ and } (\|\boldsymbol{\zeta}\| \geq \frac{\gamma\,\|\mathbf{w}^*\|}{\lambda_Q^{min}} + \frac{2\Phi_0}{\lambda_{min;Q}})\}$$

If the generalized state variable $[\boldsymbol{\zeta}, \tilde{\mathbf{w}}]^T$ belongs to $\mathcal{M}$, it is easy to see that

(i) the network inputs are bounded provided that the temporal derivatives up to order d_i of the reference r_i are bounded, and
(ii) the weights of the network are bounded.

Consequently, the outputs of the network are certainly bounded (at least for a multi-layer perceptron or a radial basis function network). Therefore, if $\boldsymbol{\nu}(t)$ is bounded, $\boldsymbol{\phi}(t)$ is bounded if the generalized state variable $[\boldsymbol{\zeta}, \tilde{\mathbf{w}}]^T$ does not escape from $\mathcal{M}$. Let us suppose that, at time $t{=}t_0$, $[\boldsymbol{\zeta}_0, \tilde{\mathbf{w}}_0]^T$ belongs to $\mathcal{M}$, but not to $\mathcal{Q}$. Then $[\boldsymbol{\zeta}, \tilde{\mathbf{w}}]^T \in \mathcal{M}$, for all $t \geq t_0$, because it has been shown that the function $(\eta\, \boldsymbol{\zeta}^T\mathbf{P}\boldsymbol{\zeta} + \tilde{\mathbf{w}}^T\tilde{\mathbf{w}})$ is decreasing in $\mathcal{M}\backslash\mathcal{Q}$ and $\mathcal{Q} \subseteq \mathcal{M}$. Since $[\boldsymbol{\zeta}, \tilde{\mathbf{w}}]^T$ does not exit from $\mathcal{M}$, the boundedness of $\boldsymbol{\phi}(t)$ is proved.

3.2.2. Bounded disturbances due to imperfect modelling of the process

We now consider the impact of imperfect modelling of the plant that results in errors in the estimation of the sensitivity derivatives. The error equation is

$$\boldsymbol{\delta}(t) = -(s\mathbf{I} + \mathbf{K})^{-1} (\mathbf{v}(t)\, \tilde{\mathbf{w}}(t)) \tag{32}$$

but the adjustment law is subject to disturbances:

$$\dot{\mathbf{w}} = \eta\, (\mathbf{v} + \boldsymbol{\Phi}(t))^T \boldsymbol{\delta}(t) - \eta\, \gamma\, \|\boldsymbol{\delta}\|\, \mathbf{w}, \qquad \text{with } \eta, \gamma > 0 \tag{33}$$

where the matrix $\boldsymbol{\Phi}(t)$ represents the error in the sensitivity derivatives, and is supposed to be bounded: $\|\boldsymbol{\Phi}(t)\, \boldsymbol{\delta}\| \leq \Phi_0\, \|\boldsymbol{\delta}\|$ for any $\boldsymbol{\delta} \in \mathfrak{R}^n$.

As in previous section, we compute

$$\dot{V} = \dot{\boldsymbol{\zeta}}^T \mathbf{P}\, \boldsymbol{\zeta} + \boldsymbol{\zeta}^T \mathbf{P}\, \dot{\boldsymbol{\zeta}} + 2\, \frac{1}{\eta}\, \tilde{\mathbf{w}}^T \dot{\tilde{\mathbf{w}}}$$

$$= -\boldsymbol{\zeta}^T \mathbf{Q}\, \boldsymbol{\zeta} + 2\, \tilde{\mathbf{w}}^T \boldsymbol{\Phi}^T \boldsymbol{\delta}(t) - 2\, \tilde{\mathbf{w}}^T \gamma\, \|\boldsymbol{\delta}\|\, \mathbf{w}$$

$$= -\boldsymbol{\zeta}^{T} \mathbf{Q} \boldsymbol{\zeta} - 2\gamma \|\boldsymbol{\delta}\| \tilde{\mathbf{w}}^{T} \tilde{\mathbf{w}} + 2\gamma \|\boldsymbol{\delta}\| \tilde{\mathbf{w}}^{T} [\frac{\boldsymbol{\Phi}^{T} \boldsymbol{\delta}(t)}{\gamma \|\boldsymbol{\delta}\|} - \mathbf{w}^{*}]$$

and we consider two different conditions **(a)** and **(b)**:

(a) $\|\tilde{\mathbf{w}}\| \geq \|\mathbf{w}^{*}\| + \frac{\Phi_0}{\gamma}$ (34)

Proof of boundedness for condition (a). We have

$$\dot{V} \leq -\boldsymbol{\zeta}^{T} \mathbf{Q} \boldsymbol{\zeta} - 2\gamma \|\boldsymbol{\delta}\| \|\tilde{\mathbf{w}}\|^2 + 2\gamma \|\boldsymbol{\delta}\| \|\tilde{\mathbf{w}}\| [\frac{\|\boldsymbol{\Phi}^{T} \boldsymbol{\delta}(t)\|}{\gamma \|\boldsymbol{\delta}\|} + \|\mathbf{w}^{*}\|]$$

$$\leq -\boldsymbol{\zeta}^{T} \mathbf{Q} \boldsymbol{\zeta} - 2\gamma \|\boldsymbol{\delta}\| \|\tilde{\mathbf{w}}\|^2 + 2\gamma \|\boldsymbol{\delta}\| \|\tilde{\mathbf{w}}\| [\frac{\Phi_0}{\gamma} + \|\mathbf{w}^{*}\|]$$

$$\leq -\boldsymbol{\zeta}^{T} \mathbf{Q} \boldsymbol{\zeta}$$

which implies that $\dot{V} \leq 0$.

(b) $\|\boldsymbol{\zeta}\| \geq \frac{\gamma}{2\lambda_{Q}^{min}} (\|\mathbf{w}^{*}\| + \frac{\Phi_0}{\gamma})^2$ (35)

Proof of boundedness for condition (b). We have

$$\dot{V} = -\lambda_{Q}^{min} \boldsymbol{\zeta}^{T} \boldsymbol{\zeta} - 2\gamma \|\boldsymbol{\delta}\| [\tilde{\mathbf{w}}^{T} \tilde{\mathbf{w}} + (\mathbf{w}^{*} - \frac{\boldsymbol{\Phi}^{T} \boldsymbol{\delta}(t)}{\gamma \|\boldsymbol{\delta}\|})^{T} \tilde{\mathbf{w}}]$$

$$\leq -\lambda_{Q}^{min} \boldsymbol{\zeta}^{T} \boldsymbol{\zeta} + \frac{\gamma \|\boldsymbol{\delta}\|}{2} \|\mathbf{w}^{*} - \frac{\boldsymbol{\Phi}^{T} \boldsymbol{\delta}(t)}{\gamma \|\boldsymbol{\delta}\|}\|^2$$

The last inequality come from the direct application of the preliminary lemma of section 3.2.1**(b)**.

Moreover, we have

$$\dot{V} \leq -\lambda_{Q}^{min} \boldsymbol{\zeta}^{T} \boldsymbol{\zeta} + \frac{\gamma \|\boldsymbol{\delta}\|}{2} (\|\mathbf{w}^{*}\| + \frac{\|\boldsymbol{\Phi}^{T} \boldsymbol{\delta}(t)\|}{\gamma \|\boldsymbol{\delta}\|})^2$$

$$\leq -\lambda_{Q}^{min} \boldsymbol{\zeta}^{T} \boldsymbol{\zeta} + \frac{\gamma \|\boldsymbol{\delta}\|}{2} (\|\mathbf{w}^{*}\| + \frac{\Phi_0}{\gamma})^2$$

$$\leq -\lambda_{Q}^{min} \|\boldsymbol{\zeta}\|^2 + \frac{\gamma \|\boldsymbol{\zeta}\|}{2} (\|\mathbf{w}^{*}\| + \frac{\Phi_0}{\gamma})^2$$

$$\leq 0$$

This implies that the signals in the adaptive loop are bounded.

For the sake of simplicity, we applied a different treatment for the case of disturbances due to noise and neural controller modelling errors (3.2.1) and the case of disturbances due to imperfect modelling of the process. One can easily show that if the two sources of disturbances are present, we come to the same conclusion, that is, the boundedness of the signals.

On the other hand, Narendra & Annaswamy (1987 1989) showed that in the ideal case, that is, in the absence of disturbances and modelling errors, the error equations are exponentially stable when the reference input is persistently exciting with a sufficiently large amplitude.

4. Weights initialization procedure

Now, the result is valid only when the weights are not too far from their optimal values (local stability). We do not expect that this hypothesis could be easily removed because there is always a risk of falling into a local minimum when dealing with nonlinear systems such as, i.e., feedforward neural networks. This suggests a preliminary initialization of the weight values in order to be in the bassin of attraction of the optimal values. One way to initialize the weights is to consider *direct linear connections* from the inputs to the outputs of the net. The weights of these connections can be trained by linear techniques, or initialized to values that solve the linear problem (Kawato 1990; Gomi & Kawato 1993; Scott, Shavlik & Ray 1992). The other weight values should be initialized near zero in order to have negligible effect on the control law. Tuning of these weights can then be started, while maintaining the direct weights from input to output constant. In other words, we have

$$\mathbf{u}(t) = \mathbf{u}_1(t) + \mathbf{u}_2(t) \tag{36}$$

with $\mathbf{u}_1(t)$ being the result of the linear transform from the input to the output, and $\mathbf{u}_2(t)$ being the output of the neural net, without the direct linear connections.

5. Simulation results

In this section, we apply the robust adjustment law to a nonlinear process: a flow mixer with two inputs (the flow of the two fluids at the input), and two outputs (the concentration of one fluid, and the level of the mixture in the tank; see Figure 2).

General description. The process equations are

$$\dot{x}_1 = \frac{2p_1 - x_1}{206x_2} u_1 + \frac{x_1}{206} u_2 \tag{37a}$$

$$\dot{x}_2 = \frac{-2p_2}{206} + \frac{u_1}{206} + \frac{u_2}{206} \tag{37b}$$

$$y_1 = x_1 \; ; \; y_2 = x_2 \tag{37c}$$

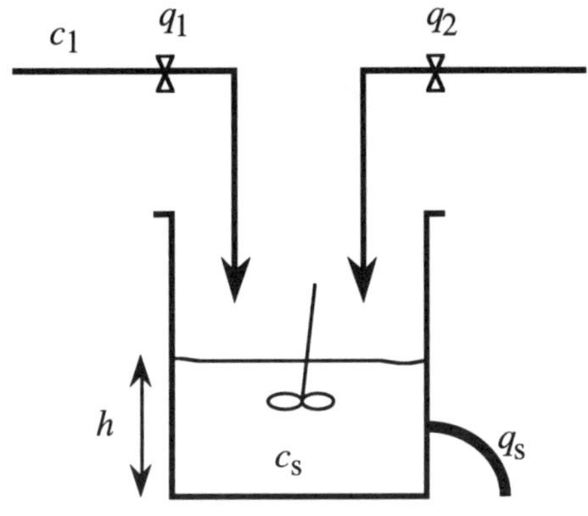

Figure 2. The controlled flow mixer.

with

u_1 being the normalized flow rate of the first fluid at the first input (q_1: see Figure 2);
u_2 being the normalized flow rate of the second fluid (water) at the second input (q_2: see Figure 2);
y_1 being the concentration of the mixture in the tank (c_s: see Figure 2);
y_2 being the normalized level of the fluid mixture in the tank (h: see Figure 2);
p_1 being the concentration of the first fluid at the first input (c_1: see Figure 2), the fluid at the second input beeing water (concentration is zero);
p_2 being the normalized flow rate of the fluid mixture at the output of the tank (q_s: see Figure 2).

All the variables are normalized in order to have $y_1, y_2, p_1, p_2 \in [0, 1]$. The inputs are constrained to stay in the interval $u_1, u_2 \in [0, 1.5]$. The aim is to achieve a given mixture concentration in the tank and a given fluid level in the tank. The parameters of the flow mixer are set to p_1=1, p_2=0.5.

The error model is given in the following form (from equation (17)) :

$$\delta_1 = \varepsilon_1 + \alpha_1 \int \varepsilon_1 \, dt$$

$$\delta_2 = \varepsilon_2 + \alpha_2 \int \varepsilon_2 \, dt$$

with $\alpha_1 = \alpha_2 = 0.04\ sec^{-1}$. This implies that the reference filter is given by (from equations (2) and (16) where the terms k_i are chosen equal to α_i)

$$\dot{\tilde{y}}_1 = \dot{r}_1 + \lambda_1^0 \varepsilon_1 + \lambda_1^I \int \varepsilon_1\, dt$$

$$\dot{\tilde{y}}_2 = \dot{r}_2 + \lambda_2^0 \varepsilon_2 + \lambda_2^I \int \varepsilon_2\, dt$$

with:

$$\lambda_1^0 = 0.08;\ \lambda_1^I = 0.0016$$

$$\lambda_2^0 = 0.08;\ \lambda_2^I = 0.0016$$

This means that the error transient follows:

$$\ddot{\varepsilon}_1 + 0.08\,\dot{\varepsilon}_1 + 0.0016\,\varepsilon_1 = 0$$

$$\ddot{\varepsilon}_2 + 0.08\,\dot{\varepsilon}_2 + 0.0016\,\varepsilon_2 = 0$$

Network architecture. The neural controller consists of a Radial Basis Function Network (RBF), with 15 hidden units (i.e. 15 centers), 6 inputs ($x_1, x_2, \rho_1, \rho_2, u_1, u_2$) and two outputs u_1, u_2. Only the weights from the Gaussian hidden units to the outputs are modified. The learning rate η and the parameter γ are chosen as $\eta = 2000$, and $\eta\gamma = 2$.

The reference signals $r_1(t)$ and $r_2(t)$ (corresponding to the concentration of the mixture in the tank and the heigh of the fluid mixture in the tank) are trapezoidal periodical signals, with $r_1(t) \in [0.4, 0.7]$, $r_2(t) \in [0.3, 0.6]$. The outputs of the network (i.e. the process inputs) are constrained to stay in the interval $[0, 1.5]$; this is realized by putting a hard-limiter at the outputs of the neural network.

For the computation of the partial derivatives $\partial\mathbf{F}[\mathbf{x}(t); \mathbf{u}(t)]/\partial\mathbf{u}(t)$ needed in the adaptation law (27), we used a rough linearized decoupled model of the flow mixer:

$$\dot{x}_1 = \frac{1}{k_1} u_1 \tag{38a}$$

$$\dot{x}_2 = \frac{1}{k_2} u_2 \tag{38b}$$

$$y_1 = x_1\ ;\ y_2 = x_2 \tag{38c}$$

with $k_1 = k_2 = 50$. Alternatively, we could also have used a neural network that identifies the process; this is the approach proposed in Jordan (1989) or Narendra & Parthasarathy (1990). We will now describe the different simulation conditions.

First simulation. In this first simulation, we use an initial linear proportional controller for the linear connections, while the weights from the Gaussian hidden units to the outputs are set to zero. For the initial linear controller, we used the same rough linearized decoupled model of the flow mixer, which leads to the following controller

$$u_1 = k_1\, \rho_1 \tag{39a}$$
$$u_2 = k_2\, \rho_2 \tag{39b}$$

with $k_1 = k_2 = 50$ and $\boldsymbol{\rho} = [\rho_1, \rho_2]^T$ is defined as in equation (5). This corresponds to a PD controller with a feedforward term.

The Figure 3 shows the tracking trajectories by using the linear PD controller alone (equations (39ab)), without adaptation. Thereafter, the robust adaptation law given by equation (27) is applied. Figures 4a and 4b respectively show tracking trajectories for time steps 0 to 2000 *sec*, and for steps 4000 to 6000 *sec*., as well as the control efforts provided by the network (no noise is introduced). Figures 5a and 5b show tracking trajectories in the presence of measurement noise. The measurement noise is uniformly distributed in the interval $y_1(t)\pm 0.01$ and $y_2(t)\pm 0.01$. Notice that after training, we do not observe any parameter adjustment any more.

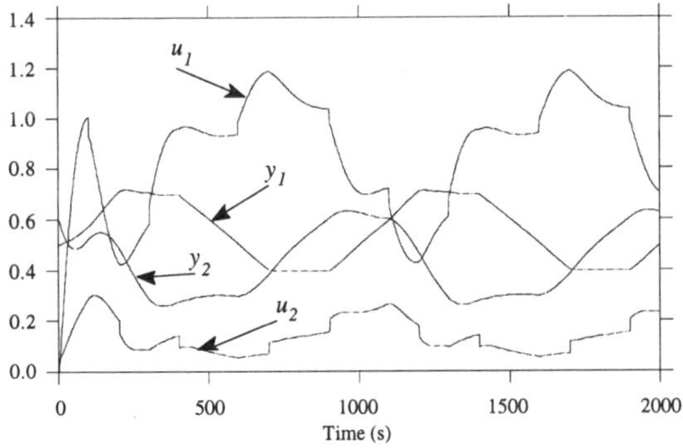

Figure 3. Time evolution of the two process inputs as well as the two outputs during the first 2000 seconds when using the initial linear controller alone (no weights adaptation).

Second simulation. In this case, we do not use any initial controller: the weights from the Gaussian hidden units to the outputs are initialized to small random values. As for the previous simulation, Figures 6a and 6b respectively show tracking trajectories for time step 0 to 2000 *sec*, and for step 4000 to 6000 *sec*., as well as the control efforts provided by the network.

Table 1 shows the mean absolute error, averaged over the last 1000 *sec*, for the different conditions.

Time (sec)	With initial linear controller (without noise)	Without initial linear controller (without noise)	With initial linear controller (with noise)
1000	0.0056	0.0148	0.0130
2000	0.0024	0.0060	0.0113
3000	0.0022	0.0050	0.0113

Table 1. Averaged absolute error for the different simulation conditions.

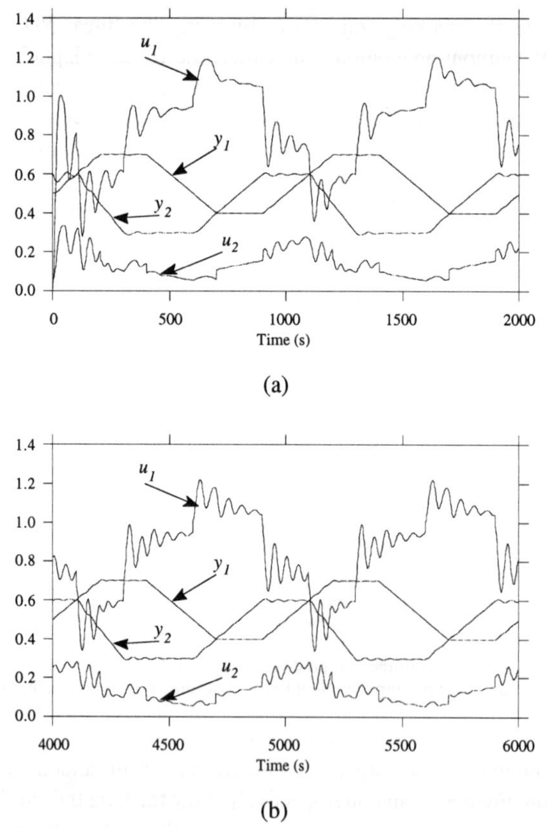

Figure 4. (a) Time evolution of the two process inputs as well as the two outputs during the first 2000 seconds when using the robust adjustment law. (b) The same plot, after 4000 seconds. A linearized proportional controller is used for weights initialization.

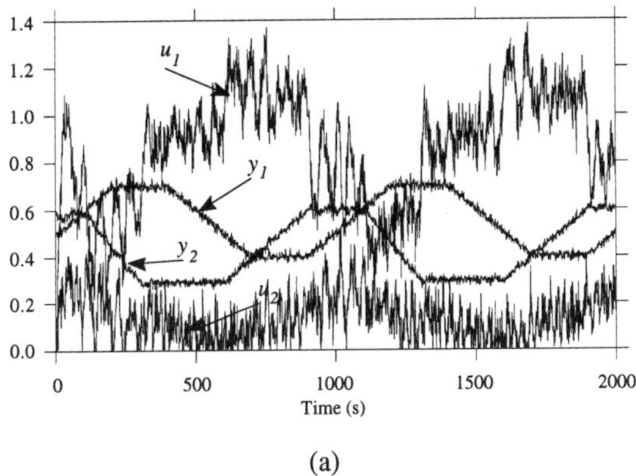

(a)

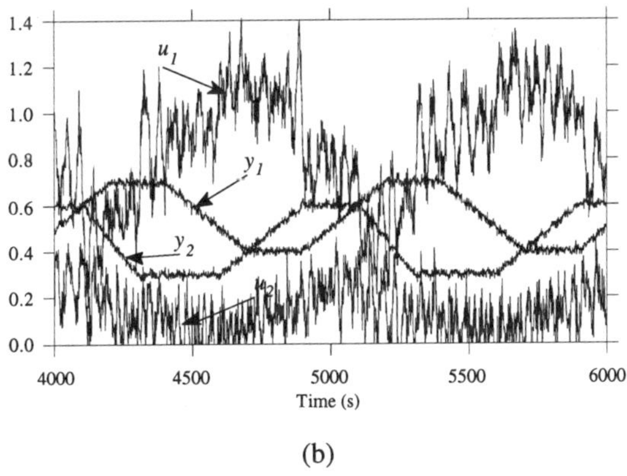

(b)

Figure 5. (a) Time evolution of the two process inputs as well as the two outputs during the first 2000 seconds when using the robust adjustment law. (b) The same plot, after 4000 seconds. A linearized proportional controller is used for weights initialization; observation noise is introduced.

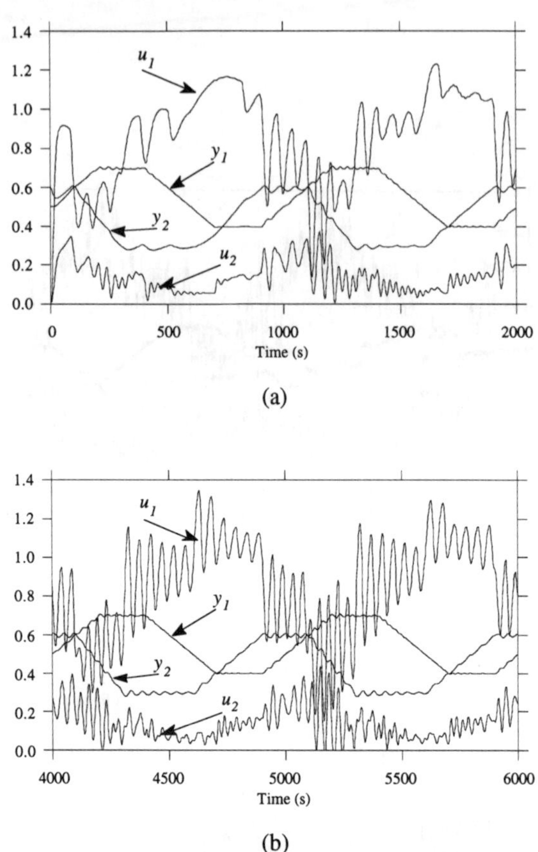

Figure 6. (a) Time evolution of the two process inputs as well as the two outputs during the first 2000 seconds when using the robust adjustment law. (b) The same plot, after 4000 seconds. The weights are initialized with small random values.

Third simulation. In order to show the benefit of using the robust adaptation law (27) versus the simple one (21), we examined the behaviour of the system in non-persistently exciting conditions, and in the presence of measurement noise. After 5000 seconds, the references were maintained at constant values. Measurement noise is uniformly distributed in the interval $y_1(t)\pm 0.05$ and $y_2(t)\pm 0.05$. Unstability appears after about 80000 seconds when using the simple adaptation law, while the system remains stable with the robust one. This is an example of the well-known parameter drift phenomenon.

Discussion. The first and second simulations clearly show the benefit of starting with predefined weight values. When initializing the weights with random values, there is always a risk to fall in a local minimum. This is what happened in the second simulation. The first simulation show the effectiveness of the robust adaptation law in the presence of measurement noise. The third simulation illustrates the problem of parameter drift in the presence of measurement noise and nonpersistent excitation.

6. Conclusions

In the first part of the paper, we proved the stability of a nonlinear MIMO (Multi-Input Multi-Output) system controlled by a neural network with a simple weight adaptation strategy. This extends our previous work (Renders 1993; Saerens, Renders & Bersini 1995), where single-input single-output (SISO) systems were considered. The proof is based on the Lyapunov formalism. In the SISO case, for radial basis function nets, the stability result was strictly valid: the weights do not have to be initialized around the perfectly tuned values. We were not able to prove a similar result in the MIMO case: the stability statement is only valid if the initial weight values are not too far from their optimal values that allow perfect model matching (local stability). We therefore propose to initialize the weights with values that solve the linear problem.

In the second part of the paper, after the treatment of the ideal case, we derived a robust adjustment law that takes account of the disturbances. Indeed, in real practical situations, the perfect model matching condition cannot be verified, and it is not be possible to ensure that the error converges to zero, but only that it remains bounded. Moreover, a perfect model of the process will certainly not be available, so that the sensitivity model needed for the adaptation strategy is only imperfectly known. We therefore use a modified adjustment law known as the "e_1-modification" scheme, originally introduced by Narendra & Annaswamy (1987) for the adaptive control of linear systems. Finally, we show the effectiveness of the robust adaptation law by applying it to the control of a simulated two inputs - two outputs flow-mixer.

Further work will be devoted to the design of algorithms that do not require the a priori knowledge of process delays and orders. In particular, the hypothesis that the process is linearizable and decouplable by static state-feedback was implicitly assumed in this work. The use of recurrent neural networks should allow to overcome this limitation by automatically constructing a general dynamic state-feedback law if necessary.

Appendix

Stability lemma. Let us consider two signals related by the following dynamic equation

$$\mathbf{y}(t) = \mathbf{H}(s)\,\lambda\,\mathbf{v}\boldsymbol{\Psi} = \mathbf{H}(s)\,\boldsymbol{\mu}(t) \qquad \text{(A1)}$$

where $\mathbf{y}(t)$ is an output signal vector, $\boldsymbol{\mu}(t) = (\lambda\ \mathbf{v}\boldsymbol{\psi})$ the input vector, and $\mathbf{H}(s)$ is a Strictly Positive Real transfer function matrix. $\boldsymbol{\mu}(t)$ is computed from λ, an unknown constant with known sign, $\boldsymbol{\psi}$, a vector function of time, and $\mathbf{v}$, a measurable vector.

If the vector $\boldsymbol{\psi}$ varies according to

$$\dot{\boldsymbol{\psi}} = -\eta\ \mathrm{sgn}(\lambda)\ \mathbf{v}^T\ \mathbf{y}(t) \tag{A2}$$

with η being a positive constant, then, if $\mathbf{v}$ is bounded, $\mathbf{y}(t) \rightarrow 0$ asymptotically (see, for instance, Slotine & Li 1991).

Proof of the stability lemma. Let us define a positive definite function V of the form

$$V = \boldsymbol{\zeta}^T \mathbf{P}\, \boldsymbol{\zeta} + \frac{|\lambda|}{\eta}\, \boldsymbol{\psi}^T \boldsymbol{\psi} \tag{A3}$$

with $\mathbf{P}$ positive definite, and $\boldsymbol{\zeta}$ being a state vector of the system described by (A1). Indeed, since $\mathbf{H}(s)$ is SPR, we know that it admits a state-space representation

$$\dot{\boldsymbol{\zeta}} = \mathbf{A}\,\boldsymbol{\zeta} + \mathbf{B}\,\boldsymbol{\mu}(t) \tag{A3a}$$

$$\mathbf{y}(t) = \mathbf{C}^T \boldsymbol{\zeta} \tag{A4b}$$

such that there exists two symmetric positive definite matrices $\mathbf{P}$ and $\mathbf{Q}$ which satisfy

$$\mathbf{A}^T \mathbf{P} + \mathbf{P}\,\mathbf{A} = -\mathbf{Q} \tag{A5a}$$

$$\mathbf{P}\,\mathbf{B} = \mathbf{C} \tag{A5b}$$

Now, let us compute

$$\begin{aligned}
\dot{V} &= \dot{\boldsymbol{\zeta}}^T \mathbf{P}\,\boldsymbol{\zeta} + \boldsymbol{\zeta}^T \mathbf{P}\,\dot{\boldsymbol{\zeta}} + 2\,\frac{|\lambda|}{\eta}\,\boldsymbol{\psi}^T \dot{\boldsymbol{\psi}} \\
&= -\boldsymbol{\zeta}^T \mathbf{Q}\,\boldsymbol{\zeta} + 2\boldsymbol{\zeta}^T \mathbf{P}\,\mathbf{B}\,\boldsymbol{\mu}(t) - 2\,\lambda\,\boldsymbol{\psi}^T \mathbf{v}^T \mathbf{y}(t) \\
&= -\boldsymbol{\zeta}^T \mathbf{Q}\,\boldsymbol{\zeta} + 2\,\boldsymbol{\mu}(t)^T \mathbf{y}(t) - 2\,\boldsymbol{\mu}(t)^T \mathbf{y}(t) \\
&= -\boldsymbol{\zeta}^T \mathbf{Q}\,\boldsymbol{\zeta} \\
&\leq 0
\end{aligned}$$

If the signal $\mathbf{y}$ is bounded; \O(ζ,\S\UP6(•)) is also bounded (equation (A1) and (A3a)). This implies that $\ddot{V}$ is also bounded, and, from Barbalat's lemma, $\dot{V} \to 0$, so that $\mathbf{y}(t) \to 0$ (see, for instance, Slotine & Li 1991).

Acknowledgements

This work was partially supported by the ARC 92/97-160 (BELON) project from the "Communauté Française de Belgique", and the FALCON (6017) Basic Research ESPRIT project from the European Communities.

References

ÅSTRÖM K.J. & WITTENMARK B., 1989, "Adaptive control". Addison-Wesley Publishing Company.

BARTO A., 1990, "Connectionist learning for control: an overview". In Neural networks for control, W Thomas Miller, R. Sutton & P. Werbos (editors), The MIT Press.

CARDALIAGUET P. & EUVRARD G., 1992, "Approximation of a function and its derivative with a neural network". Neural Networks, 5, pp. 207-220.

CHEN F. & KHALIL H., 1991, "Adaptive control of nonlinear systems using neural networks – A dead-zone approach". Proceedings of the American Control Conference, pp. 667-672.

CHEN F. & LIU C., 1992, "Adaptively controlling nonlinear continuous-time systems using neural networks". Proceedings of the American Control Conference, pp. 46-50.

GALLANT A. & WHITE H., 1992, "On learning the derivatives of an unknown mapping with a multilayer feedforward network". Neural Networks, 5, pp. 129-138.

GOMI H. & KAWATO M, 1993, "Neural network control for a closed-loop system using feedback-error-learning". Neural Networks, 6, pp. 933-946.

GOODWIN G. & SIN K.S., 1984, "Adaptive filtering, prediction and control". Prentice-Hall, Englewood Cliffs.

GUPTA M. & SINHA N. (EDITORS), 1994, "Intelligent Control Systems: Theory and applications". IEEE Press, under press.

HORNIK K., STINCHOMBE M. & WHITE H., 1989, "Multilayer feedforward networks are universal approximators". Neural Networks, 2, pp. 359-366. Reprinted in "Artificial Neural Networks: Concepts and Control Applications" by Rao Vemuri (editor), IEEE Computer Society Press, 1992.

HORNIK K., STINCHOMBE M. & WHITE H., 1990, "Universal approximation of an unknown mapping and its derivatives using multilayer feedforward networks". Neural Networks, 3, pp. 551-560.

HUNT K. & SBARBARO D., 1991, "Neural networks for nonlinear internal control". IEE Proceedings-D, 138 (5), pp. 431-438.

HUNT K., SBARBARO D., ZBIKOWSKI R. & GAWTHROP P., 1992, "Neural networks for control systems – A survey". Automatica, 28 (6), pp. 1083-1112.

IOANNOU & DATTA, 1991, "Robust adaptive control: A unified approach". Proceedings of the IEEE, 79, pp. 1736-1768.

JIN L., NIKIFORUK P. & GUPTA M., 1993, "Direct adaptive output tracking control using multilayer neural networks". IEE Proceedings-D, 140 (6), pp. 393-398.

JIN Y., PIPE T. & WINFIELD A., 1993, "Stable neural adaptive control for discrete systems". Proceedings of the World Congres on Neural Networks, Portlend, Vol III, pp. 277-280.

JOHANSEN T.A., 1994, "Robust adaptive control of slowly-varying discrete-time non-linear systems". Technical report of the Norwegian Institute of Technology, Department of Engineering Cybernetics.

JORDAN M.I. , 1989, "Generic constraints on underspecified target trajectories". Proceedings of International Joint Conference on Neural Networks, Washington, Vol I, pp. 217-225.

JORDAN M.I. & RUMELHART D., 1991, "Internal world models and supervised learning". Proceedings of the eighth International Workshop on Machine Learning, Illinois, pp. 70-74.

KAWATO M., 1990, "Feedback-error-learning neural network for supervised learning". In "Advanced neural computers", Eckmiller R. (editor), Elsevier Publishers, pp. 365-372.

LANDAU Y., 1979, "Adaptive control. The model reference approach". Marcel Dekker.

LIU C. & CHEN F., 1993, "Adaptive control of nonlinear continuous-time systems using neural networks - General relative degree and MIMO cases". International Journal of Control, 58 (2), pp. 317-335.

MILLER W., SUTTON R. & WERBOS P., 1990, "Neural networks for control". The MIT Press.

NARENDRA K.S., 1992, "Adaptive control of dynamical systems using neural networks". In "Handbook of intelligent control", White D. & Sofge D. (editors), pp. 141-184. Van Nostand Reinhold.

NARENDRA K.S. & ANNASWAMY A., 1987, "A new adaptive law for robust adaptation without persistent excitation". IEEE Transactions on Automatic Control, 32 (2), pp. 134-145.

NARENDRA K.S. & ANNASWAMY A., 1989, "Stable adaptive systems". Prentice-Hall.

NARENDRA K.S. & PARTHASARATHY K., 1990, "Identification and control of dynamic systems using neural networks". IEEE Transactions on Neural Networks, 1 (1), pp. 4-27. Reprinted in "Artificial Neural Networks: Concepts and Control Applications" by Rao Vemuri (editor), IEEE Computer Society Press, 1992.

NARENDRA K.S. & PARTHASARATHY K., 1991, "Gradient methods for the optimization of dynamical systems containing neural networks". IEEE Transactions on Neural Networks, 2 (2), pp. 252-262. Reprinted in "Artificial Neural Networks: Concepts and Control Applications" by Rao Vemuri (editor), IEEE Computer Society Press, 1992.

NARENDRA K.S. & PARTHASARATHY K., 1992, "Neural networks and dynamical systems". International Journal of Approximate Reasoning, 6, pp. 109-131.

NGUYEN D. & WIDROW B., 1989, "The truck backer-upper: an example of self-learning in neural networks". Proceedings of International Joint Conference on Neural Networks, Washington, Vol II, pp. 357-353.

NGUYEN D. & WIDROW B., 1990, "Neural networks for self-learning control systems". IEEE Control Systems Magazine, 10 (3), pp. 18-23.

ORTEGA R. & TANG Y., 1989, "Robustness of adaptive controllers – A survey". Automatica, 25 (5), pp. 651-677.

PARTHASARATHY K. & NARENDRA K., 1991, "Stable adaptive control of a class of discrete-time nonlinear systems using radial basis networks". Report No. 9103, Yale University, Center for Systems Science, New Haven, Connecticut.

POGGIO T. & GIROSI F., 1990, "Networks for approximation and learning". Proceedings of the IEEE, 78 (9), pp. 1481-1497. Reprinted in "Artificial Neural Networks: Concepts and Control Applications" by Rao Vemuri (editor), IEEE Computer Society Press, 1992.

PSALTIS D., SIDERIS A. & YAMAMURA A., 1987, "Neural controllers". Proceedings of IEEE first International Conference on Neural Networks, San Diego, Vol 4, pp. 551-558.

PSALTIS D., SIDERIS A. & YAMAMURA A., 1988, "A multilayered neural network controller". IEEE Control Systems Magazine, 8 (2), pp. 17-21. Reprinted in "Artificial Neural Networks: Concepts and Control Applications" by Rao Vemuri (editor), IEEE Computer Society Press, 1992.

RAO VEMURI V. (EDITOR), 1992, "Artificial neural networks. Concepts and control applications". IEEE Computer Society Press.

RENDERS J.-M., 1991, "A new approach of adaptive neural controller design with application to robotics control". In IMACS Annals on Computing and Applied Mathematics, 10: Mathematical and Intelligent Models in System Simulation, J.C. Baltzer AG, Scientific Publishing Co., pp. 361-366.

RENDERS J.-M., 1993, "Biological metaphors for process control (in french)". Ph.D Thesis, Université Libre de Bruxelles, Faculté Polytechnique, Belgium.

RENDERS J.-M., BERSINI H. & SAERENS M., 1993, "Adaptive neurocontrol: How black-box and simple can it be ?". Proceedings of the tenth International Workshop on Machine Learning, Amherst, pp. 260-267.

RENDERS J.-M., SAERENS M. & BERSINI H. (1995), "Adaptive neurocontrol of a certain class of discrete-time MIMO processes based on stability theory". Chapter 3 of "*Neural Network Engineering in Dynamic Control Systems*", K. Hunt, G. Irwin & K. Warwick (editors), pp. 43-60. Springer-Verlag.

RICHARD M. & LIPPMANN R., 1991, "Neural network classifiers estimate Bayesian a posteriori probabilities". Neural Computation, 3, pp. 461-483.

RIMON E. & NARENDRA K., 1992, "A new adaptive estimator for linear systems". IEEE Transactions on Automatic Control, 37 (3), pp. 410-412.

ROHRS C., VALAVANI L., ATHANS M. & STEIN G., 1985, "Robustness of continuous-time adaptive control algorithms in the presence of unmodelled dynamics". IEEE Transactions on Automatic Control, 30 (9), pp. 881-889.

ROVITHAKIS G. & CHRISTODOULOU M., 1994, "Adaptive control of unknown plants using dynamical neural networks". IEEE Transactions on Systems, Man, and Cybernetics, 24 (3), pp. 400-412.

RUCK D., ROGERS S., KABRISKY M., OXLEY M. & SUTER B., 1990, "The multilayer perceptron as an approximation to a Bayes optimal discriminant function". IEEE Transactions on Neural Networks, 4 (1), pp. 296-298.

RUMELHART D.E., HINTON G.E. & WILLIAMS R.J., 1986, "Learning internal representation by error propagation". In Distributed Parallel Processing, Vol. 1, MIT Press, Cambridge.

SAERENS M., RENDERS J.-M. and BERSINI H., 1995, "Neural controllers based on backpropagation algorithm". Chapter 12 in the "*IEEE-Press Book on Intelligent Control Systems: Theory and applications*", Gupta M. & Sinha N. (editors), pp. 292-327. IEEE Press.

SAERENS M. & SOQUET A. (1991), "Neural Controller Based on Back-Propagation Algorithm". IEE Proceedings–F, 138 (1), pp. 55-62. Reprinted in "Artificial Neural Networks: Concepts and Control Applications" by Rao Vemuri (editor), IEEE Computer Society Press, 1992.

SAERENS M., SOQUET A., RENDERS J-M. & BERSINI H., 1992, "Some preliminary comparisons between a neural adaptive controller and a model reference adaptive controller". In "Neural Networks in Robotics" by Bekey G. & Goldberg K. (editors), pp. 131-146. Kluwer Academic Press.

SANNER R. & SLOTINE J.-J., 1992, "Gaussian networks for direct adaptive control". IEEE Transactions on Neural Networks, 3 (6), pp. 837-863.

SCOTT G., SHAVLIK J. & HARMON RAY W., 1992, "Refining PID controllers using neural networks". Neural Computation, 4, pp. 746-757.

SEUBE N., 1990, "Construction of learning rules in neural networks that can find viable regulation laws to control problems by self-organization". Proceedings of the International Neural Networks Conference, Paris, pp. 209-212.

SEUBE N. & MACIAS J.-C., 1991, "Design of neural network learning rules for viable feedback laws". Proceedings of the European Control Conference, Grenoble, pp. 1241-1246.

SLOTINE J.-J. & LI W., 1991, "Applied nonlinear control". Prentice-Hall.

SONTAG E., 1992, "Feedback stabilization using two-hidden-layer nets". IEEE Transactions on Neural Networks, 3 (6), pp. 981-990.

STINCHCOMBE M. & WHITE H., 1989, "Universal approximation using feedforward networks with non-sigmoid hidden layer activation functions". Proceedings of the International Joint Conference on Neural Networks, Washington D.C., pp. 613-617.

TZIRKEL-HANCOCK E. & FALLSIDE F., 1992, "Stable control of nonlinear systems using neural networks". International Journal of Robust and Nonlinear Control, 2, pp. 63-86.

WHITE D. & SOFGE D. (EDITORS), 1992, "Handbook of intelligent control". Van Nostand Reinhold.

WHITE H., 1989, "Learning in artificial neural networks: A statistical perspective". Neural Computation, 1, pp. 425-464.

WHITE H., 1992, "Artificial neural networks: Approximation and learning theory". Blackwell.

WILLIAMS R.J. & ZIPSER D., 1989, "A learning algorithm for continually running fully recurrent neural networks". Neural Computation, 1 (2), pp. 270-280.

BLACK-BOX MODELING WITH STATE-SPACE NEURAL NETWORKS

ISABELLE RIVALS and LÉON PERSONNAZ

ESPCI, Laboratoire d'Électronique, 10 rue Vauquelin
75231 Paris Cedex 05, France.
E-mail: rivals@neurones.espci.fr, personna@neurones.espci.fr

ABSTRACT

Neural network black-box modeling is usually performed using nonlinear input-output models. The goal of this paper is to show that there are advantages in using nonlinear state-space models, which constitute a larger class of nonlinear dynamical models, and their corresponding state-space neural predictors. We recall the fundamentals of both input-output and state-space black-box modeling, and show the state-space neural networks to be potentially more efficient and more parsimonious than their conventional input-output counterparts. This is examplified on simulated processes as well as on a real one, the hydraulic actuator of a robot arm.

1. Introduction

During the past few years, several authors [Narendra and Parthasarathy 1990, Nerrand et al. 1994] have suggested neural networks for nonlinear dynamical black-box modeling. The problem of designing a mathematical model of a process using only observed data has attracted much attention, both from an academic and an industrial point of view. Neural models can be used either as simulators (for fault detection, personnel training, etc., see for example [Ploix et al. 1994]), as models for the training of a neural controller, and/or as models to be used within control systems [see for instance the robust "neural" internal model control of a 4WD Mercedes vehicle in Rivals et al. 1994].

To date, most of the work in neural black-box modeling has been performed making the assumption that the process to be modeled can be described accurately by input-output models, and using the corresponding input-output neural predictors. We show in the present paper that, since state-space models constitute a larger class of nonlinear dynamical models, there is an advantage in making the assumption of a state-space description, and in using the corresponding state-space neural predictors. In section 2, we give some useful definitions, and show state-space neural networks to be potentially more efficient than their conventional input-output counterparts. In section 3, in order to enlighten the links and the differences between state-space and input-output black-box modeling, we recall some results in linear modeling. In section 4.1, we present a family of nonlinear input-output models and their associated predictors, and in section 4.2, a family of nonlinear state-space models is presented in the same manner. In section 5, we describe the training of the input-output and state-space neural predictors in a unified fashion. State-space modeling is then illustrated and compared to input-output modeling on simulated processes in section 6, and on the hydraulic actuator of a robot arm in section 7. In the latter case, our results are compared to those obtained by other groups on the same data.

2. Some Definitions

The problem of modeling a process is to find a mathematical model which is able to describe its dynamic behavior, given some prior knowledge about the process and input-output measurements. Two types of models can be chosen according to the prior knowledge one has about the internal behavior of the process: knowledge-based models and black-box models. We are interested in the latter type.

2.1. Knowledge-Based versus Black-Box Modeling

A model is termed knowledge-based if it has been possible to construct it entirely from prior knowledge and physical insight, or if the physical insight suggests a model structure where the only unknowns are parameters which will be estimated from measurements. Knowledge-based models are usually in the state-space form, their state variables and the relationships between them having a physical meaning.

If no or unsufficient physical insight is available, but only observed data, no model structure can be defined *a priori*. In this case, one must resort to black-box models. A discrete-time black-box model is a recursive filter whose outputs are functions of its past external inputs and state variables; these functions are defined by a set of parameters (polynomial functions, or neural networks, radial basis networks, wavelet networks...). For nonlinear modeling, the family of parameterized functions should be as flexible as possible, in order to be able to describe any complex dynamical process. Black-box models are usually chosen to be input-output models; our aim is precisely to show that it may be advantageous to use state-space models as black-box models.

2.2. Input-Output versus State-Space Black-Box Models

We are interested in black-box state-space models, for the following reasons:

- First, state-space models can describe a broader class of dynamical systems than input-output models: whereas it is always possible to rewrite a nonlinear input-output model in a state-space representation, conversely, an input-output model globally equivalent to a given state-space model might not exist [Leontaritis and Billings 1985];

- Second, even when an input-output representation does exist, it may have a higher order. Let us consider the following deterministic SISO state-space model:

$$\begin{cases} x(k+1) = f\left(x(k),\, u(k)\right) & \textit{(state equation)} \\ y(k) = g\left(x(k)\right) & \textit{(output equation)} \end{cases} \qquad (2.2.1)$$

where $u(k)$ is the scalar external input, $y(k)$ is the scalar output, and $x(k)$ is the n-state vector of the model at time k; f and g are nonlinear functions. It has been shown that, under fairly general conditions on the observability of (2.2.1), an equivalent input-output model does exist, and that it is given by:

$$y(k) = h\left(y(k\text{-}1),\, \ldots,\, y(k\text{-}r),\, u(k\text{-}1),\, \ldots,\, u(k\text{-}r)\right) \qquad (2.2.2)$$

with $n \leq r \leq 2\,n + 1$ [Levin 1992, Levin and Narendra 1995].

Therefore, state-space models are likely to have a lower order and to require a smaller number of past inputs (i.e. a smaller number of regressors) than input-output models, and hopefully a smaller number of parameters. This is of importance especially when a limited amount of data is available.

2.3. Black-Box Modeling

The black-box modeling of a process is achieved in three steps: 1) choice of a set of candidate models; 2) derivation of the associated predictors; 3) selection of the best model.

Step 1: choice of a set of candidate models.

The analysis of the process behavior leads to a set of control inputs and measured disturbances - the external inputs - acting on the observed outputs. In this paper, all unmeasured disturbances are supposed to be stochastic processes. This analysis also gives possible values for the order of a black-box model, either input-output or state-space. Candidate models are thus characterized by their input-output or state-space structure, their order and number of external inputs, and by the functional relationships between their inputs and their output.

In this paper, we will consider time-invariant stochastic SISO models (the extension to MISO models is straightforward):

- Input-output candidate models:

$$y_p(k) = h\left(y_p(k\text{-}1),\, \ldots,\, y_p(k\text{-}n),\, u(k\text{-}1),\, \ldots,\, u(k\text{-}m),\, w(k),\, \ldots,\, w(k\text{-}p)\right) \qquad (2.3.1)$$

where $y_p(k)$ denotes the output of the process and $u(k)$ the known external input at time k; $\{w(k)\}$ is a sequence of zero mean, independent and identically distributed (i.i.d.) random variables; h is a nonlinear function.

- State-space candidate models:

$$\begin{cases} x_p(k+1) = f\left(x_p(k),\ u(k),\ v_1(k)\right) \\ y_p(k) = g\left(x_p(k),\ v_2(k)\right) \end{cases} \tag{2.3.2}$$

where $y_p(k)$ denotes the output of the process, $u(k)$ the known external input, and $x_p(k)$ the n-state vector at time k; $\{v_1(k)\}$ is a sequence of zero mean, i.i.d. random vectors (state noise), and $\{v_2(k)\}$ is a sequence of zero mean, i.i.d. random variables (output noise); f and g are nonlinear functions. Since we deal with black-box modeling, the state x_p is not measured.

A candidate model, built upon a set of assumptions about the process, is called an *assumed model.*

Step 2: derivation of the associated predictors.

We define the *theoretical predictor associated to a given assumed model* as the recursion giving the conditional expectation $E\left(y_p(k+1) \mid k\right)$ of the output $y_p(k+1)$ given the past observations $\{y_p(k),\ y_p(k\text{-}1),\ \dots,\ y_p(0)\}$ if the assumed model is true.

Theoretical predictor associated to an input-output assumed model:

$$y(k+1) = h_{pred}\left(y_p^k,\ u^k,\ y^k,\ k\right) \tag{2.3.3}$$

where y is the predictor output, and y_p^k denotes a finite set of past outputs of the process (the same notation is used for u, y, x). h_{pred} is a nonlinear function, generally time-varying, depending on h in (2.3.1), and whose expression will be given in section 4 for some particular input-output assumed models.

Theoretical predictor associated to a state-space assumed model:

$$\begin{cases} x(k+1) = f_{pred}\left(x^k,\ u^k,\ y_p^k,\ k\right) \\ y(k+1) = g_{pred}\left(x^k,\ u^k,\ y_p^k,\ k\right) \end{cases} \tag{2.3.4}$$

where y is the predictor output, and x its state vector. f_{pred} and g_{pred} are nonlinear functions depending on f and g in (2.3.2), whose expressions will be given in section 5 for some particular models.

The theoretical predictor associated to the true model is thus the *minimal variance* - or *optimal - predictor*. At this step of the modeling procedure, the goal is to find an *empirical predictor* as close as possible to the optimal predictor.

Once candidate assumed models have been chosen, empirical predictors giving an estimate of $E\left(y_p(k+1) \mid k\right)$ are then designed, using the theoretical predictors associated to the candidate models: the functions h_{pred} , or f_{pred} and g_{pred}, of each theoretical predictor are replaced by *functions parameterized by a set of parameters* θ. Any universal function approximator can be used. In the modeling examples of sections 6 and 7, we use neural networks with sigmoidal hidden neurons: these networks are known to be universal [Hornik et al. 89] and parsimonious [Hornik et al. 94] approximators. For a review of other possible approximators, see [Sontag 93, Sjöberg et al. 1995]. The training of a candidate neural predictor is achieved using input-output sequences. Its weights can be estimated with a *prediction error* (P.E.) method, or possibly an *extended Kalman filter* (E.K.F.) method.

Step 3: selection of the best model.

Eventually, the performances of the candidate predictors are estimated with the system the model is being designed for (predictor, simulator, control system...). According to these performances, the "best" predictor is chosen, and hence the best model.

Parts 3 and 4 are devoted to the theoretical predictors associated to various assumed models, which will be needed for the design of neural empirical predictors.

3. Linear Modeling

In order to show the links and the differences between state-space and input-output modeling, some results in linear modeling are recalled. These results are mostly of tutorial nature, and we will generally refer to [Goodwin and Sin 1984].

3.1. The Linear State-Space Assumed Model

The general SISO linear time-invariant stochastic model is:

$$\begin{cases} x_p(k+1) = F\, x_p(k) + G\, u(k) + v_1(k) \\ y_p(k) = H\, x_p(k) + v_2(k) \end{cases} \tag{3.1.1}$$

with dim $(F) = n \times n$, dim $(G) = n \times 1$, and dim $(H) = 1 \times n$.

Let us assume that the initial state $x_p(0)$ is gaussian with mean μ_0, and that the noises are also gaussian. The conditional expectation of the state $x_p(k+1)$ and of the output $y_p(k+1)$ of model (3.1.1) given observations $y_p(0)$ up to $y_p(k)$ can be estimated by the *Kalman predictor*. Let $x(k+1)$ and $y(k+1)$ denote these predictions. They satisfy the following recursion:

$$\begin{cases} x(0) = \mu_0 \\ x(k+1) = F\, x(k) + G\, u(k) + K(k)\left(y_p(k) - H\, x(k)\right) \\ y(k+1) = H\, x(k+1) \end{cases} \tag{3.1.2}$$

where $K(k)$ is the time-varying Kalman gain. Under certain stabilizability and observability conditions, the error covariance and the Kalman gain $K(k)$ converge to *steady-state* values as $k \to \infty$.

If the gaussian assumption is removed, (3.1.2) is the minimum variance *linear* predictor.

In the modeling problem, the matrices F, G, H, K are parameterized by an unknown vector θ, which can be estimated using a P.E. or an E.K.F. method:

- Prediction Error method: the steady-state form of predictor (3.1.2) ($K(k) = K \ \forall \ k$) is used. The estimation of θ is obtained by minimizing the mean square prediction error (MSPE):

$$J(\theta) = \frac{1}{N}\sum_{k=0}^{N-1} e(k+1)^2 = \frac{1}{N}\sum_{k=0}^{N-1} (y_p(k+1) - y(k+1))^2 \qquad (3.1.3)$$

- Extended Kalman Filter method: the unknown parameter vector θ is included in an augmented state vector. The E.K.F. gives a suboptimal estimation of the state vector (and therefore of θ) using a linearization of the model around the current state estimate.

3.2. *The Innovations Model*

Using the *innovations* $\varepsilon(k)$, defined as $\varepsilon(k) = y_p(k) - H\,x(k) = y_p(k) - y(k)$, the Kalman predictor can be rewritten as what is called the innovations model, disturbed by the random innovations sequence only:

$$\begin{cases} x(k+1) = F\,x(k) + G\,u(k) + K(k)\,\varepsilon(k) \\ y_p(k) = H\,x(k) + \varepsilon(k) \end{cases} \qquad (3.2.1)$$

The latter is equivalent to the following *time-varying input-output model*:

$$A(q^{-1})\,y_p(k) = B(q^{-1})\,u(k) + C(k, q^{-1})\,\varepsilon(k) \qquad (3.2.2)$$

where q^{-1} is the backward shift operator, A, B and C are polynomials of degree n, A and C being monic, and with $B(q^{-1}) = b_1\,q^{-1} + \dots + b_n\,q^{-n}$. The time-varying nature of C in (3.2.2) arises from the fact that the Kalman gain is time-varying. However, if the Kalman gain is asymptotically time-invariant, C is also asymptotically time-invariant. We can thus replace the time-varying model (3.2.2) by its steady-state form:

$$A(q^{-1})\,y_p(k) = B(q^{-1})\,u(k) + C(q^{-1})\,\varepsilon(k) \qquad (3.2.3)$$

3.3. *The ARMAX Assumed Model*

The equivalence between the state-space model (3.1.1) and the input-output model (3.2.3) in the steady state motivates the choice of time-invariant input-output assumed models:

$$A(q^{-1})\,y_p(k) = B(q^{-1})\,u(k) + C(q^{-1})\,w(k) \qquad (3.3.1)$$

where A, B and C are polynomials of degree n, m and p, A and C being monic, and where $B(q^{-1}) = b_1\,q^{-1} + \dots + b_m\,q^{-m}$; the noise $\{w(k)\}$ is a sequence of zero mean, i.i.d. random variables. This model is called the ARMAX (AutoRegressive Moving Average with eXogenous input) model.

The designer thus has the choice between two assumed models which are equivalent in the steady state, the model in the state-space form (3.1.1) and the ARMAX model (3.3.1). The latter is often chosen for the simplicity of the modeling procedure. Consider the following one-step-ahead predictor:

$$C(q^{-1})\, y(k+1) = \left(C(q^{-1}) - A(q^{-1})\right) y_p(k+1) + B(q^{-1})\, u(k+1) \qquad (3.3.2)$$

or equivalently:

$$y(k+1) = \left(1 - A(q^{-1})\right) y_p(k+1) + B(q^{-1})\, u(k+1) + \left(C(q^{-1}) - 1\right) e(k+1) \quad (3.3.3)$$

where $e(k+1) = y_p(k+1) - y(k+1)$ is the prediction error*. Provided that the assumed model is true, and that the polynomial C has no zeros on or outside the unit circle, the effect of arbitrary initial conditions will diminish exponentially, and the prediction error e becomes equal to the noise w: predictor (3.3.3) is *asymptotically* optimal. Therefore, we will consider (3.3.3) as the theoretical predictor associated to the assumed model (3.3.1).

The process identification can thus be performed by minimizing the MSPE of a predictor parameterized by θ:

$$\begin{aligned} y(k+1) = &\left(1 - A(\theta, q^{-1})\right) y_p(k+1) + B(\theta, q^{-1})\, u(k+1) \\ &+ \left(C(\theta, q^{-1}) - 1\right) e(k+1) \end{aligned} \qquad (3.3.4)$$

Several particular cases of the ARMAX model are well known:

The ARX (AutoRegressive with eXogenous input) model.

It is also called the "*Equation-Error*" model. It is obtained by taking $C(q^{-1}) = 1$:

$$A(q^{-1})\, y_p(k) = B(q^{-1})\, u(k) + w(k) \qquad (3.3.5)$$

The theoretical predictor associated to this ARX model is:

$$y(k+1) = \left(1 - A(q^{-1})\right) y_p(k+1) + B(q^{-1})\, u(k+1) \qquad (3.3.6)$$

The "Output-Error" model.

It corresponds to the assumption of *additive output noise* (i.e. $C(q^{-1}) = A(q^{-1})$):

$$A(q^{-1})\, y_p(k) = B(q^{-1})\, u(k) + A(q^{-1})\, w(k) \qquad (3.3.7)$$

If the assumption is true, and if A has no zeros on or outside the unit circle, the following theoretical predictor is asymptotically optimal:

$$y(k+1) = \left(1 - A(q^{-1})\right) y(k+1) + B(q^{-1})\, u(k+1) \qquad (3.3.8)$$

* It is important to distinguish between the noise sequence occuring in a given assumed model ($\{v_1(k)\}$, $\{v_2(k)\}$, or $\{w(k)\}$), the prediction error sequence $\{e(k)\}$ of some predictor, and the innovations sequence $\{\varepsilon(k)\}$, which is the prediction error sequence of the optimal predictor.

4. Nonlinear Modeling

As recalled in section 2 for deterministic models, in contradistinction to the linear case, there is no simple equivalence between nonlinear input-output and state-space models. Since the latter build a larger class of nonlinear models, it would be too restrictive to make only input-output assumptions. Furthermore, even when there exists an input-output model equivalent to a given state-space assumed model, its function h_{pred} in (2.3.3) might require many more arguments than the functions f_{pred} and g_{pred} of the equivalent state-space model (2.3.4). Therefore, if input-output modeling is unsatifactory (typically if the training becomes uneasy because too many inputs and/or neurons are needed), one should resort to neural state-space models. Their state variables not being imposed to be delayed values of their output, these are more flexible.

As in the linear case, a possible presentation would start with state-space modeling, and deal with input-output modeling as a particular case. We choose first to recall the input-output approach, which should be tried in the first place, and then to emphasize the new aspects of neural state-space modeling.

4.1. Input-Output Modeling

The input-output modeling approach for neural networks has been partly introduced and illustrated in [Narendra and Partasarathy 1990, Narendra and Partasarathy 1991, Nerrand et al. 1994]. In this subsection, we summarize its basic principles.

The NARMAX assumed model.

The most general input-output model whose associated predictor can be easily expressed is the NARMAX model, which is an extension of the linear ARMAX model to the nonlinear case [Leontaritis and Billings 1985, Chen and Billings 1989]. It is given by:

$$y_p(k) = h\left(y_p(k\text{-}1), \ldots, y_p(k\text{-}n), u(k\text{-}1), \ldots, u(k\text{-}m), w(k\text{-}1), \ldots, w(k\text{-}p)\right) + w(k) \tag{4.1.1}$$

Let us consider the following theoretical feedback predictor of order p (its state variables are the p past prediction errors $e = y_p - y$):

$$y(k+1) = h\left(y_p(k), \ldots, y_p(k\text{-}n+1), u(k), \ldots, u(k\text{-}m+1), e(k), \ldots, e(k\text{-}p+1)\right) \tag{4.1.2}$$

If the assumed model is true, and if the p first prediction errors are equal to the noise, $e(k+1) = w(k+1)\ \forall\ k$; therefore, the variance of the prediction error e is minimal. The effect of arbitrary initial conditions will vanish depending on the unknown function h. We will consider (4.1.2) as the theoretical predictor associated to the assumed model (4.1.1). Note that the function h_{pred} defining the associated predictor is the function h of the assumed model.

In order to implement the associated predictor, one must therefore train a feedback neural network of the form:

$$y(k+1) = \varphi\left(y_p(k), ..., y_p(k\text{-}n+1), u(k), \ldots, u(k\text{-}m+1), e(k), \ldots, e(k\text{-}p+1); \theta\right) \tag{4.1.3}$$

where φ is the nonlinear function implemented by the feedforward part of the network with weights θ. If the assumed model is true, and if φ approximates h with an arbitrary precision, then predictor (4.1.3) is arbitrarily close to the asymptotically optimal one.

As in the linear case, it is usual in nonlinear modeling to first consider particular cases of the NARMAX model leading to more simple predictors.

The NARMAX assumed model with additive correlated (ARMA) noise.

This model of intermediate complexity is discussed in [Goodwin and Sin 1984]:

$$A(q^{-1})\, y_p(k) = h\left(y_p(k\text{-}1), \ldots, y_p(k\text{-}n), u(k\text{-}1), \ldots, u(k\text{-}m)\right) + C(q^{-1})\, w(k) \tag{4.1.4}$$

where A and C are monic polynomials with degree n and p respectively, and h is a nonlinear function. The theoretical predictor associated to (4.1.4) is given by:

$$\begin{aligned} y(k+1) = h\left(y_p(k), \ldots, y_p(k\text{-}n+1), u(k), \ldots, u(k\text{-}m+1)\right) \\ + \left(1 - A(q^{-1})\right) y_p(k+1) + \left(C(q^{-1}) - 1\right) e(k+1) \end{aligned} \tag{4.1.5}$$

If C has no zeros outside or on the unit circle, the effect of arbitrary initial conditions will die away.

The neural network used for training thus imposes a linear dependency between its output and the past prediction errors:

$$y(k+1) = \varphi\left(y_p(k), \ldots, y_p(k\text{-}n+1), u(k), \ldots, u(k\text{-}m+1); \theta\right) + \Pi\,(q^{-1})\, e(k+1) \tag{4.1.6}$$

where Π is a polynomial with adjustable coefficients.

The NARX assumed model.

The NARX model assumes *pseudo-white additive state noise*, and is given by:

$$y_p(k) = h\left(y_p(k\text{-}1), ..., y_p(k\text{-}n), u(k\text{-}1), \ldots, u(k\text{-}m)\right) + w(k) \tag{4.1.7}$$

The associated theoretical predictor is feedforward:

$$y(k+1) = h\left(y_p(k), ..., y_p(k\text{-}n+1), u(k), \ldots, u(k\text{-}m+1)\right) \tag{4.1.8}$$

If the assumed model is true, the prediction error $e(k+1) = y_p(k+1) - y(k+1)$ is equal to the noise $w(k+1)$, and its variance is minimal.

Thus, the neural predictor must be made of a feedforward network:

$$y(k+1) = \varphi\left(y_p(k), ..., y_p(k\text{-}n+1), u(k), \ldots, u(k\text{-}m+1); \theta\right) \tag{4.1.9}$$

The Output-Error assumed model.

As in the linear case, the Output-Error model assumes *pseudo-white additive output noise*. It is given by:

$$\begin{cases} z_p(k) = h\,(z_p(k\text{-}1), \ldots, z_p(k\text{-}n), u(k\text{-}1), \ldots, u(k\text{-}m)) \\ y_p(k) = z_p(k) + w(k) \end{cases} \quad (4.1.10)$$

Let us consider the following theoretical feedback predictor of order n:

$$y(k+1) = h\,(y(k), \ldots, y(k\text{-}n+1), u(k), \ldots, u(k\text{-}m+1)) \quad (4.1.11)$$

If the assumed model is true, and if the n first prediction errors are equal to the noise, then the prediction error is also equal to the noise and the prediction error variance is minimal. The effect of arbitrary initial conditions will vanish depending on the unknown function h.

In order to implement the associated predictor (4.1.11), one must train a feedback neural network of the form:

$$y(k+1) = \varphi\left(y(k), \ldots, y(k\text{-}n+1), u(k), \ldots, u(k\text{-}m+1); \theta\right) \quad (4.1.12)$$

4.2. *State-Space Modeling*

State-space models can be used either as *knowledge-based* models if enough prior knowledge on the physics of the process is available, or as *black-box* models when input-output models prove to be inefficient; we are concerned with the latter case [for a presentation including knowledge-based models, see Rivals et al. 1995].

The specificity of black-box state-space modeling arises from the fact that only the outputs of the state-space predictors have desired values $\{y_p(k)\}$, while the internal state variables are not imposed. This gives state-space predictors more flexibility (i.e. the ability to have a more complex input-output behavior), but the price to be paid is that different state-space representations might display the same input-output behavior. To present this specificity, we start with deterministic state-space assumed models; we then consider stochastic models.

4.2.1. *Determistic State-Space Models*

The assumed model is:

$$\begin{cases} x_p(k+1) = f\,(x_p(k), u(k)) \\ y_p(k) = g\,(x_p(k)) \end{cases} \quad (4.2.1.1)$$

where x_p denotes the n-state vector of the assumed model. Consider the predictor:

$$\begin{cases} x(k+1) = f\,(x(k), u(k)) \\ y(k+1) = g\,(x(k+1)) \end{cases} \quad (4.2.1.2)$$

where the n-vector x is the prediction of the state x_p. If the assumed model is true, and if $x(k) = x_p(k)$, the prediction error $e(k+1) = y_p(k+1) - y(k+1)$ is zero.

If the functions f and g are known, the extended Kalman predictor gives a *suboptimal* solution using a linearization of the model around the current state estimate. The one-step-ahead predictions of the state and of the output are computed using the following recursion:

$$\begin{cases} x(k+1) = f\left(x(k),\, u(k),\, 0\right) + K(k,\, x(k))\left(y_p(k) - g\left(x(k),\, 0\right)\right) \\ y(k+1) = g\left(x(k+1),\, 0\right) \end{cases} \tag{4.2.2.2}$$

where $K(k, x(k))$ is the time-varying and state-dependent extended Kalman gain, which is computed using the linearizations of f and g [Goodwin and Sin 1984]. But, in the modeling problem, the functions f and g are partially or completely unknown. There is thus no reason to restrict the complexity of the associated predictor structure to the form (4.2.2.2). We propose a time-invariant associated predictor with the same arguments, but of the more general form:

$$\begin{cases} x(k+1) = f_{pred}\left(x(k),\, u(k),\, y_p(k)\right) \\ y(k+1) = g_{pred}\left(x(k+1)\right) = g\left(x(k+1),\, 0\right) \end{cases} \tag{4.2.2.3}$$

A state-space neural predictor corresponding to predictor (4.2.2.3):

$$\begin{cases} \xi(k+1) = \varphi\left(\xi(k),\, u(k),\, y_p(k);\, \theta_\varphi\right) \\ y(k+1) = \omega\left(\xi(k+1);\, \theta_\omega\right) \end{cases} \tag{4.2.2.4}$$

with n-state vector ξ will be trained to minimize the MSPE on the training set.

Remark.

As in the deterministic case, the following state-space network can also be used:

$$\begin{cases} \xi(k+1) = \varphi\left(\xi(k),\, u(k),\, y_p(k);\, \theta\right) \\ y(k+1) = \psi\left(\xi(k),\, u(k),\, y_p(k);\, \theta\right) \end{cases} \tag{4.2.2.5}$$

where φ and ψ are nonlinear functions implemented by a single feedforward network with weights θ.

As in the input-output case, one would like to consider particular cases of the general model that lead to less complex predictors. This can be done essentially when making the assumption of additive output noise:

"Additive Output Noise" state-space assumed model.

In the case where additive output noise only is assumed:

$$\begin{cases} x_p(k+1) = f\left(x_p(k),\, u(k)\right) \\ y_p(k) = g\left(x_p(k)\right) + v_2(k) \end{cases} \tag{4.2.2.6}$$

an associated predictor takes a more simple form (it does not use the observations $\{y_p(k)\}$):

$$\begin{cases} x(k+1) = f\left(x(k),\, u(k)\right) \\ y(k+1) = g\left(x(k+1)\right) \end{cases} \tag{4.2.2.7}$$

If the assumed model is true, and if it is stable, the prediction error of predictor (4.2.2.7) is asymptotically equal to the noise.

The following state-space neural predictor will be trained:

$$\begin{cases} \xi(k+1) = \varphi\left(\xi(k),\, u(k);\, \theta_\varphi\right) \\ y(k+1) = \omega\left(\xi(k+1);\, \theta_\omega\right) \end{cases} \qquad (4.2.2.8)$$

where ξ is the n-state vector of the predictor, and where φ and ω are nonlinear functions implemented by cascaded subnetworks with weights θ_φ and θ_ω.

Remark.

Again, the following state-space neural network can also be used:

$$\begin{cases} \xi(k+1) = \varphi\left(\xi(k),\, u(k);\, \theta\right) \\ y(k+1) = \psi\left(\xi(k),\, u(k);\, \theta\right) \end{cases} \qquad (4.2.2.9)$$

where φ and ψ are nonlinear functions implemented by a single feedforward network with weights θ.

5. Training of the Neural Predictors

Let us consider a set of candidate neural predictors, each candidate being characterized by the assumed model it is associated to, and by its architecture (connectivity and number of neurons). The goal is now to obtain the best neural predictor among the candidates. As in the linear case considered in section 3.1, the training of each candidate (the estimation of its weights θ) is achieved using a P.E. method, or possibly an E.K.F. method in the case of a state-space neural network; the candidate with the best performance is then selected.

The E.K.F. training method consists in including the weights of the neural network in an augmented state vector, and to estimate this state recursively using the extended Kalman predictor associated to the augmented model. But it should be noted that the E.K.F. method requires the knowledge of the noise covariance matrices, which is unrealistic in black-box modeling. The E.K.F. algorithm might thus diverge, whereas the P.E. method will always provide the best predictor of a given arbitrary structure. As a consequence, P.E. methods are preferred for the training of neural networks. For a presentation of E.K.F. methods applied to neural network training, see [Singhal and Wu 1989, Matthews and Moschytz 90].

In the framework of a P.E. method, a candidate network is trained by minimizing its MSPE on input-output *training sequences*. Since we are concerned with time-invariant models, the MSPE on the training sequences can be minimized in a non recursive, iterative fashion ("batch" training). The best candidate is the predictor with the smallest MSPE on appropriate *test sequences*.

More precisely, the training and the selection use:
- sequences applied to the external inputs of the neural predictor $\{I(k)\}$, and sequences of the corresponding desired values $\{y_p(k+1)\}$ for the outputs (the training and test sequences);
- a cost function defined as the MSPE on the training sequence of size N, which is to be minimized iteratively. At iteration i, its value is:

$$J(\theta^i) = \frac{1}{N} \sum_{k=0}^{N-1} \left(e^i(k+1)\right)^2 = \frac{1}{N} \sum_{k=0}^{N-1} \left(y_p(k+1) - y^i(k+1)\right)^2 \qquad (5.1)$$

where $y^i(k+1)$ is the output of the predictor at time k and iteration i.
- the MSPE on the test sequence.

Training of a neural candidate with a given architecture.

At each iteration, the computation of the cost function $J(\theta^i)$ is performed using N copies of the feedforward part of the neural predictor (see Figure 5.1). In the following, we will omit the index i of the current iteration.

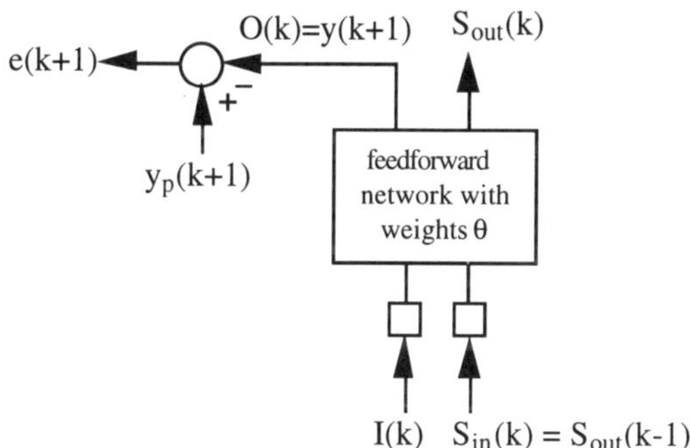

Figure 5.1.
Copy k of a trained neural predictor.

In the general case of a feedback predictor, the inputs of the copy k are:
- $I(k)$, the nonfeedback inputs of the predictor (external inputs u and possibly the measured outputs y_p);
- $S_{in}(k)$, the input state variables of the predictor at time k (past outputs y, or state ξ, or prediction errors e), which verify $S_{in}(k) = S_{out}(k-1)$ for k > 0; for copy k=0, the initial state must be fixed arbitrarily.

The outputs of the copy k are:
- $O(k)$, the output of the predictor $y(k+1)$;
- $S_{out}(k)$, the output state variables of the predictor at time k+1.

The cost function $J(\theta^i)$ can be minimized with any gradient method (the gradient being computed by the Backpropagation algorithm); quasi-Newtonian methods are particularly efficient. Feedforward predictors are trained in a directed fashion [Nerrand et al. 1993], i.e. using the Teacher Forcing algorithm [Jordan 1985], and feedback predictors are trained in a semi-directed fashion, i.e. using the Backpropagation-Through-Time algorithm [Rumelhart et al. 1986]. For further details about the training of state-space neural networks, see [Rivals 1995].

Example 1.

Let us make the NARX assumption for a given process; the associated predictor is given by (4.1.8). A feedforward neural predictor given by (4.1.9) will therefore be used:

$$y(k+1) = \varphi\left(y_p(k), ..., y_p(k-n+1), u(k), \ldots, u(k-m+1); \theta\right) \tag{5.2}$$

The external inputs $I(k)$ of copy k used for training are:

$$I(k) = \left[y_p(k), \ldots, y_p(k-n+1), u(k), \ldots, u(k-m+1)\right] \tag{5.3}$$

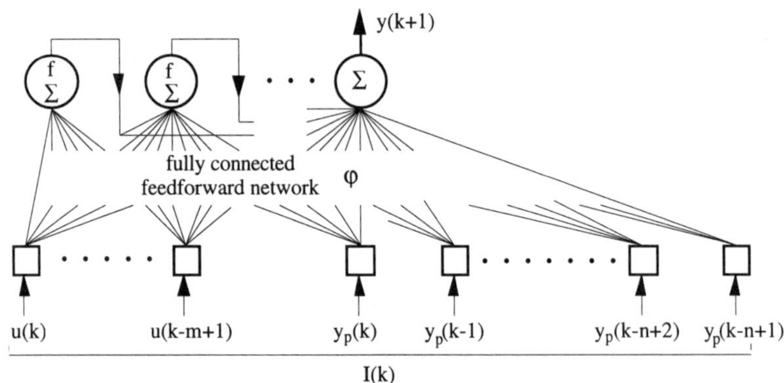

Figure 5.2.
Copy k of a neural NARX predictor.

Figure 5.2 shows a possible implementation of predictor (5.2) where the function φ is computed by a fully connected feedforward network with weights θ.

Example 2.

The predictor associated to a NARMAX assumed model is given by (4.1.2). A feedback neural predictor given by (4.1.3) must therefore be used:

$$y(k+1) = \varphi\left(y_p(k), ..., y_p(k-n+1), u(k), \ldots, u(k-m+1), e(k), \ldots, e(k-p+1); \theta\right) \tag{5.4}$$

The external and state inputs, and the state outputs of copy k, are:

$$\begin{aligned} I(k) &= [y_p(k),\ \ldots,\ y_p(k\text{-}n\text{+}1),\ u(k),\ \ldots,\ u(k\text{-}m\text{+}1)] \\ S_{in}(k) &= S_{out}(k\text{-}1) = [e(k),\ \ldots,\ e(k\text{-}p\text{+}1)] \end{aligned} \tag{5.5}$$

The initial state can for instance be taken equal to: $S_{in}(0) = [0, \ldots, 0]$. Figure 5.3 shows a possible implementation of predictor (5.4), the function φ being computed by a fully connected feedforward network with weights θ.

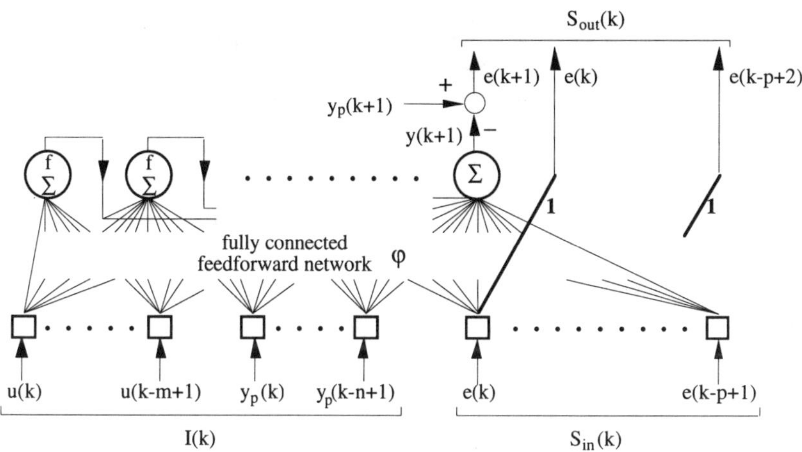

Figure 5.3.
Copy k of a neural NARMAX predictor.

Example 3.

A predictor associated to a stochastic state-space assumed model of order n is given by (4.2.2.3). A state-space neural network is used, possibly (4.2.2.4):

$$\begin{cases} \xi_1(k+1) = \varphi_1\left(\xi_1(k),\ \ldots,\ \xi_n(k),\ u(k),\ y_p(k);\ \theta_{\varphi_1}\right) \\ \quad \ldots \\ \xi_n(k+1) = \varphi_n\left(\xi_1(k),\ \ldots,\ \xi_n(k),\ u(k),\ y_p(k);\ \theta_{\varphi_n}\right) \\ y(k+1) = \omega\left(\xi_1(k+1),\ \ldots,\ \xi_n(k+1);\ \theta_\omega\right) \end{cases} \tag{5.6}$$

The external and state inputs, and the state outputs of copy k, are:

$$\begin{aligned} I(k) &= [u(k),\ y_p(k)] \\ S_{in}(k) &= S_{out}(k\text{-}1) = [\xi_1(k),\ \ldots,\ \xi_n(k)] \end{aligned} \tag{5.7}$$

The state may be initialized with: $S_{in}(0) = [0, \ldots, 0]$.

(5.6) is represented in Figure 5.4. The n functions $\varphi_1, \ldots, \varphi_n$ and the output function ω are implemented by n+1 one-layer feedforward subnetworks.

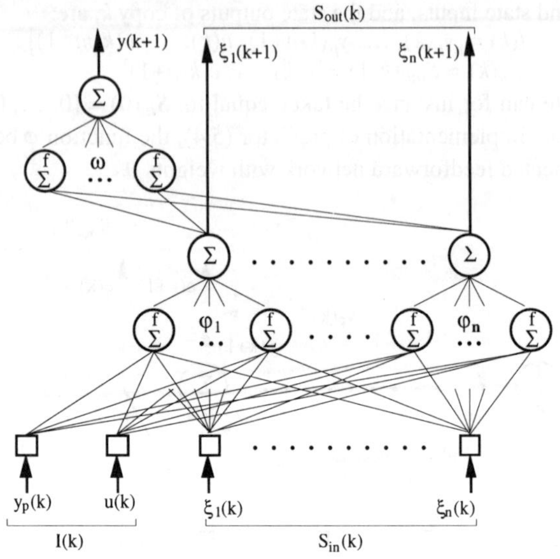

Figure 5.4.
Copy *k* of a predictor (4.2.2.4) associated to a stochastic state-space model.

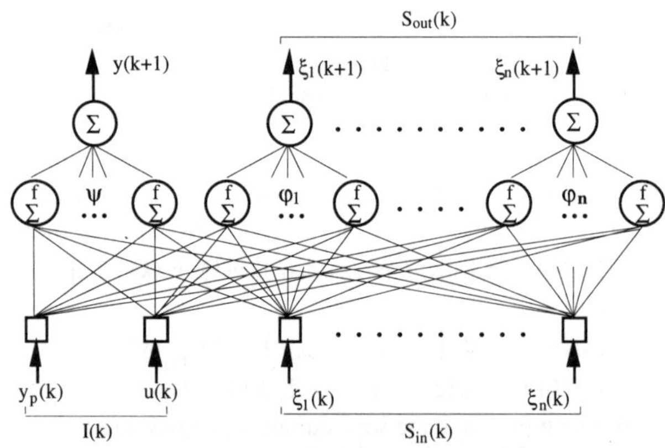

Figure 5.5.
Copy *k* of a predictor (4.2.2.5) associated to a stochastic state-space model.

Another possible neural predictor for (4.2.2.3) is of form (4.2.2.5):

$$\begin{cases} \xi_1(k+1) = \varphi_1\left(\xi_1(k),\ \ldots,\ \xi_n(k),\ u(k),\ y_p(k);\ \theta_{\varphi_1}\right) \\ \qquad\cdots \\ \xi_n(k+1) = \varphi_n\left(\xi_1(k),\ \ldots,\ \xi_n(k),\ u(k),\ y_p(k);\ \theta_{\varphi_n}\right) \\ y(k+1) = \psi\left(\xi_1(k),\ \ldots,\ \xi_n(k),\ u(k),\ y_p(k);\ \theta_{\psi}\right) \end{cases} \tag{5.8}$$

The external and state inputs, and the state outputs of copy k, are also given by (5.7). A possible implementation of predictor (5.8) is shown in Figure 5.5.

Remark.

The state-space networks proposed here are minimal state-space representations, as opposed to those used in [Elman 1990], or more recently in [Lo 1994, Connor et al. 1994], which have also been shown to be universal identification models [Sontag 1993]. In the present paper, the state variables are computed by a neural network with as many nonlinear hidden neurons as required, whereas in Elman's networks, each function φ_i is constrained to be computed with a *single* hidden neuron. This constraint leads generally to a non minimal state-space representation.

To summarize:

Let the training and test sequences be chosen according to the future use of the model (type of inputs, sample size, etc.): if *the assumptions underlying the selected neural predictor are true,* if *the family of functions defined by the feedforward part of the network is rich enough with respect to the complexity of the process behavior (i.e.* if *the number of neurons is sufficient), and* if *the training algorithm is efficient,* then *the trained predictor will be arbitrarily close to the optimal predictor in the state domain covered by the training and test sequences.*

6. Modeling Simulated Processes

To illustrate the neural black-box modeling approach we have outlined, we first model processes simulated by deterministic and stochastic state-space systems.

6.1. Modeling a Deterministic Process

We consider a process simulated by the deterministic second-order state-space system:

$$\begin{cases} x_{1p}(k+1) = 1.145\ x_{1p}(k) - 0.549\ x_{2p}(k) + 0.584\ u(k) \\ x_{2p}(k+1) = \dfrac{x_{1p}(k)}{1 + 0.01\ x_{2p}(k)^2} + 0.330\ u(k) \\ \quad y_p(k) = 4\ tanh\left(\dfrac{x_{1p}(k)}{4}\right) \end{cases} \tag{6.1.1}$$

The behavior of this process is oscillatory with unity static gain around zero; the denominator of the second state equation, together with the *tanh* in the output equation, increases the damping and decreases the static gain in the large-amplitude regime. The training and test input sequences consist of steps of random amplitude in [-5, +5] with a duration of 20 sampling periods; the total length of the sequences is

1200. Figure 6.1 shows the first third of the test sequence (for clarity, the graphs will always display the first third of the sequences only).

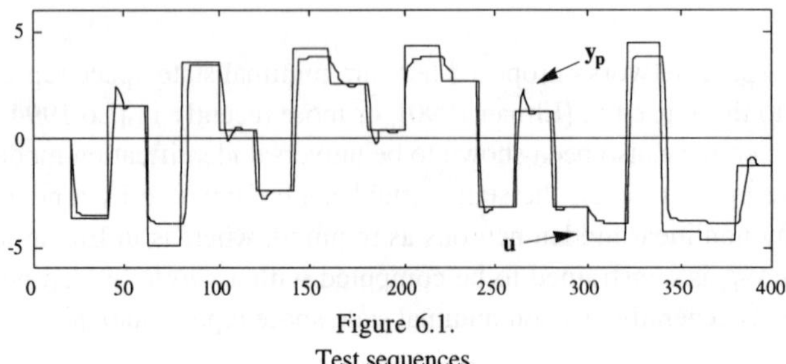

Figure 6.1.
Test sequences.

A predictor associated to the true model is given by Eq. (4.2.1.2). We use a second-order fully connected network of form (4.2.1.5):

$$\begin{cases} \xi_1(k+1) = \varphi_1\left(\xi_1(k),\, \xi_2(k),\, u(k);\, \theta\right) \\ \xi_2(k+1) = \varphi_2\left(\xi_1(k),\, \xi_2(k),\, u(k);\, \theta\right) \\ y(k+1) = \psi\left(\xi_1(k),\, \xi_2(k),\, u(k);\, \theta\right) \end{cases} \tag{6.1.2}$$

With only 2 hidden neurons and 27 weights (see Figure 6.2), the MSPE equals 1.6×10^{-7} on the training sequence, and 3.8×10^{-7} on the test sequence.

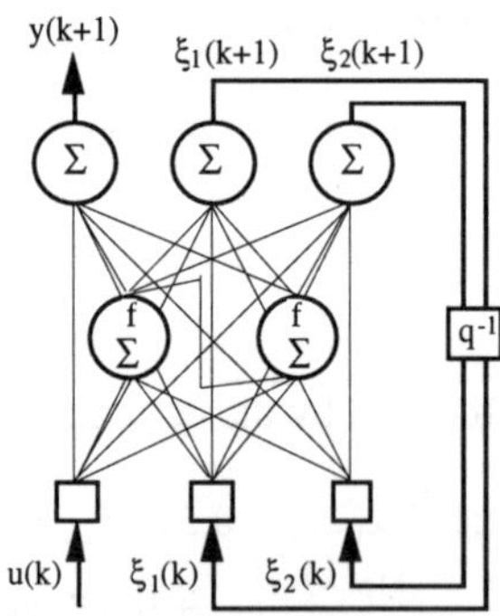

Figure 6.2.
Neural state-space predictor associated to deterministic or to "Additive Output Noise" state-space assumed models (*f* is the *tanh* function).

Figure 6.3a shows the outputs of the process and of the neural predictor; the prediction error is displayed on Figure 6.3b. Figures 6.3c and 6.3d show the state variables of the process and of the model. As expected, there is no straightforward relationship between them.

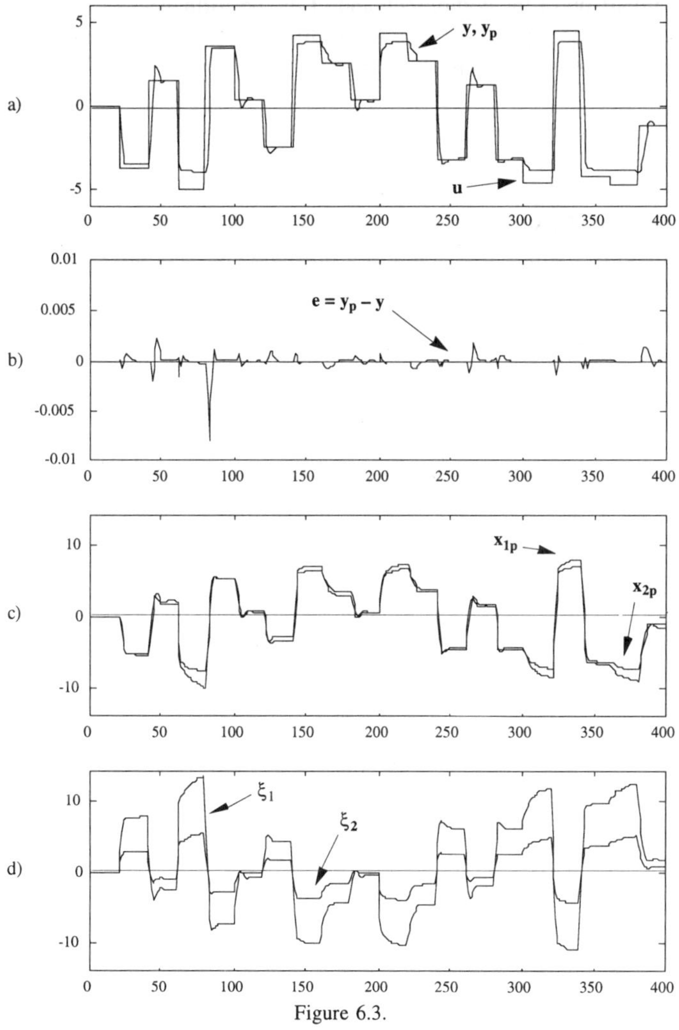

Figure 6.3.
State-space modeling of a deterministic process.

Input-output candidate predictors of the following form were also trained:

$$y(k+1) = \varphi\big(y(k),\ \ldots,\ y(k\text{-}n+1),\ u(k),\ \ldots,\ u(k\text{-}m+1);\ \theta\big) \qquad (6.1.3)$$

For n=m=2, 10 hidden neurons are necessary to obtain a MSPE of 10^{-4} on the test sequence; higher values of n and m do not lead to better results.

The comparison of the complexity and of the performances of the neural input-output and state-space predictors shows the difficulty that may arise when modeling an unknown process with an input-output predictor.

6.2. *Modeling Stochastic Processes*

Let $\{v_1(k)\}$ and $\{v_2(k)\}$ be gaussian white noise sequences with zero mean and variance 8,3 10^{-2}. The input sequences $\{u(k)\}$ for training and testing are the same as without noise. Since the predictors presented in section 3 and 4 are only *asymptotically* optimal, and since, for general stochastic models, their error variance is state-dependent, the error variance on the test sequence will be estimated using 100 realizations of the noise.

6.2.1. *Modeling a Process with Additive Output Noise*

The process is simulated by the following stochastic system:

$$\begin{cases} x_{1p}(k+1) = 1.145\ x_{1p}(k) - 0.549\ x_{2p}(k) + 0.584\ u(k) \\ x_{2p}(k+1) = \dfrac{x_{1p}(k)}{1 + 0.01\ x_{2p}(k)^2} + 0.330\ u(k) \\ y_p(k) = 4\ tanh\left(\dfrac{x_{1p}(k)}{4}\right) + v_2(k) \end{cases} \qquad (6.2.1.1)$$

A predictor associated to the true model is given by (4.2.2.7). Network (6.1.2) was thus trained to be a predictor for process (6.2.1.1). The best performance (i.e. the smallest MSPE on the test set) is obtained with 2 hidden neurons (27 weights). The MSPE equals 8.2x10^{-2} on the training sequence, and 8.4x10^{-2} on the test sequence, while the variance of $v_2(k)$ is 8.3x10^{-2}. The variance of the prediction error on the test, estimated with 100 realizations of the noise, is shown on Figure 6.4. In the steady state, the variance estimation is 8.3x10^{-2}, which is equal to the noise variance.

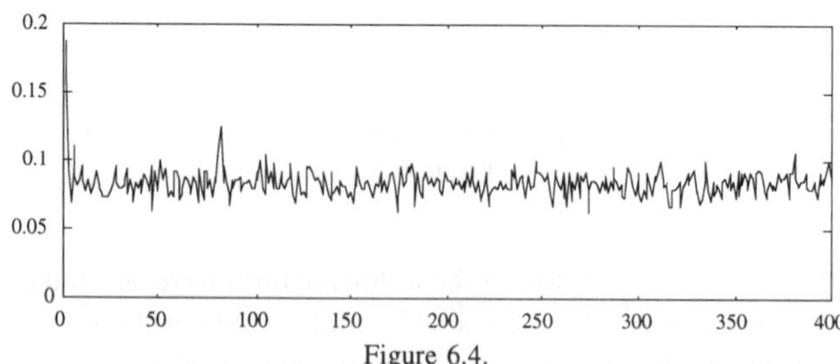

Figure 6.4.
Prediction error variance of a neural state-space predictor associated to an "Additive Output Noise" state-space assumed model.

6.2.2. Modeling a Process with Additive State and Output Noise

The simulated process has also additive noise on the first state variable:

$$\begin{cases} x_{1p}(k+1) = 1.145\, x_{1p}(k) - 0.549\, x_{2p}(k) + 0.584\, u(k) + v_1(k) \\ x_{2p}(k+1) = \dfrac{x_{1p}(k)}{1 + 0.01\, x_{2p}(k)^2} + 0.330\, u(k) \\ y_p(k) = 4\, tanh\left(\dfrac{x_{1p}(k)}{4}\right) + v_2(k) \end{cases} \quad (6.2.2.1)$$

Modeling with a predictor associated to a stochastic state-space assumed model.

As shown in section 4.2.2, the theoretical predictor (4.2.2.3) will lead to minimum variance prediction. The best MSPE on the test is obtained with a second-order fully connected network of form (4.2.2.5):

$$\begin{cases} \xi_1(k+1) = \varphi_1\left(\xi_1(k),\, \xi_2(k),\, u(k),\, y_p(k);\, \theta\right) \\ \xi_2(k+1) = \varphi_2\left(\xi_1(k),\, \xi_2(k),\, u(k),\, y_p(k);\, \theta\right) \\ y(k+1) = \psi\left(\xi_1(k),\, \xi_2(k),\, u(k),\, y_p(k);\, \theta\right) \end{cases} \quad (6.2.2.2)$$

with 2 hidden neurons (32 weights), which is shown on Figure 6.5.

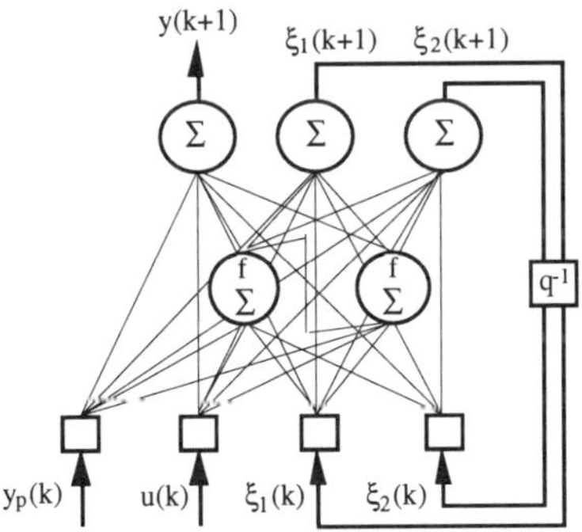

Figure 6.5.

Neural state-space predictor associated to a stochastic state-space assumed model.

The variance of the prediction error on the test sequence, estimated with 100 realizations of the noise, is shown on Figure 6.6. As opposed to the case of additive output noise, the variance is state-dependent. The MSPE equals 1.2×10^{-1} on the training sequence and 1.3×10^{-1} on the test sequence.

The value of the minimal variance is unknown; nevertheless, in order to evaluate the optimality of the neural predictor, we computed the MSPE of the extended Kalman predictor using the exact model (6.2.1.1): it is equal to 1.4×10^{-1}. This demonstrates the ability of the trained state-space neural predictor (*without knowledge of the exact model of the process*) to perform at least as well as the extended Kalman predictor.

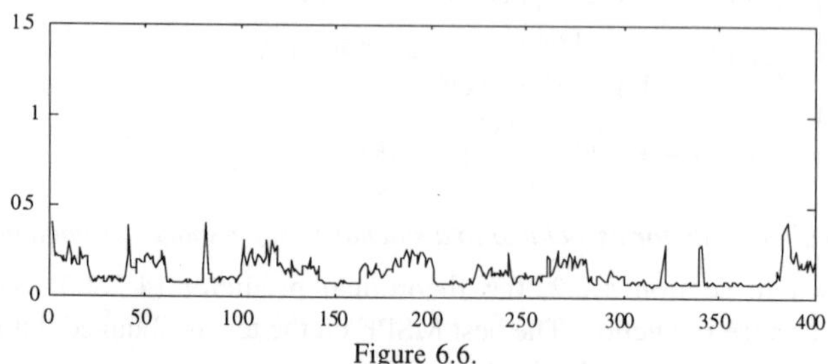

Figure 6.6.
Prediction error variance of a neural state-space predictor associated to a stochastic state-space assumed model.

Modeling with a NARMAX predictor.

We also tried to model the process (6.2.2.1) with neural NARMAX predictors, with $n=m=2$, of order $p=1$:

$$y(k+1) = \varphi\left(y_p(k),\ y_p(k\text{-}1),\ u(k),\ u(k\text{-}1),\ e(k);\ \theta\right) \qquad (6.2.2.3)$$

and order $p=2$:

$$y(k+1) = \varphi\left(y_p(k),\ y_p(k\text{-}1),\ u(k),\ u(k\text{-}1),\ e(k),\ e(k\text{-}1);\ \theta\right) \qquad (6.2.2.4)$$

The best results were obtained with a NARMAX predictor of order 2 with 5 hidden neurons (51 weights). The MSPE equals 1.5×10^{-1} on the training sequence, and 1.6×10^{-1} on the test sequence. The prediction error variance on the test sequence, estimated with 100 realizations of the noise, is shown on Figure 6.7.

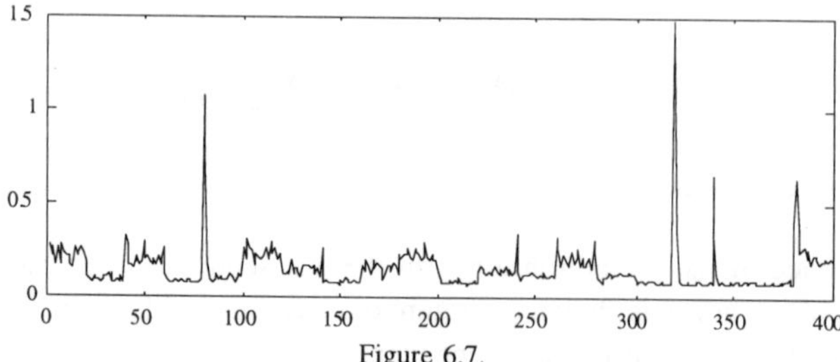

Figure 6.7.
Prediction error variance of a neural input-output predictor associated to a NARMAX assumed model.

This result confirms the fact that a NARMAX predictor might require many more arguments, neurons and weights than a state-space predictor to achieve a comparable (though not as good) performance.

7. Modeling a Hydraulic Actuator

We consider the hydraulic actuator of a robot arm, whose position depends on the valve-controlled oil pressure. The available input (valve position u) and output (oil pressure y_p) sequences were split in two for training and testing, and are shown on Figure 7.1. The training and test sequences have 512 samples each.

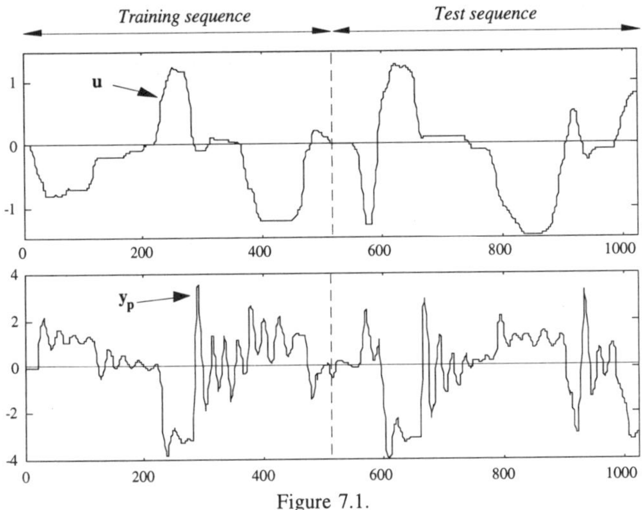

Figure 7.1.
Training and test sequences for the modeling of the hydraulic actuator.

The control and output sequences have roughly the same amplitude in the training sequence as in the test sequence, but the maximum amplitude of the control sequence is larger in the latter than in the former. Furthermore, the response of the system seems to undergo more sustained oscillations, for similar control sequences, in the training sequence than in the test sequence.

The goal of the modeling procedure is here to design a simulator of the process. The simulation model corresponding to a given predictor is a feedback model, which is obtained by replacing the process output y_p (if any) acting as external input on the predictor by the predictor output y. The performance (the MSPE on the test sequences) will be estimated with the simulation models corresponding to the candidate predictors used for training.

7.1. Input-Output Modeling of the Actuator

Under the Output-Error assumption, the input-output predictor giving the best performance without overfitting is a fully connected second-order network:

$$y(k+1) = \varphi\big(y(k), y(k\text{-}1), u(k); \theta\big) \qquad (7.1.1)$$

with 2 hidden neurons (15 weights), which is shown on Figure 7.2.

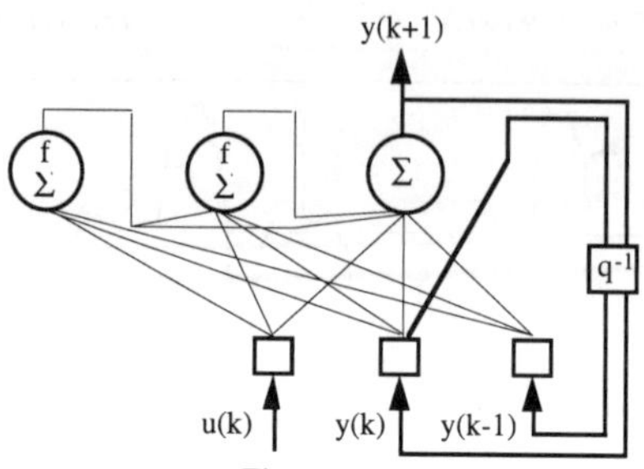

Figure 7.2.
Input-output neural predictor associated to an Output-Error assumed model.

Since the predictor (7.1.1) does not use the output of the process y_p, the corresponding simulation model is identical to the predictor. The result obtained on the test sequence is shown on Figure 7.3. The MSPE are 0.085 on the training sequence and 0.30 on the test sequence. Adding hidden neurons or extra inputs leads to overfitting.

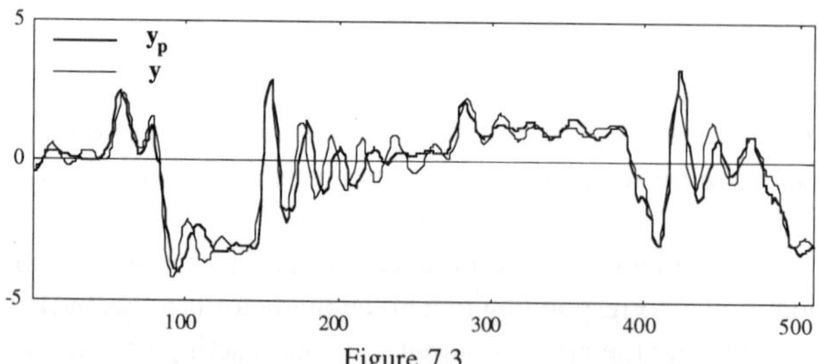

Figure 7.3.
Input-output modeling of the actuator (Output-Error assumption).

Making the NARX assumption leads to lower quality results (the NARX predictors being tested as simulation models, i.e. as feedback models). NARMAX predictors do not give better results than Output-Error predictors.

7.2. State-Space Modeling of the Actuator

Under the "Additive Output Noise" assumption, the best results were obtained with a second-order network:

$$\begin{cases} \xi_1(k+1) = \varphi_1\left(\xi_1(k),\ \xi_2(k),\ u(k);\ \theta\right) \\ \xi_2(k+1) = \varphi_2\left(\xi_1(k),\ \xi_2(k),\ u(k);\ \theta\right) \\ y(k+1) = \psi\left(\xi_1(k),\ \xi_2(k),\ u(k);\ \theta\right) \end{cases} \tag{7.2.1}$$

with 2 hidden neurons (27 weights), which is the same network as in Figure 6.2.

Again, since the predictor (7.2.1) does not use the output of the process y_p, the corresponding simulation model is identical to the predictor. The result on the test sequence is shown on Figure 7.4. The MSPE is 0.10 on the training sequence, and 0.11 on the test sequence, which is much better than under input-output assumptions. This clearly illustrates the advantage of using state-space predictors when small training sets only are available: since they require a smaller number of inputs, they are more parsimonious, hence less prone to overfitting.

A neural state-space predictor associated to a general stochastic assumed model, i.e. given by (4.2.2.4) or (4.2.2.5), did not give a better performance.

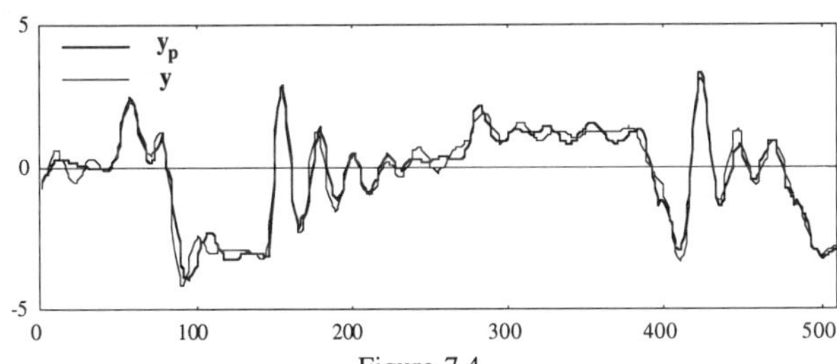

Figure 7.4.
State-space modeling of the actuator (state-space "Additive Output Noise" assumption).

Nevertheless, the results presented above compare very favorably to those obtained on the same data by other groups: in [Sjöberg 1993] a MSPE on the test sequence of 0.46 is obtained with a NARX second-order input-output multilayer Perceptron with 8 hidden neurons; similar results are obtained using wavelets in [Benveniste et al. 1994].

8. Conclusion

We have given an overview of the neural network approach to nonlinear modeling, from the choice of the candidate models to the estimation of the parameters of their corresponding neural predictors. We have emphasized the importance of state-space candidate models, which are likely to need less arguments than input-output models: this is clearly an advantage when small data sets only are available. The trained state-space neural predictors were shown to be able to perform at least as well as the extended Kalman predictor, but without the knowledge of a model of the process. The modeling approach has been illustrated on several simulated examples and on a real process.

In the case where an accurate nonlinear model of the process is available, another possible application of neural state-space predictors is the synthesis of an optimal filter using training sequences generated by computer simulations with the model: a trained neural predictor is likely to outperform the extended Kalman filter, or even the iterated extended Kalman filter, as shown in [Lo 1994].

An additional attractive feature of state-space neural modeling is the fact that it is well suited to the introduction of domain knowledge, when the latter is in the form of state equations; this is illustrated in [Ploix et al. 1994] by the modeling of a large industrial process, a distillation column, described by more than one hundred state equations.

9. Acknowledgements

The authors are deeply indebted to Pierre-Yves Glorennec for providing the data on the hydraulic actuator.

10. References

Benveniste A., Juditsky A., Delyon B., Zhang Q., Glorennec P.-Y. Wavelets in identification, *10th IFAC Symposium on Identification* (Copenhagen, 1994).

Chen S., Billings S. A. Recursive prediction error parameter estimation for non linear models, *Int. J. Control* **49** (1989), pp. 569-594.

Connor J. T., Martin R. D., Atlas L. E. Recurrent neural networks and robust time series prediction, *IEEE Trans. on Neural Networks* **5** (1994), pp. 240-254.

Elman J. L. Finding structure in time, *Cognitive Science* **14** (1990), pp. 179-221.

Goodwin G. C., Sin K. S. *Adaptive filtering prediction and control* (Prentice-Hall, New Jersey, 1984).

Hornik K., Stinchcombe M., White H. Multilayer feedforward networks are universal approximators, *Neural Networks* **2** (1989), pp. 359-366.

Hornik. K., Stinchcombe M., White H., Auer P. Degree of approximation results for feedforward networks approximating unknown mappings and their derivatives, *Neural Computation* **6** (1994), pp. 1262-1275.

Jordan M. I. *The learning of representations for sequential performance*, Doctoral Dissertation, University of California, San Diego (1985).

Leontaritis I. J., Billings S. A. Input-output parametric models for non-linear systems. Part I: deterministic non-linear systems. Part II: stochastic non-linear systems, *Int. J. Control* **41** (1985), pp. 303-344.

Levin A. U. *Neural networks in dynamical systems; a system theoretic approach*, PhD Thesis, Yale University (1992).

Levin A. U., Narendra K. S. Identification using feedforward networks, *Neural Computation* **7** (1995), pp. 349-357.

Lo J. T.-H. Synthetic approach in optimal filtering, *IEEE Trans. on Neural Networks* **5** (1994), pp. 803-811.

Matthews M. B., Moschytz G. S. Neural-network nonlinear adaptive filtering using the extended Kalman filter algorithm, *International Neural Network Conference*, (Paris, 1990), pp. 115-118.

Narendra K. S., Parthasarathy K. Identification and control of dynamical systems using neural networks, *IEEE Trans. on Neural Networks* **1** (1990), pp. 4-27.

Narendra K. S., Parthasarathy K. Gradient methods for the optimization of dynamical systems containing neural networks, *IEEE Trans. on Neural Networks* **2** (1991), pp. 252-262.

Nerrand O., Roussel-Ragot P., Personnaz L., Dreyfus G. Neural networks and non-linear adaptive filtering: unifying concepts and new algorithms, *Neural Computation* **5** (1993), pp. 165-199.

Nerrand O., Roussel-Ragot P., Urbani D., Personnaz L., Dreyfus G. Training recurrent neural networks: why and how? An Illustration in Process Modeling, *IEEE Trans. on Neural Networks* **5** (1994), pp. 178-184.

Ploix J.-L., Dreyfus G., Corriou J.-P., Pascal D. From knowledge-based models to recurrent networks: an application to an industrial distillation process, *Neural Networks and their Applications*, J. Hérault ed.(1994).

Rivals I., Canas D., Personnaz L., Dreyfus G. Modeling and control of mobile robots and intelligent vehicles by neural networks, *IEEE Conference on Intelligent Vehicles* (Paris, 1994), pp. 137-142.

Rivals I. *Modélisation et commande de processus par réseaux de neurones; application au pilotage d'un véhicule autonome*, Thèse de Doctorat de l'Université Paris 6 (1995).

Rivals I., Personnaz L., Dreyfus G., Ploix J.-L. Modélisation, classification et commande par réseaux de neurones: principes fondamentaux, méthodologie de conception, et illustrations industrielles, in *Récents progrès en génie des procédés* **9** (Lavoisier technique et documentation, Paris, 1995).

Rumelhart D. E., Hinton G. E., Williams R. J. Learning internal representations by error back-propagation, in *Parallel Distributed Processing: explorations in the microstructure of cognition. Vol.1 : Foundations*, D. E. Rumelhart, J. L.

McLelland and the PDP Research Group eds. (MIT Press, Cambridge MA, 1986), pp. 318-362.

Singhal S., Wu L. Training multilayer perceptrons with the extended Kalman algorithm, *Advances in Neural Information Processing Systems* **I** (Morgan Kaufmann, San Mateo 1989), pp. 133-140.

Sjöberg J. *Regularization issues in neural network models of dynamical systems*, LiU-TEK-LIC-1993:08, ISBN 91-7871-072-3, ISSN 0280-7971 (1993).

Sjöberg J., Zhang Q., Benveniste A., Deylon B., Glorennec P.-Y., Hjalmarsson H., Juditsky A., Ljung L. Nonlinear black-box modelling in system identification: model structures and algorithms, submitted to *Automatica* (1995).

Sontag E. D. Neural networks for control, in *Essays on control: perspectives in the theory and its applications*, Trentelman H. L. and Willems J. C. eds. (Birkhäuser , Boston 1993), pp. 339-380.

AN APPROACH TO INTELLIGENT IDENTIFICATION AND CONTROL OF NONLINEAR DYNAMICAL SYSTEMS

Donald A. Sofge and David L. Elliott
NeuroDyne, Inc.
775 Gateway Dr. SE, #514, Leesburg, VA 22075, U.S.A.
E-mail: sofge@ai.mit.edu, delliott@src.umd.edu

ABSTRACT

We have developed an approach to modeling nonlinear dynamical systems (with inputs) by an identifier consisting of tapped delay lines used as the input layer to a feedforward neural network with a bottleneck architecture. After training, the identifier is used in externally recurrent mode to approximate the dynamic behavior of the given system. The adequacy of approximation is evaluated by commanding both the given system and the identifier with the same control signal and comparing their responses. We demonstrate the effectiveness of this approach on three challenging dynamic systems: a Saturn engine model (with model retrained and verified on an actual Saturn engine plant) a van der Pol oscillator, and a lead-acid storage battery. Further, we have developed an approach to neural adaptive control of the engine plant based upon dynamic programming approaches of optimal control theory. The goal of the auto engine controller is to optimize multiple engine parameters consisting of drivability, emissions performance, and fuel economy.

1. Introduction

Although controlling a given dynamical system (plant) can be approached directly by neural network methods, many of the more popular control methods based on dynamic programming ideas require that the plant be *identified*—that is, we need a computable model of the plant, capable of predicting near-future outputs from trial controls.

System identification is one of the most basic problems in control system theory. In the case of linear plants a usual approach to identification is to construct an approximate transfer function representing the plant's behaviour in discrete or continuous time; using the superposition theorem, the initial state and output can be assumed to be zero. (For a good survey of the subject see Ljung[1].) For nonlinear plants identification is more difficult because transfer functions cannot be used and because the input-output relation depends on the initial state. (There may be many states for which the output is constant or zero.)

Several representations of nonlinear systems have been used in identification. The ideal method may be to have a first-principles model of the plant, with only the parameters to be chosen, as in mechanics, circuit theory, enzyme kinetics and many other disciplines where a lumped parameter model of low dimension is available.

Then the identification is done by parameter-fitting to minimize a suitable cost, such as the sum over time of squared differences of plant and model outputs. This technique is useful when the plant is well understood and systematic experimentation can be used to calibrate subsystems, as Hendricks and Sorenson[2] did for their mean-value model for spark-ignition internal combustion engines discussed in Section 4.

A second approach assumes that the plant is a black box that can only be known by its input-output relation in the neighborhood of a (single) equilibrium and for small values of the plant's input commands. In the 1960's this approach was implemented with Volterra series and other functional expansions involving multiple integrals polynomial in the inputs and initial states. It is usually taken for granted that beyond the third term (cubic in the inputs) these expansions are experimentally hard to evaluate, and since the series are rarely known to be convergent one stops with whatever accuracy one can obtain at that point. More recently there have been interesting nonlinear identification techniques involving canonical variate analysis[3], and delay coordinate methods[4] inspired by Chaos theory, for instance.

Neural network approaches to identification are used when a homogeneous computational architecture (capable of parallel computation) is desirable, little is known about the plant, the plant structure is strongly nonlinear, and only input-output data are available. Requirements include great quantities of data, preferably with sampling of large regions in the state space and many inputs. One payoff is speed of evaluation of the identified model, which can make good control possible even in the presence of inaccuracy.

Recently there have been many attempts at using neural methods with plant input-output data in the hope of representing the input-output relations. One approach is to construct a neural network with time-delays between nodes, and structure that imitates that of the plant. In contrast to this approach, we use a canonical architecture[5] that uses time-delayed samples of system controls and outputs as first layer, upon which sits a bottlenecked feedforward network in which the top nodes each are given by a subnet built over the bottleneck layer.

2. Identification with Linear ARMA Filters

An nth-order discrete-time autoregressive moving-average (ARMA) filter is given by its transfer function

$$H(z) = \frac{(B_1 + \cdots + B_n z^n)}{(1 + A_1 z + \cdots + A_n z^n)}.$$

If its weights are adapted by the Widrow-Hoff LMS algorithm[6] as discussed below, one may approximately identify a linear system P given by transfer function $P(z)$. We discuss only the single-input single-output case, but the generalization to the multi-input multi-output version is straightforward. (In the continuous time case, the plant input and output are sampled at some clock interval τ.)

An input which is "persistently exciting" (a spectral property guaranteeing uniqueness of the filter H, see Chapter 14 of Ljung[1]) is presented as an input to the system P; past values of the input and output, respectively u_k and y_k, are used as inputs to two delay-lines of length n. The delay-line values are related by the plant transfer function: $\hat{y} = P(z)\hat{u}$, where

$$P(z) = \frac{(B_1 + \cdots + B_n z^n)}{(1 + A_1 z + \cdots + A_n z^n)}.$$

The two delay-lines are used to construct an adaptive *transverse filter* as follows. The delay-line taps are provided with adjustable weights A_k, B_k and the outputs summed to give

$$Y_t = \Sigma_{k=1}^{n}(A_k y_{t-k\tau} + B_k u_{t-k\tau}).$$

The most recent sample $y(t)$ of the plant output is compared with the value Y_t predicted by the filter, and the squared error minimized by adjusting the A_k and B_k on-line by the LMS rules for $1 \leq k \leq L$:

$$\begin{aligned} \Delta A_k &= \alpha(Y_t - y_t) y_{t-k\tau}, \\ \Delta B_k &= \alpha(Y_t - y_t) u_{t-k\tau} \end{aligned} \tag{1}$$

or by similar rules (called normalized LMS) in which the right-hand sides of Equation (1) are normalized by the variances of y and u respectively.

Note that these difference equations (1) are linear in the weights. Finding the best L and α to achieve rapid convergence is difficult, and those problems are of course no easier in the nonlinear generalization to be discussed below.

3. Neural Networks for Nonlinear Systems

The advent of neural network technology has raised the prospect of highly parallel computational approaches to identification, based on theorems showing that several of the function systems used in neural networks can be used to uniformly approximate multivariable functions that are sufficiently smooth. The transition maps for nonlinear systems appear to be an appropriate application. The first type of system we have in mind has the control theoretic model

$$\begin{aligned} \dot{x} &= f(x) + g(x)c(t), \\ y(t) &= h(x(t)) \end{aligned}$$

with the output sampled at some clock interval τ. It is assumed for simplicity that τ is small enough so that the control $c(t)$ can be regarded as constant between samples; in that case, the transition map is $(x(t-\tau), c(t)) \to x(t)$. It is, of course, easy enough to specify the control in a linear or polynomial way in each interval $(t-\tau, t)$ if τ must be large for some reason. In the Identifier structure, the additional parameters of the control would need corresponding delay lines. A second type of system is discussed in Section 6.

The identification task may be attacked through the use of *recurrent* neural networks[7]—in the above case of the linear adaptive filter this means $\Delta A_k = \alpha(Y_t - y_t)Y_{t-k\tau}$, which is quadratic in the weights, so that the dynamics in weight-space may become chaotic. Recurrent networks, like adaptive control systems, rely on slow variations of the trained parameters (weights) to ensure stability. However, the training can be performed by minimizing prediction error of a feedforward network[8,9,10]; in this case recursion is used in the *trained* network but *not during training*. That is the approach we adopt here, but with an architecture based on a state-space approach, related to control theory notions of state, observability and controllability. The approach of Chen and Billings[8] also uses a general feedforward network; our Identifier will be described in Section 5.

4. Lean-Burn Engine Application

Under a three year program sponsored by the National Science Foundation and Clean Car Program, Grant ECS-9216530, NeuroDyne is leading a research team in the development of lean-burn engine controls using advanced methods in intelligent sensor processing and controls. A challenge facing the intelligent control community is the development of technology which allows for the production of automobiles that can meet stringent EPA standards and can satisfy performance requirements of drivers. The new EPA fuel economy and emissions requirements at 100,000 miles inevitably require tuning the automobile control algorithms over time. The goal of this effort is to develop intelligent lean-burn engine controls using existing sensor technology to meet forthcoming EPA regulations. We are using mean value internal combustion engine models based on Ford and Saturn engines to get preliminary neural network specifications.

Once the architecture is roughed out by simulation data, it is necessary to train with experimental data from many actual engine runs; the simulation, therefore, need not be accurate at all time-scales to be valuable in the preliminary design of an Identifier.

A first-principles model of an engine would involve not only its mechanical and electrical components but also nonlinear partial differential equations with moving boundaries to describe the aerodynamics of the inflowing air/fuel mixture and the behavior of flame fronts. The control elements are multiplicative and saturate. However, the overall input-output behavior of an engine is repeatable, given the parameters of fuel composition, torque load, throttle settings, etc.; and it is quite reasonable to identify parametric and nonparametric models directly from experimental data under controlled conditions.

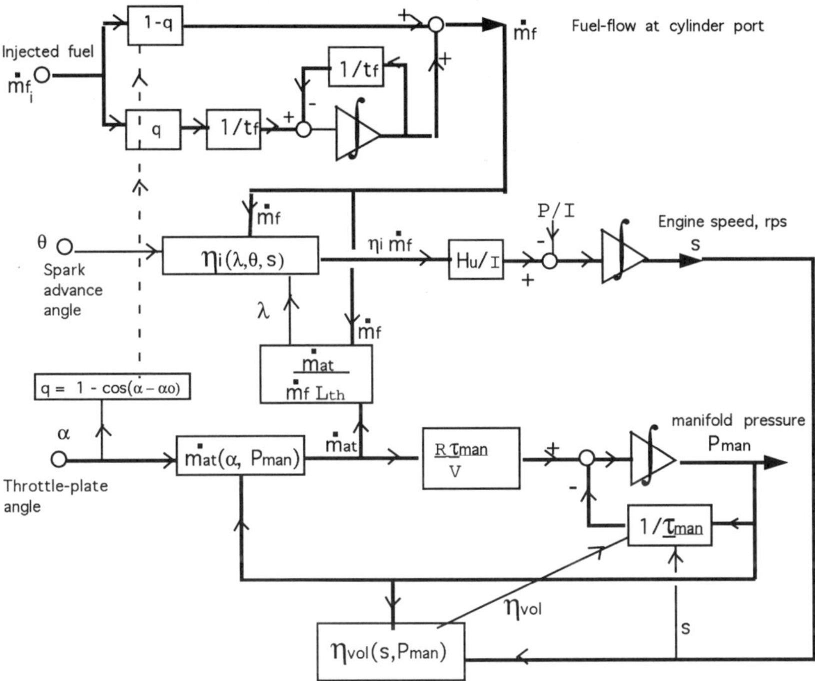

Mean value model of Spark Ignition Engine
simplified from Hendricks - Sorenson, SAE Paper 900616

Figure 1: Mean-value model, internal combustion engine.

Hendricks and Sorenson[2] gave a third-order mean-value engine model, which under our Grant was modified and adapted to a Ford engine and later a Saturn engine in 1994. The state variables in this model are fuel film mass flow, crank-shaft speed, and manifold pressure. This simplicity is achieved by making several assumptions about the engine, e.g. that processes that reach equilibrium within one or two engine cycles may be neglected.

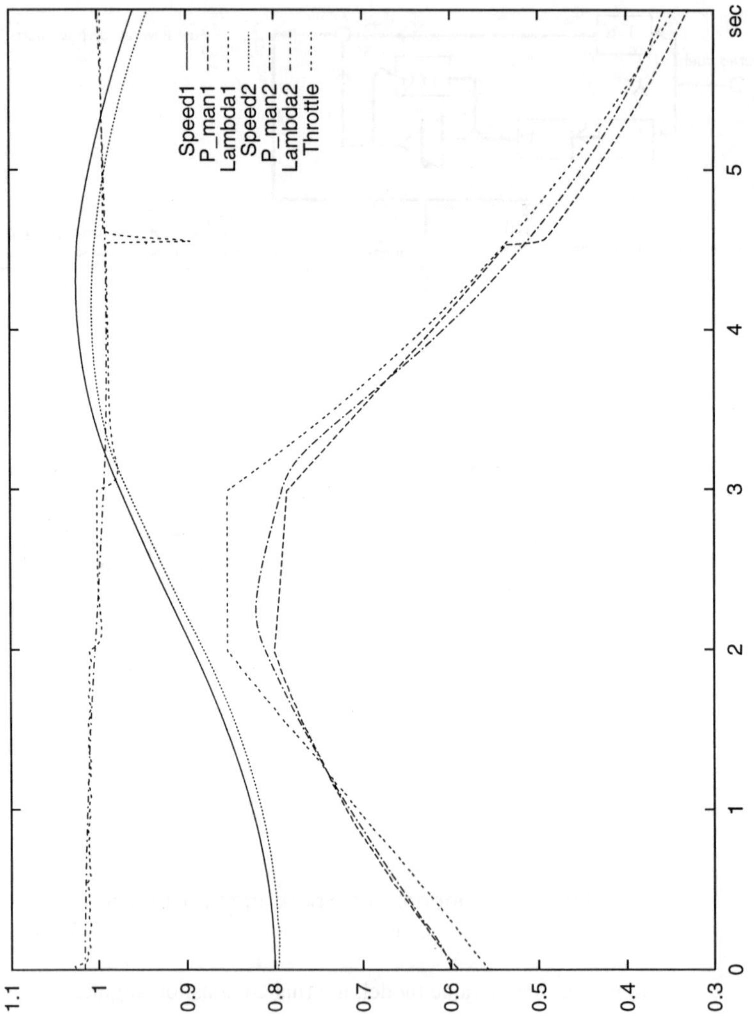

Figure 2: Normalized Engine (1) and Reconstructed (2) Dynamics.

The neural network weights are being improved by the use of Saturn engine data from West Virginia University. Initial results show that starting with a network trained to within 99% accuracy on the simulation, and then retraining on actual Saturn engine data, the network quickly converges to within 98% accuracy.

In the current automobile engine identifier c has two components: the throttle-plate angle, and fuel-injection rate computed from the fuel/air ratio λ (via an engine sensor in experimental data, via an algorithm in simulations). The three outputs y_1, y_2, y_3 are respectively engine speed, P_{man} (manifold pressure), and λ (stoichiometry). We kept 6 samples in each delay line; run length is 6 seconds with time-step 0.05 sec. In Figure 2 the response of the engine simulation (Speed1, ...) and the Identifier's response to the same Throttle command (Speed2, ...) are shown in normalized units.

5. NeuroDyne Identifier Structure

A NeuroDyne Identifier has a structure comprising a tapped delay line and a feedforward neural network of bottleneck type with segmented subnet for each output.

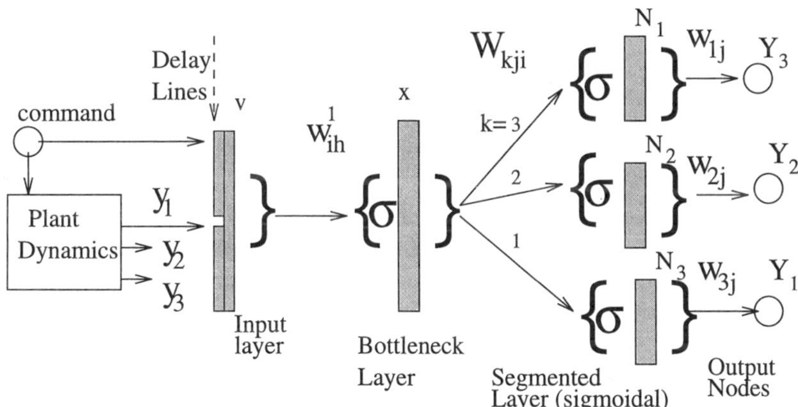

Figure 3: Training mode.

It is trained by a fast backpropagation algorithm developed by NeuroDyne so that several outputs approximate, in mean square, a sequence of plant outputs stored in the delay line. For the purpose of illustration, we will assume that the nonlinear plant in Figure 3 is described by a nonlinear dynamical system with vector output signal $[y_1(t), y_2(t), y_3(t)]$ and commanded by the control $c(t)$, as in the automobile engine discussed above.

Both $y(t)$ and $c(t)$ are sampled periodically. The delay lines in the example hold N sampled values of the control and each of one or more of the outputs, say $y_1(t)$ which (stacked into one vector) together constitute the vector of input nodes

$$v = [y_1(t-N), \cdots, y_1(t), c(t-N), \cdots, c(t)].$$

(If the plant is observable with respect to output y_1 we do not need other outputs, but in general it is best to incorporate as many independent plant outputs as are readily measured.)

In the feedforward network we use the fast Elliott sigmoid[11]

$$\sigma(u) = \frac{u}{(1+|u|)};$$

its derivative is given by

$$\sigma' = (1-|\sigma|)^2.$$

Often the hidden layer or bottleneck layer, node vector (x), is obtained by sigmoidal functions of linear combinations of input node values v; a good initial guess for the size of vector x is the dimension n of a good lumped-parameter approximation of the plant:

$$x_i = \sigma(\sum_1^{2N}(w_{ih}^1 v_h) + w_0^1), \quad 1 \leq i \leq n.$$

Here the values of x lie in a box in $\mathbb{R}^n$. For the automobile engine $n = 3$; for this engine, in our studies giving rise to the data shown in Figure 2, the sigmoid nonlinearity is not needed for this layer.

The bottleneck layer may be thought of as an internal state space for the Identifier. Finally, each of the outputs (Y_1, Y_2, Y_3) is a linear combination of the sigmoidal outputs of its own independent subnet; the sigmoidal hidden layer segments are labelled (N_1, N_2, N_3), each with some number Q of nodes, to be determined experimentally to achieve the desired accuracy.

Training, using a large set of randomly chosen control inputs $c(t)$, is via the well-known back-propagation algorithm, to predict the next plant outputs by minimizing the error $\sum_1^3 (Y_i(t) - y_i(t))^2)$.

This projection of the sampled inputs down to the bottleneck layer is an attempt to make x a true state vector, and of the same dimension as the plant. This point is made plausible by considering local linearization of the net (see Baldi and Hornik[12] for similar ideas). In some applications there is a single scalar output y and we would choose the y_i to be a vector of samples, say $[y_(t-1), y(t-2), y(t-3)]$ in order to force the prediction of (by observability theory[13]) a version of the plant state.

Using independently segmented subnets for the outputs has been shown in earlier NeuroDyne work on a Pratt & Whitney jet engine study[14] to result in more rapid training, since the weight corrections for the different outputs do not interfere with each other. To train a new choice of output, once the system dynamics are well captured, one need only add a new segment N and output Y, and set the learning rate at 0 for the weights between the input and bottleneck. The number of nodes is larger than usual than in a fully connected three-layer net, but the number of connections and weights is smaller.

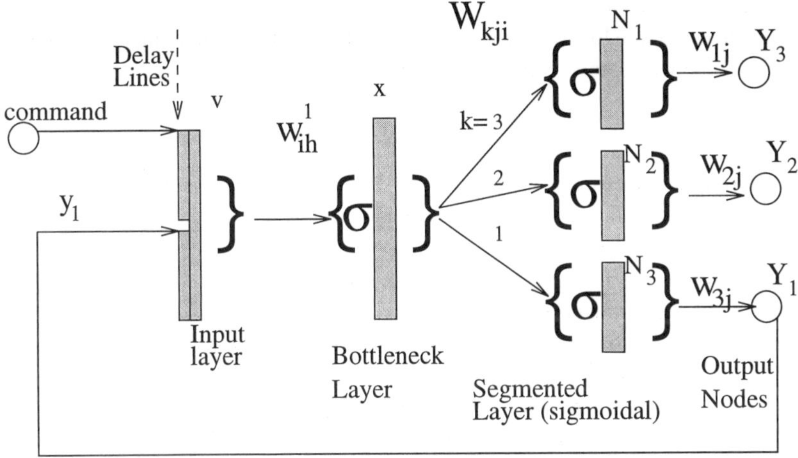

Figure 4: Reconstruction mode.

We are trying to capture both plant dynamical structure and plant state; structure and parameters are stored as weights, states as bottleneck-layer node values. State information stored in the delay lines is projected down to the bottleneck layer.

Figure 3 shows the Identifier in training; it is strictly feedforward. The trained net of Figure 4 provides a system model by replacing $y(t)$ from the plant with Y_3. (For autonomous systems the net can be initialized by choosing values for the v nodes and using all of the Y_k to load the delay line, with c = 0.) Once initialized, only the top output Y_3 is fed back to the beginning of the delay line. For controlled behavior, $c(t)$ arbitrary, it is necessary only to start at steady-state values of control and output to initialize, so Y_1, Y_2 can be suppressed from the trained net if unneeded.

The use of more samples than states may be needed for nonlinear observability; this is well known in the Chaos community from the work of F. Takens, and in the control community from the Aeyels theorem[13] on generic observability. A.U. Levin[9] has pointed out the importance of observability of sampled systems, in the context of nonlinear prediction with neural nets. Stability of the plant is useful since only a finite region in its state space can be learned. Under such conditions, the Identifier approximates state transition dynamics and output maps.

The Identifier, once trained, can be used to reproduce the dynamic behavior of the plant either in autonomous or commanded mode. For instance Figure 2 shows sample results for the same varying throttle command given to our computer simulation of a Saturn engine and to an Identifier trained as in Figure 3, starting at steady-state values so that the delay-lines can be initialized successfully. In the trained Identifier, outputs from the neural network are fed back to the delay-line inputs as shown in Figure 4.

We have used the Identifier with a wide variety of systems, including a commanded version of the well-known van der Pol oscillator:

$$\ddot{y} + A\dot{y}(y^2 - 1) + y = c(t).$$

The output waveforms for the van der Pol oscillator are accurate to within 99% accuracy in amplitude and frequency as it approaches its own stable limit-cycle (see Figure 5).

It should be emphasized that this figure shows reconstructed dynamics, not just prediction based on the past of signals as is often shown in the ANN literature. Other methods of prediction are available in the Chaos literature involving algorithms to create local approximations to lag-coordinate transition maps for chaotic attractors; they cannot capture the transient dynamics nor input-output behavior.

The trained network of Figure 4, which is now recurrent, can be used to reproduce the dynamic behavior of the plant either in autonomous or commanded mode. Using our automotive engine model as plant, its output variables are reconstructed by the Identifier to within 1% of those of the simulation. We have examined a problem important for real-time uses of neural networks in learning system dynamics: *Constant training in a limited part of the state space can distort response of the network for states which are rarely visited.*

A solution to this problem of "overtraining" was found by looking at systems with transient behavior leading into a limit-cycle, where the transient behavior is reconstructed less accurately. The solution is to provide a dead-band for updating; that is, choose a number ϵ (between 0 and 1) and use the rule:

If $|Y(t) - y(t)| > \epsilon$ update the weights.

This technique has previously been found to be necessary[16] in the use of adaptive critic controllers which train on-line to avoid overtraining one area of the network mapping while simultaneously degrading another area.

Figure 5 shows, for an input pulse chosen at random in the training set, the response of the van der Pol system and the response of its neural net simulation; C is the control pulse. The error for both transient and steady state is uniform; without the dead-band the first (transient) cycle would not have been adequately learned.

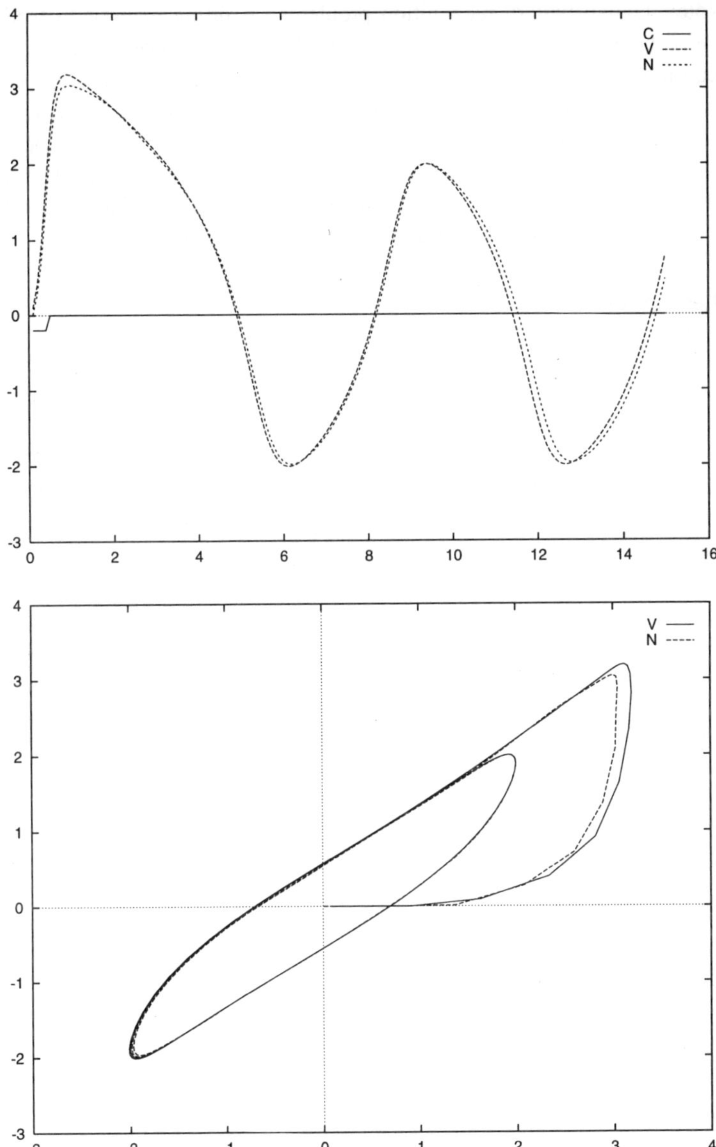

Figure 5: Time and phase-plane responses of van der Pol oscillator and Identifier to input pulse C.

Note: Unlike pattern recognition problems, in fitting maps with feedforward networks the error does not approach zero. As training proceeds (and one adjusts the learning rates to reduce error variance) it is essential to check the procedure by running the Identifier, with typical inputs c, in recurrent mode and comparing its output graphically with the corresponding plant output. We have developed a suite of software that produces output files which make this easy; the freeware program gnuplot[17] is convenient for examining these records visually and looking at the error history; gnuplot was used to produce the graphs in this paper.

6. Lead-Acid Battery

For recent work on electric vehicles it was necessary to develop an Identifier for a lead-acid storage battery; to choose a good architecture (sampling intervals, number of samples, number of nodes in each layer) we first need to develop a computer simulation of a lead-acid battery with which to try the Identifier. The dynamics of such a secondary battery is given by electrochemists[18] in the form of nonlinear partial differential equations with nonlinear boundary conditions, describing transport and diffusion of ions (Hydrogen, Sulfate, etc.)

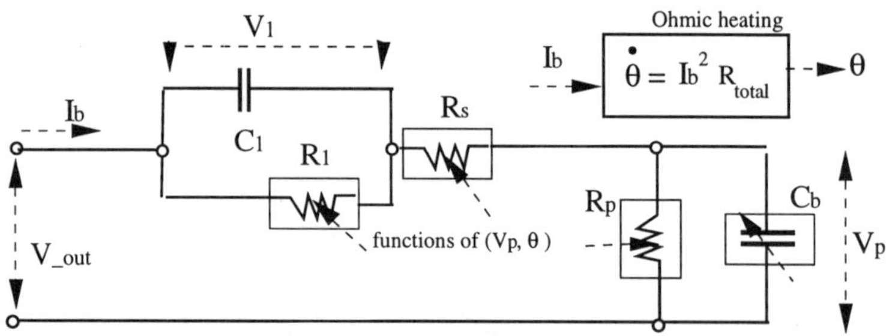

Figure 6: Lead-acid battery equivalent circuit.

For control engineering purposes, lumped parameter models have evolved[19]. For this model, an equivalent circuit is shown in Figure 6. The corresponding simulation diagram and a dynamic neural network identifier that has been found to approximate the dynamic response of the simulation (and of available battery data from the literature[20] or manufacturers) are shown in Figure 7. The simulation schematizes integral equations describing the evolution of the voltages V_1, V_p on the capacitors and the battery temperature θ. $V_o = V_1 + I_b R_s + V_p$ is the nominal battery voltage as measured across its terminals. V_p is related[20] to the state of charge (SOC) of the battery, the remaining fraction of ampere-hour capacity.

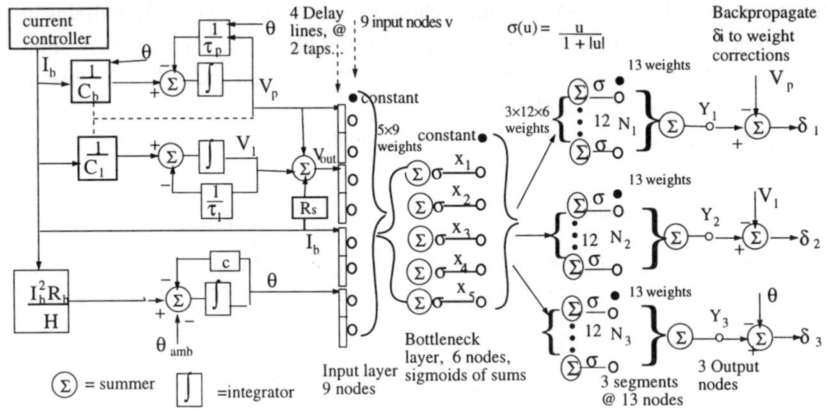

Figure 7: Learning Battery Dynamics.

The battery resistances and capacitances are highly nonlinear (exponential of power-law) functions of temperature θ, V_p and input current I_b, so the number of sigmoidal nodes may be expected to be large. The neural network shown in Figure 7 was obtained after extensive experimentation. The battery circuit with which we work is of the form

$$\dot{z} = f(z, c), \quad y = h_1(z) + ch_2(z)$$

where

$$z = (V1, V_p, \theta), \quad y = (V_{out}, V_p, \theta)$$

(a very nonlinear lag-lead network in which the control directly influences the output) and the simulation[19] has three-dimensional state z, the bottleneck layer must therefore have more than three nodes; in this study five nodes gave good results. The explicit formula for the bottleneck layer is

$$N_{kq} = \sigma(W_{kq0} + \sum(W_{kqj}x_j)$$

and the following subnets, one for each of the three outputs, are given by

$$N_{kq} = \sigma(W_{kq0} + \sum(W_{kqj}x_j).$$

The three outputs are given by $Y_k = w^3_{k0} + \sum_{q=1}^{12} w^3_{kq}N_{kq}$.The number of nodes, $Q = 12$, in each of the segments (N_1, N_2, N_3) was determined experimentally to achieve the desired accuracy of 0.1 volt in the dynamic reconstruction. The error $\delta_i = Y_i - y_i$, $1 \leq i \leq 3$; training to minimize expected prediction error $J = E\sum_1^3(\delta_i^2)$ is by a fast backpropagation algorithm developed by NeuroDyne—a stochastic approximation method using a large set of randomly chosen current inputs $c(t)$. Initializing the system near an equilibrium, always required for the Identifier, is particularly simple here since with zero current the internal state changes (self-discharge) are microscopic for low-maintenance lead-acid batteries.

Figure 7 shows the Identifier in training mode; the neural network is strictly feedforward.

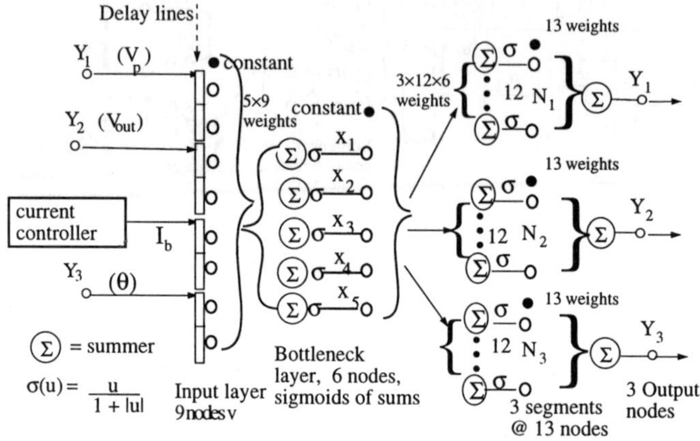

Figure 8: Battery Dynamics Reconstruction.

After training, the net—as in Figure 8—provides a system model by replacing $y_i(t)$ from the plant with Y_i. (For autonomous systems the net can be initialized by choosing values for the v nodes and using all of the Y_k to load the delay line, with $c(t) \equiv 0$.)

The network was trained by presenting random-amplitude step inputs and triangle-wave inputs (with random period and amplitude) to achieve battery discharge, with current cutoff when the voltage V_{out} reaches a lower limit of 10.2 volts. The results are shown in Figure 9 which shows (in volts) the nominal battery output V_{out} and Identifier reconstruction of it, V_N; Figure 10 shows the voltage V_p across C_1 (proxy for SOC) as well as its reconstruction from the Identifier V_{pN}. Also shown is the commanded current $c = I_b$, in units of 10 amperes.

7. Controller Development

We will apply a neural adaptive control technique to perform real-time planning and optimization of auto engine performance trajectories and measures of emissions. We hope to employ them also to optimize battery charging and energy management in electric vehicles.

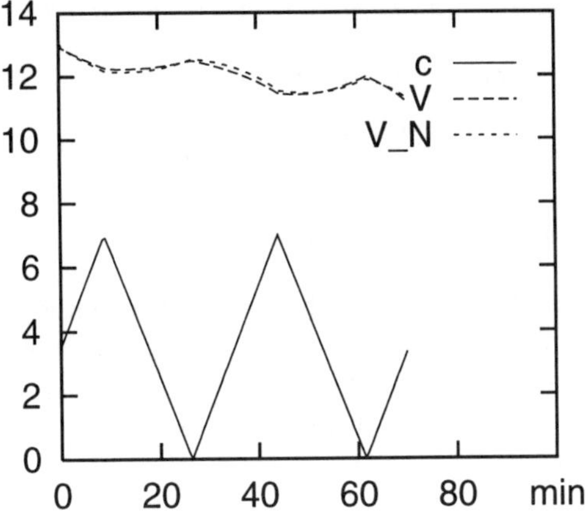

Figure 9: V_{out} and Identifier reconstruction V_N.

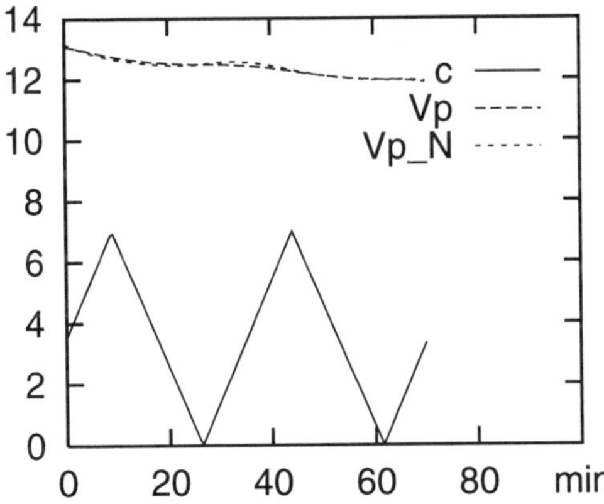

Figure 10: SOC V_p and Identifier reconstruction V_{pN}.

Dynamic programming is based upon Bellman's principle of optimality. Heuristic Dynamic Programming (HDP) techniques which may provide a potential solution to the curse of dimensionality and nonlinear system identification problems traditionally experienced with dynamic programming approaches through the use of supervised learning methods to approximate the functional equation. The method used to solve classical nonlinear optimization problems is to determine the optimal control $u*$ for a nonlinear system shown in Equation 2 that will minimize the performance measure in Equation 3.

$$\dot{x}(t) = f(x(t), u(t), t), \tag{2}$$

$$J = h(x(t_f), t_f) + \int_{t_0}^{t_f} g(x(t), u(t), t)dt. \tag{3}$$

7.1. *Approximate Dynamic Programming*

To demonstrate how our approach utilizes the nonlinear system identification of neural nets and robustness characteristics of optimal control theory, we will begin by describing our implementation of dynamic programming and then show our extensions for nonlinear control using neural network based on approximate dynamic programming (or adaptive critics).

The adaptive critic provides a compact, computationally efficient structure for calculating the cost to go in real-time, thus removing the "curse of dimensionality". The critic network also implicitly provides a nonlinear model for calculating the Jacobian for adapting the controller. The controller objective function at time stage k over a horizon N is:

$$J = \sum_{i=1}^{N} \lambda_i \frac{X_{k+i}^2}{2},$$

where λ_i is a weighting factor. It can be either a forgetting factor with $\lambda_i = e^i$, where $0 < e < 1$, or as an alternative λ_i can be set as $1/N$. J may be decomposed as:

$$J = \sum_{i=1}^{N} \lambda_i J_i,$$

where

$$J_i = \frac{X_{k+i}^2}{2}.$$

We can then compute the derivative with respect to the control (u_k) at stage k for each of the individual J_i.

7.2. *Optimal Control Computation for a Horizon of N*

The optimal control at stage k is computed in two steps. At the first step, a trial optimal control based on horizon of one is computed for each step within the horizon N. In the second step, an iterative search on trial control seeks the optimal control.

At time stage k, the optimal u_k at a horizon of one can be computed by a search through the forward dynamics model using the one step objective function J_1. The search may be conducted using a gradient search. The algorithm is almost identical to the back propagation method for computing the optimal weights except that here the free parameter is the control u and the weights are clamped. If the control is a one dimensional quantity, a direct line search to obtain the u that minimizes J is preferred. The state at $k+1$ can be estimated using the forward model, the windowed state input and the trial control u_k. This same procedure is repeated at subsequent time stages $k+i$ within the horizon to produce a trial u_{k+i} and an estimated state at $k+i+1$. Using gradient search, the optimal control u_k for a horizon of N is:

$$u_k = u_k(\text{trial}) - \partial J/\partial u_k.$$

The term $\partial J/\partial u$ can be decomposed as

$$\frac{\partial J}{\partial u_k} = \lambda_1 \frac{\partial J_1}{\partial u_k} + \lambda_2 \frac{\partial J_2}{\partial u_k} + \cdots + \lambda_N \frac{J_N}{\partial u_k},$$

where $\frac{\partial J_1}{\partial u_k}$ is the gradient for horizon of one.

As u is updated, each u_{k+i} within the horizon is also updated based on $\frac{\partial J}{\partial u_{k+i}}$. The new sequence of u is applied to the forward dynamics model to produce a new sequence of predicted state and a new value for the objective function. The above procedure of correcting for the control is repeated until either the value of u_k converges or a specified number of iterations is exceeded.

8. Concluding Remarks

The system theoretic study of identification via neural networks and delay-lines has many open problems, some of which have been alluded to here. The chief of these is to gain a better understanding of the dynamics of weight-training itself, which even in the case of linear adaptive filters has been the subject of serious research[22].

The Identifier architecture described above lends itself to prediction as much as to reconstruction, like many previous neural network methods, and we have recently begun using it for that purpose with the automobile engine, toward the goal of maintaining undesirable emissions below stated levels. Such prediction methods may lead to new general approaches to nonlinear filtering once the stochastic aspects have been addressed. Such problems may be attractive to the Statistics and Control Theory Communities, which have both been taking an active interest in feedforward networks and their training algorithms. Besides the optimization schemes we have addressed, there are possibilities of using Identifiers in system stabilization as well as constraint satisfaction.

We have not attempted to discuss observability and observers here, but refer the interested reader to the work of Aeyels[13], Levin[9] and very recently Moraal[21] who use lag coordinate methods.

9. Acknowledgements

A portion of this research was supported by NSF Grant ECS 92-16530 to a team headed by Emil Hanzevack of the University of South Carolina and NSF Grant DMII 94-61400 to NeuroDyne, Inc.. Two members of the first team, Douglas Foster of Adrenaline Inc. and Michael Sampson of MIT (now at General Motors Research), did most of the programming for the engine simulation, especially in mapping Ford and Saturn engines to the Hendricks-Sorenson[2] mean-value model. Early versions of some sections of this paper have been given at the American Control Conference, the Institute for Systems Research and the NACT I Workshop; the questions and comments from those audiences have contributed considerably to its development.

10. References

1. L. Ljung, *System Identification: Theory for the User* Englewood Cliffs, NJ : Prentice-Hall, 1987.

2. E. Hendricks and S. C. Sorenson, "Mean value modelling of spark ignition systems," SAE Tech. Paper 900616 Int. Cong. & Expo., Detroit, MI, February 26–March 2, 1990.

3. W. E. Larimore, "Identification and Filtering of Nonlinear Systems Using Canonical Variate Analysis," in *Nonlinear Modeling and Forecasting* (M. Casdagli and S. Eubank, eds.), New York: Addison-Wesley, 1992.

4. T. Sauer, "Time Series Prediction Using Delay Coordinate Embedding," in *Predicting the Future and Understanding the Past: A Comparison of Approaches* (A. Weigend and N. Gershenfeld, eds.), Reading, MA: Addison-Wesley Pub. Co., 1994.

5. D. L. Elliott, "Reconstruction of Nonlinear Systems Using Delay Lines and FeedForward Networks," *Proceedings of the 1995 American Control Conference*, Seattle, Washington June 21–23, 1995; Paper WM-13.5.

6. P. M. Clarkson, *Optimal and Adaptive Signal Processing* Boca Raton, Florida, CRC Press, 1993.

7. P. Werbos, T. McAvoy and H. T. Su, "Neural Networks, System Identification, and Control in the Chemical Process Industries," in *Handbook of Intelligent Control* (D. A. Sofge and D. A. White, eds.), New York: Van Nostrand Reinhold, 1992.

8. S. Chen, S. A. Billings, "Neural networks for nonlinear dynamic system modeling and identification," *International Journal Control* **56:** No.2, 319-346, 1992.

9. A. U. Levin, *Neural Networks in Dynamical Systems: a System Theoretic Approach*, Ph.D. Thesis, Yale University, November 1992.

10. A. U. Levin, K. S. Narendra, "Identification of nonlinear dynamical systems using neural networks," to appear in *Neural Networks for Control* (D. L. Elliott, ed.), Ablex Publishing Corporation, Norwood, NJ, 1996.

11. D. L. Elliott, "A Better Activation Function for Artificial Neural Networks," ISR Report 93-8, University of Maryland, Jan. 1993.

12. P. Baldi and K. Hornik, "Learning in Linear Neural Networks: A Survey," *IEEE Transactions on Neural Networks* **6**, No. 4, 837-858, 1995.

13. D. Aeyels, "Generic observability of differentiable systems," *SIAM Journal of Control and Optimization* **19** 595-603, 1981.

14. T. Long, D. A. White and D. L. Elliott, "Enhanced Performance Seeking Control using Neural Networks", Final Report on Contract No. NAS3-26911, NASA Lewis Research Center, 1993.

15. F. Takens, "Detecting strange attractors in turbulence," in *Dynamic Systems: Warwick, 1980* (D. Rand and L. S. Young, eds.), Berlin: Springer Lec. Notes Math. **898**, 366–381 (1981).

16. D. A. Sofge and D. A. White, "Applied Learning-Optimal Control for Manufacturing," Chapter 9, pp. 259-279, *Handbook of Intelligent Control: Neural, Fuzzy, and Adaptive Approaches* (D. A. White and D. A. Sofge, eds.), Van Nostrand Reinhold, New York, 1992.

17. T. Williams and C. Kelley, *gnuplot* for Sun, Macintosh, etc.; for information e-mail to info-gnuplotdartmouth.edu.

18. H. Gu, T. V. Nguyen and R. E. White, "A Mathematical Model of a Lead-Acid Cell," *Journal of the Electrochemical Society* **134:** 2953-2960 (Dec. 1987).

19. Z. H. Salameh, M. A. Casacca and W. A. Lynch, "A Mathematical Model for Lead-Acid Batteries," *IEEE Transactions on Energy Conversion* **7:** (1) 93-97, March 1992.

20. Bode, Hans, *Lead-Acid Batteries,* New York: John Wiley & Sons, 1977.

21. P. E. Moraal and J. W. Grizzle, "Observer design for nonlinear systems with discrete-time measurements," *IEEE Transactions on Automatic Control* **40**, No. 3, 395-404, 1995.

22. V. Solo, "The Limiting Behavior of LMS," *IEEE Transactions on Acoustics, Speech, and Signal Processing* **37**:1909-1922, 1989.

HOW TO ADAPT IN NEUROCONTROL
A DECISION FOR CMAC

W. S. Mischo
*Fachgebiet Regelsystemtheorie und Robotik**
Technische Hochschule Darmstadt†
Landgraf-Georg-Str. 4
D-64283 Darmstadt, Germany
E-mail: wsm@rt.e-technik.th-darmstadt.de

ABSTRACT

CMAC is one of the first neural networks successfully applied to real world control problems. Its ability to locally "generalise" an input/output behaviour based on a non-linear input point processing and a linear algorithm for modifying internal states guarantees fast convergence to an implicit model. Thus it is a good candidate for learning control. In this Chapter some criteria for learning elements in control are first presented, and various approaches are judged against these criteria. Then the features of CMAC (in a sense a winner of this comparison) are described together with some improvements, one of them probably being a new result. Finally, guidelines for a CMAC hardware realisation are discussed, as they were used for the implementation of an ASIC, which now is available in a first version.

1. Introduction

1.1. Neurocontrol Is Learning Control

The term *Neurocontrol* has been established for control systems, which use neural networks as a device for implementation of a process model (or other process related modelling tasks), a controller, or even both. That usually involves solving of a given control problem, for which Control Theory has to provide the methods and not only the elements of a system, but also the system as a whole has to be taken into account. From this point of view, each system, which controls a process without *a priori* process knowledge, like e.g. a physical model, can be considered a Neurocontrol system. That is because such systems are in a situation similar to a living thing, which "initially" (after being born) knows nothing about its environment. It must learn to control its environment only by observing the input/output behaviour. An excellent aid for this task is its ability to generalise, i.e. to recognise similarities of situations in the environment. Thus Neurocontrol is also *Learning Control*. And it seems to be possible to implement a learning control loop with "non-neural" methods as well.

*member of the Control Systems Theory and Robotics group
†Darmstadt University of Technology

1.2. Why Learning Control?

Learning control is a method to deal with processes, which are *unknown in most aspects*, except for the process order and a sufficient set of inputs and outputs (controllability and observability). In the linear case there exists a powerful theory for both analysis and synthesis of control loops, but in general processes behave *non-linearly* and they may be *time-varying*.

An additional motivation for applying learning control can also be that mathematically oriented modelling needs a great effort in terms of engineering know-how, prototyping of processes etc. Anyway, there are processes, where conventional modelling approaches fail (e.g. in Biotechnology). Such processes are normally driven by a human operator. Sometimes people try rule based control approaches, or even learning control using the human operator's control strategy (which has been learned) as an initial solution. Even if a mathematical model from a process has been obtained, it can be so complex (or the process so fast), that calculation time of the modelling algorithm becomes critical. Finally, a model for the prototype of a process can become worthless due to ageing or the variation of components.

By identifying the modelling strategy with elements used in implementing it, several basic methods for "learning elements" can be given:

- multi-variable polynomial approximation,
- artificial neural networks with sigmoid-type neurons, adapted by gradient descent methods, commonly called *error backpropagation*, which is the reason, why people simply call it *backprop(agation) networks*,
- adaptive fuzzy rule bases, which facilitate incorporation of qualitative process knowledge,
- radial basis function neural networks,
- locally interpolating neural networks, such as CMAC and its derivations,
- interpolation memories following mathematical interpolation methods (n-order surfaces, B-splines, etc.).

Also the structure in the control loop is important as a criterion of difference:

- a simple control loop with an off-line adapted "intelligent" (i.e. mainly non-linear) controller,
- pilot control by an inverse process model—on-line learned or off-line trained with disturbance suppression and basic stabilization by a linear controller; Figure 1 shows the on-line variant, which has been used in [1],

- model predictive control, called LERNAS by the author of [2]: an on-line learned model of the unknown process is used to optimize controller outputs, which are then trained into a learning controller, Figure 2.

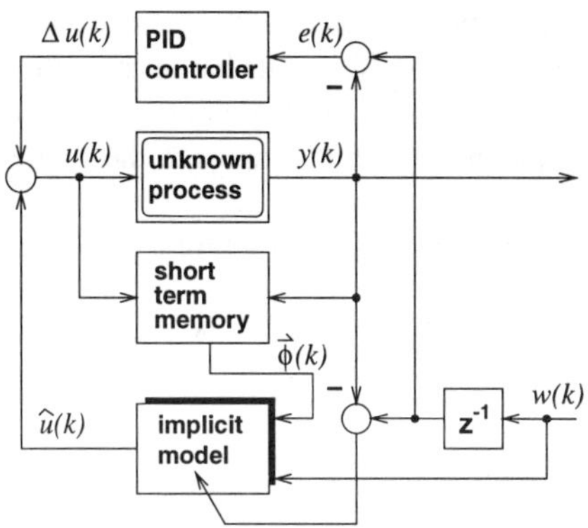

Figure 1: Intelligent control loop using an inverse process model.

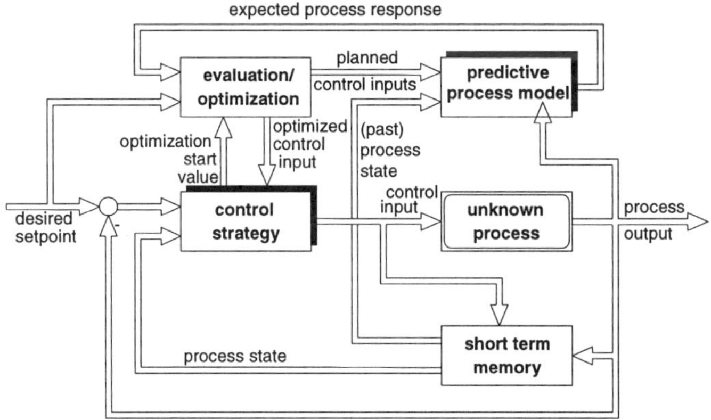

Figure 2: The learning control loop LERNAS.

2. Intelligent Components and Their Principles

2.1. *Prerequisites*

In this Chapter only discrete-time processes are considered. Also, due to implementation, the inputs and outputs are assumed to be discrete, though this is not discussed before the section about CMAC and its hardware realisation. Only single output behaviour is analysed, because multiple output systems can be transformed into a set of single output systems (by adding inputs to the single output models to model coupling between them).

Given these assumptions, the problem is to find for a process y an estimate $\hat{y}$:

$$\begin{aligned} \hat{y} \;=\; & f(u_1(k-1), \ldots, u_1(k-n_1), \ldots u_r(k-1), \ldots, u_r(k-n_r), \\ & y(k-1), \ldots, y(k-m)) \end{aligned} \tag{1}$$

such that

$$\sum_k \|y(k) - \hat{y}(k)\| \to \min \tag{2}$$

in an already defined norm $\|\cdot\|$ (usually L_2).[a]

2.2. *Overview*

2.2.1. Mathematical Framework

Intelligence is an attribute of systems which are able to learn. In this context intelligence is not a well defined property, but a mixture of abilities, which are satisfied to different degrees by structures whose basic behaviour can be given as:

$$\hat{p}: \vec{s} \;\mapsto\; \hat{p}(\vec{s}) = f(\vec{x}, \vec{\varphi}(\vec{s}, \vec{x})), \tag{3}$$

$$\vec{\varphi}: \vec{s} \;\mapsto\; \vec{\varphi}(\vec{s}, \vec{x}), \tag{4}$$

where $\vec{x}$ are the inner states of the model and $\vec{\varphi}$ in principle a mapping of the input $\vec{s}$ to a function vector, which represents the essential features of the model; f is the output function, usually a linear combination of the functions $\vec{\varphi}$ and sometimes an additional scalar function. Training or adaptation is performed by a rule, which can be expressed as a state transition from discrete time k to $k+1$:

$$\psi: \vec{x}^{(k)} \mapsto \vec{x}^{(k+1)} \tag{5}$$

---

[a]In some structures a penalty term for the model complexity is added in (2). Because this is not appropriate for all of them (and in general complicates the training/adaptation situation), it should only be considered, where authors explicitly introduced it.

(recall that we are dealing with discrete-time, single output processes only). In general, this training function should satisfy:

$$\sum_{\vec{s}_j \in S'} \| f(\vec{x}^{(k+1)}, \vec{\varphi}(\vec{s}_j, \vec{x}^{(k+1)})) - p_j \| < \sum_{\vec{s}_j \in S'} \| f(\vec{x}^{(k)}, \vec{\varphi}(\vec{s}_j, \vec{x}^{(k)})) - p_j \| \quad (6)$$

for a given subset S' of the training set $S = \{(\vec{s}_1, p_1), \ldots, (\vec{s}_t, p_t)\}$, which can be seen as a set of t observed input/output values of the behaviour to be modelled. In principle (6) specifies an optimization problem with both sides being successive values of the quality criterion. As mentioned before this quality criterion can be extended to include model complexity.

Finally, this training procedure should satisfy the obvious requirement:

For any $\epsilon > 0$ there exists N such that

$$\| f(\vec{\varphi}(\vec{s}_j, \vec{x}^{(N)})) - p_j \| < \epsilon \quad (7)$$

for any $(\vec{s}_j, p_j) \in S$.

For this an adequate training schedule (choice of subsets S') is necessary, which depends on the application.

This general convergence behaviour cannot be guaranteed in any training situation (given a training set S and a modelling strategy) mainly for two reasons:

- disturbances of the training outputs p_j, resulting in $p_i \neq p_j$ for $\vec{s}_i = \vec{s}_j$,
- the strategy leads to an unavoidable remaining error, such that in the right side of (7) a term $e_c(S)$ has to be added, where index c represents the used modelling strategy.

In the first case it is important for the strategy to implement noise suppressing training strategies, like stochastic approximation. A simple technique employing counters for the modifying accesses to the inner states will be presented later for CMAC.

The second case arises for one or more of the following reasons:

- the number of internal states x_i is not sufficient for the concrete modelling problem;
- the functions $\vec{\varphi}$, f (especially $\vec{\varphi}$) represent "inappropriate" knowledge about the function to be modeled. In principle, the approach presented in Section 2.3.1. deals with this problem for improving the approximation result;

- the training algorithm is not able to detect a (sufficiently "deep") minimum in the error surface, especially in cases where the relationship between inputs/inner states and error is strongly non-linear, as e.g. for backpropagation networks (see Section 2.3.2.).

Given the above constructs, the modelling strategies can be distinguished by the input to states mapping $\vec{\varphi}$, the output function f, and the training method ψ and its convergence (speed, remaining error $e_c(S)$). These components are of course not independent of each other. For mapping $\vec{\varphi}$ there are two cases:

- each φ_i covers the whole input space; this can be called a global approximation approach;

- the φ_i affect only (compact) subsets of the input space, which is part of a local approximation (or interpolation) procedure.

In most cases, the input mapping is not dependent on the inner states: $\vec{\varphi} = \vec{\varphi}(\vec{s})$. In the cases, where also the inner states $\vec{x}$ are arguments of $\vec{\varphi}$, the training and the input processing are in general more complicated (see Sections 2.3.4., 2.3.6.).

The number of internal states for local modelling strategies is greater (usually much greater) than for global methods. This also has effects on the training process: due to the larger number of degrees of freedom local strategies converge more rapidly than global methods.

As for the output function f, the most important property is its linearity or otherwise, linearity obviously being an advantage.

For the training methods ψ, the first important criterion is the size of the sets S' used for a training step satisfying the convergence criterion (6). Only the training algorithms, where it is possible to choose $|S'| = 1$, guarantee optimal on-line behaviour, whereas training procedures requiring $|S'| > 1$ are slow, because process model updates are delayed. Also, the calculation effort increases with the size $|S'|$; very large sizes suggest off-line training.

2.3. *Specific Modelling Strategies*

2.3.1. Polynomial Approximation [3]

For the sake of simplicity only the SISO case is discussed.

From the Stone-Weierstraß theorem it follows that it is possible to model a process y by a Kolmogorov-Gabor polynomial, the complete q-order multi-variable polynomial of all process inputs:

$$\begin{aligned}
\hat{y}(k) &= \bar{y} \\
&+ \textstyle\sum_{i_1=0}^{n} b_{i_1} u(k-i_1) \\
&+ \textstyle\sum_{i_1=0}^{n} \sum_{i_2=i_1}^{n} b_{i_1 i_2} u(k-i_1) u(k-i_2) \\
&+ \ldots \\
&+ \textstyle\sum_{i_1=0}^{n} \cdots \sum_{i_q=i_{q-1}}^{n} b_{i_1 \ldots i_q} u(k-i_1) \cdots u(k-i_q) \\
&+ \textstyle\sum_{j_1=1}^{m} a_{j_1} y(k-j_1) \\
&+ \textstyle\sum_{j_1=1}^{m} \sum_{j_2=j_1}^{m} a_{j_1 j_2} y(k-j_1) y(k-j_2) \\
&+ \ldots \\
&+ \textstyle\sum_{j_1=1}^{m} \cdots \sum_{j_q=j_{q-1}}^{m} a_{j_1 \ldots j_q} y(k-j_1) \cdots y(k-j_q) \\
&+ \textstyle\sum_{i_1=0}^{n} \sum_{j_1=1}^{m} c_{i_1 j_1} u(k-i_1) y(k-j_1) \\
&+ \textstyle\sum_{i_1=0}^{n} \sum_{i_2=i_1}^{n} \sum_{j_1=1}^{m} c_{i_1 i_2 j_1} u(k-i_1) u(k-i_2) y(k-j_1) \\
&+ \ldots \\
&+ \textstyle\sum_{i_1=0}^{n} \cdots \sum_{i_{q-1}=i_{q-2}}^{n} \sum_{j_1=1}^{m} c_{i_1 \ldots i_{q-1} j_1} u(k-i_1) \cdots u(k-i_{q-1}) y(k-j_1) \\
&+ \ldots \\
&+ \textstyle\sum_{i_1=0}^{n} \cdots \sum_{i_g}^{n} \sum_{j_1=1}^{m} \cdots \sum_{j_h}^{m} c_{i_1 \ldots i_g j_1 \ldots j_h} \; u(k-i_1) \cdots u(k-i_g) \\
&\qquad\qquad y(k-j_1) \cdots y(k-j_h) \\
&+ \ldots \\
&+ \textstyle\sum_{i_1=0}^{n} \sum_{j_1=1}^{m} \cdots \sum_{j_{q-1}=j_{q-2}}^{m} c_{i_1 j_1 \ldots j_{q-1}} u(k-i_1) y(k-j_1) \cdots y(k-j_{q-1}) \\
&+ \ldots \\
&+ \textstyle\sum_{i_1=0}^{n} \sum_{j_1=1}^{m} \sum_{j_2=j_1}^{m} c_{i_1 j_1 j_2} u(k-i_1) y(k-j_1) y(k-j_2).
\end{aligned} \tag{8}$$

The first problem arising with this approach is that the order q of the polynomial has to be estimated *a priori*. Then the coefficients of the polynomial can be estimated using the LMS algorithm. However, another problem emerges: the number of coefficients, γ, grows with process orders n, m and the order of the polynomial q by

$$\gamma = \frac{(n+m+q)!}{(n+m)!q!}. \tag{9}$$

Some values of γ are given in Table 1 from which it can be seen that even for moderate values of n, m and q the parameter estimation becomes a time consuming computing process, because it leads to the problem of calculating a Moore-Penrose inverse of a $\gamma \times \xi\gamma$ matrix (in [4] $\xi = 3 \div 5$ is suggested).

In [3] methods have been proposed to eliminate terms in (8) by means of statistical significance measures and quality criteria. Additionally, a transformation for the input signal products in (8) has been given with the goal of getting a decoupled (orthogonal) set of input products, the *signal vector*. Still, this means an enormous calculation effort for an initial model evaluation. Therefore, a transputer net was proposed in [4] to overcome this computing bottleneck.

$q \backslash n$	2	3	4	5
2	15	28	45	66
3	35	84	165	286
4	70	210	495	1001
5	126	462	1287	3003

Table 1: Some values of the number of parameters in the Kolmogorov-Gabor polynomial for process orders $n = m = 2, \ldots, 5$ and polynomial orders $q = 2, \ldots, 5$.

For parameter estimation purposes (8) can be rewritten as:

$$\hat{y}(k) = \vec{x}^T \vec{v}(k) \tag{10}$$

with $\vec{v}(k)$ being the (transformed) signal vector and $\vec{x}$ the parameter vector to be estimated (the inner states of this model). Thus the input to inner state mapping $\vec{\varphi}$ can be seen as the calculation of the components in $\vec{v}(k)$, with the products representing the basis functions of the model. Output function f is then the linear combination of the parameter vector $\vec{x}$ and the signal vector $\vec{v}(k)$.

From the form of the basis functions it is not surprising that the modelling is best for non-linear behaviour given by a polynomial.[b] Other non-linearities (e.g. a saturation function) can only be modelled by high orders q or only in a small interval of the input domain. Also, the (initially) large number of internal states[c] requires a large number of data from the process to be modelled ($\xi\gamma$ data points, see above). Due to numerical stability the data must originate from different process setpoints/states; otherwise the matrix would degenerate. It should therefore be possible to safely drive the process through several setpoints. For this reason (and also because an estimate of the polynomial order has to be given in advance) it is non-trivial to apply this strategy to a process, whose behaviour is not well known.

2.3.2. Neural Nets with Sigmoid-Type Neurons

There is a large list of references for this topic: for basics see [5]; for the control theory/signal processing context [6,7,8]; for a control-orientated discussion of approximation capabilities [9].

The basic structure is depicted in Figure 3. In contrast to the other modelling strategies presented here, it is not possible to separate functions $\vec{\varphi}$ and f for the whole

[b]Though this is not very often encountered in practice.

[c]It is worth noting that their number is varying in the model. This is expensive, because variations in the number of parameters require new adaptation of the parameters. For this reason the mentioned orthogonal transformation of the products in (8) has been introduced.

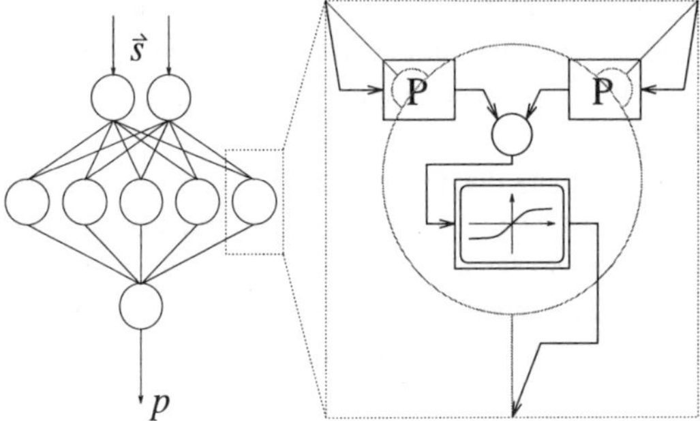

Figure 3: A two layered feedforward neural network with sigmoid-type neurons. On the left side a complete net for a 2-1 functional mapping is shown. It is considered to be two layered, because the "input layer" (top two neurons) only get the inputs to feed them into the next layer. On the right side a scheme for a single neuron with two inputs is given.

net. However, it is possible to do it for a single neuron (cf. Figure 3): $\vec{\varphi}$ simply is a linear combination of the weighted inputs. For sigmoid-type neurons f is given as:

$$f: z \mapsto \frac{1}{1+e^{-(z-\vartheta)}}, \tag{11}$$

where z is the sum of the weighted inputs and ϑ a bias value being an additional inner state for each neuron in the net.

Due to the non-linearity of the neuron outputs, the training has to be organized as a gradient descent method. For the training set S (normally just $S' = S$ is chosen) the error is determined as shown in Section 2.2.1.. The evaluation of the gradient leads to a localised training rule for each neuron's weights and bias.

There are several issues arising for this type of neural networks, especially if their use in control is planned. First, the convergence speed of the training algorithm can strongly depend on parameters used for its implementation (e.g. the learning rate, or step width, as it is called in the gradient descent algorithm). In general, even if good values for these parameters are chosen, the convergence is slower than for local modelling approaches, usually too slow for on-line applications. It is even possible (in the case of wrong parameters) for the network not to converge at all. In the terminology of Section 2.2.1.: the residual error $e_c(S)$ depends not only on the training set S, but also on (in general randomly chosen) the initial state vector $\vec{x}^{(0)}$. It is possible for the training procedure to get trapped into a local minimum of the global output error resulting in a $e_c(S)$ which is too high for the application. If this situation is detected, normally the training procedure is restarted with some other $\vec{x}^{(0)}$

(where some statistics may help to find a better initial state vector [10]). This makes this approach even less attractive for on-line applications. Remarks of this kind can also be found, e.g. in [11] for a rather simple case (control of a SISO linear third order process at few different setpoints).

As a conclusion we remark that feedforward networks are universal approximators [12], but this only is an existence result for the approximation problem; it is, however, not constructive (see [9]). So far as the author knows, no training method has been given for this type of neural networks, which leads to a certain convergence with an acceptable remaining modelling error for continuous functions with restrictions, as they normally apply for control applications. There are, however, numerous suggestions for improvement or replacement of the basic error backpropagation algorithm, e.g. [13 14].

2.3.3. CMAC

Because CMAC will be treated later in detail (Section 3.), here we only note that the inputs are transformed non-linearly: $\vec{\varphi}$ is a local distribution of the input point $\vec{s}$, whose components mostly are 0, except for a certain fixed number ϱ of them being 1. This transformation has a "local" property: the indices of the components being 1 are calculated using a partitioned representation of an environment of $\vec{s}$ in the input space. In other words: each φ_i is a two-valued function, which is 0 over nearly the whole input space, except for a small compact subrange (whose position is determined by the index i), where it evaluates to 1.

The output function f is then simply the value of the linear combination of this association vector with the vector of the inner states divided by ϱ.

The essential difference between CMAC and backpropagation is that CMAC employs a non-linear (localised) input processing and linear output calculation, whereas the neurons in the backpropagation network work in the opposite way: linear (thus global) input processing and non-linear output function.

The training in CMAC can be carried out by ordinary least mean square approaches, in single-step ($|S'| = 1$) or in multi-step ($|S'| > 1$) methods, as well. Also, stochastic approximation procedures for noisy training signals (outputs p) are possible. The capability of CMAC to model even non-linear functions using linear training methods is based on its localised input transformation, leading to a locally interpolating behaviour. Of course the number of inner states required is large compared to global modelling strategies. This can be seen as a price for the fast convergence due to the local character and the linear training algorithm.

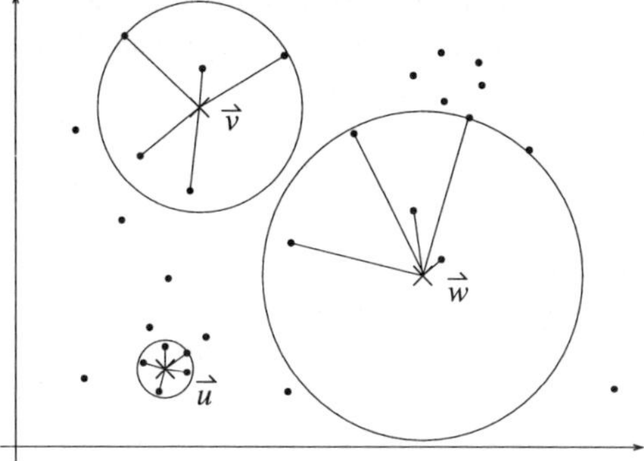

Figure 4: Interpolating Memory for two-dimensional input: the dots are the stored data points in the input domain. The interpolation areas are implicitly given by the q (here: 5) nearest neighbours of a considered input. The stored points for the requested inputs $\vec{u}$ and $\vec{v}$ are well distributed, whereas for $\vec{w}$ the memory will rather extrapolate.

2.3.4. Interpolating Memories

The problem of scattered data interpolation is well known in geodesy and graphical data processing [15]. The principle is to simply store input points with associated output values. The output for a given input point is then calculated in two stages: first a nearest neighbours search is carried out, see Figure 4. The output values y_i associated with the found input points $\vec{s}_i$ are then used to calculate the memory response. Of course the number of points q used for interpolating is at least as large as the input dimension, i.e. $q \geq n$. The simplest method for output calculation is to solve the least squares problem:

$$\begin{pmatrix} \vec{s}_1^{\,T} & 1 \\ \vdots & \vdots \\ \vec{s}_q^{\,T} & 1 \end{pmatrix} \vec{x} = \vec{y} + \vec{\epsilon} \tag{12}$$

with $\vec{\epsilon}$ being the equation system error. The output is then the projection of the presented input point $\vec{s}$ onto the hyperplane defined by the resulting vector $\vec{x}$:

$$\hat{y} = (\vec{s}^{\,T}, 1)\vec{x}. \tag{13}$$

Refinements of the basic interpolation strategy include weighting of the outputs y_i by a function of the distance between the associated input point $\vec{s}_i$ from the actually considered input $\vec{s}$. Of course also higher order interpolation surfaces are possible.

Scattered data interpolation in general is a concept well-suited for modelling (smooth) arbitrary input/output behaviour. It has the main advantage, that it automatically adapts to varying input point density.

One main problem of this approach is the dependency of the access time on the number of stored points due to the search process. The training is very simple: the observed input/output data from a process are stored in memory. However, the memory requirement for on-line application is not known *a priori* and the access times (search times) increase during operation.

In Figure 4 an additional problem is illustrated: it is possible that the data points for the interpolation are not properly distributed around the input point. In this case the output can be considered as being more likely a result of extrapolation, which may be dangerous for control applications.

In the case of disturbances acting on the training signals y_i the selection of parameter q becomes critical. If it is too small, the interpolation leads to poor modelling behaviour ("interpolation of noise"), if it is too large, the areas of influence (as depicted by the circles in Figure 4) become too large and functions with steep gradients cannot be modelled.

Methods to overcome the memory size and search time problems by input space segmentation and detailed control of the training process are given in [16]. A link between these interpolation ideas with free placement of the training inputs and the fixed grid input processing of e.g. CMAC are the spline interpolation memories/neural networks described in [17].

2.3.5. Adaptive Fuzzy Systems

A comparison of fuzzy systems with the other modelling strategies presented here may seem at first glance not entirely appropriate, but it is still possible to "translate" the fuzzy approach into the language of local interpolating strategies [17].

The fuzzy scheme is depicted in Figure 5 for the two-dimensional case. Fuzzy systems may be seen as expert systems with continuous valued logic. The basic structure is the rule base, a set of implications in the form: if input values each have a certain "fuzzy" value then the output value gets a certain fuzzy value, e.g.

`if` s_1 `is` **P** and s_2 `is` **N** `then` y `is` **BP**
`if` s_1 `is` **Z** and s_2 `is` **BN** `then` y `is` **P**

The fuzzy values for inputs and outputs are given by value range symbols, here **BN**, **N**, **Z**, **P** and **BP** have been chosen, with the *semantic* meaning[d] **b**ig **n**egative, **n**egative, **z**ero, **p**ositive and **b**ig **p**ositive, respectively.

[d] It is not obvious, what *semantic* really means. It cannot be far away from the truth, if it is seen as "understandable by a human being".

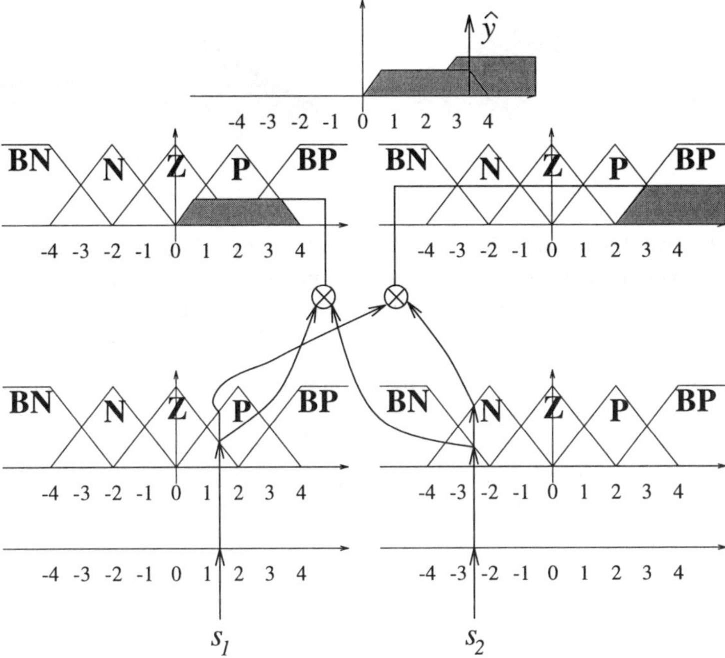

Figure 5: A simple two input fuzzy system. Two fuzzy rules are shown with their evaluation, see text.

These rules are translated into a mathematical procedure, by first introducing univariate basis functions (*membership functions*) for the above value range symbols. The responses of these basis functions on the inputs are combined by a *fuzzy relation*, i.e. the `and` in the rules above corresponds to a binary operation $\otimes$, which can be a minimum or a product of the basis function outputs. The last step is the output calculation: the basis functions for the output are scaled by the fuzzy relation response as shown in Figure 5. The areas of these scaled basis function shapes are combined to form the output, mostly by the calculation of the resulting centre of gravity.

The similarity to other local methods presented here is obvious, if the univariate basis (membership) functions are used to construct basis functions covering compact domains in the input space. Figure 6 shows possible results for the two-dimensional case.

Now, the processing of the fuzzy rule base method can be formulated as follows: each premise (the part between `if` and `then`) in the rule base corresponds to a basis function φ_i. Provided all output membership functions used in the rule base

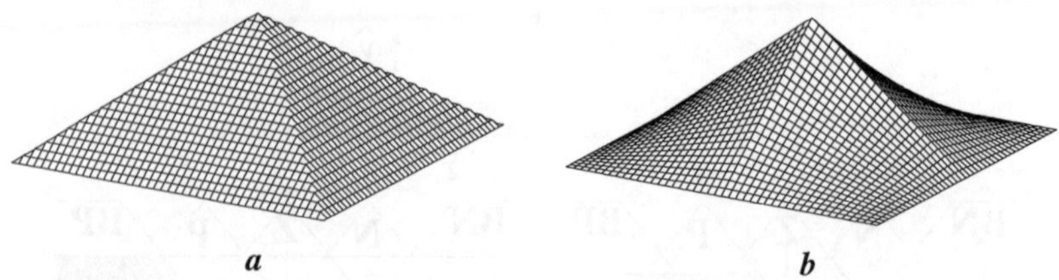

Figure 6: Two-dimensional basis functions resulting from the univariate membership functions with triangular shape: ***a*** using the minimum operator, ***b*** using the product operator for combining the outputs of the univariate functions.

are of equal width, then (also due to the linearity of the centre of gravity operation) the output $\hat{y}$ is simply a linear combination of basis function outputs[e] φ_i and the centre points of the output membership functions. Thus training or adaptation of the rule base can be carried out by moving the centre points of the output membership functions. This means that these centres are the inner states $\vec{x}$ of an adaptive fuzzy system.

There are many variations of the basic approach presented here. But the obvious advantages and disadvantages remain the same:

- Due to the semantic principle, it is possible even for persons not familiar with mathematical theory to work with it, at least on a trial-and-error base.
- As for other basis function approaches the effort grows exponentially with the input dimension. Thus adequate "generalisation" principles have to be included, to extend the influence region of a rule (a basis function), if no rules have been given for input regions around it. But this can be dangerous, if the implied hypothesis about the output surface between the given rules is not true.

2.3.6. Radial Basis Function Networks

The network structure is the same as for backpropagation networks with one hidden layer (cf. Figure 3). But, as shown in Figure 7, the function of the neurons has been changed: the non-linearity in the hidden layer now is a "radial basis function", i.e. a radially symmetric function with maximum at a certain *centre point* $\vec{c}_i$ in the input space and asymptotically decreasing to 0 with increasing distance from this centre point:

[e]More precisely: outputs of slightly modified φ_i due to non-linearity of the rescaling procedure for the output membership functions, which is normally in use.

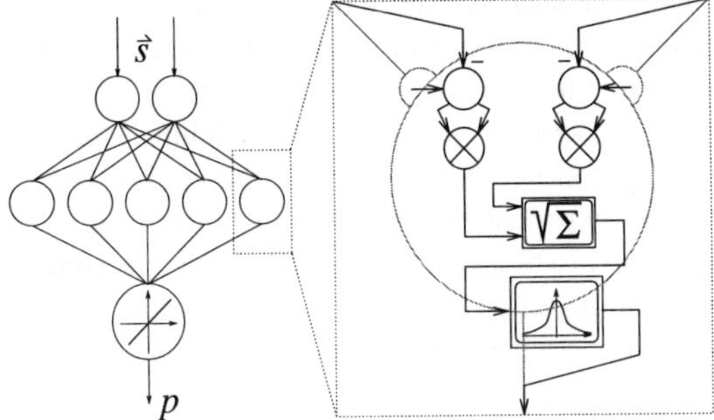

Figure 7: A radial basis function network.

$$\lim_{\|\vec{s}-\vec{c}_i\|\to\infty} \varphi_i(\vec{s}) = 0$$

as, e.g. the Gaussian "hat":

$$\varphi_i(\vec{s}) = \frac{1}{1+e^{-\|\vec{s}-\vec{c}_i\|^2}}. \tag{14}$$

As already done in the above formulation, these basis functions can be identified with the input transformation functions from Section 2.2.1.. The output neuron simply does a linear combination of the φ_i with its weight vector.

Further classification of this method depends on whether the centre points $\vec{c}_i$ are considered as being inner states (in addition to the weights of the output neuron) or not. In the latter case, the situation is comparable to other local, linear output modelling strategies (CMAC, fuzzy systems, spline interpolation).[f] But in practice the problem is to choose a set of appropriate centre points, because in general the modelling capabilities for a given training set S depend on the choice of them. Therefore an off-line evaluation is necessary to determine the $\vec{c}_i$. For this it is essential (especially for high dimensional input spaces) to have a data set, which at least has an input point distribution similar to the data during on-line operation.

In the case of varying centres $\vec{c}_i$ and if the number of the φ_i is changing, the network dynamics become complicated. For the first case [8] gives a training algorithm, whereas heuristic rules are necessary to control the number of neurons in the hidden layer.

[f]Though the basis function approach is not local in a strict sense, because the values of the Gaussian hat, (14), never reach 0. But at large distances from the centre point $\vec{c}_i$ the output is sufficiently small, such that network modifications due to training steps can be considered as having only local influence.

This is due to the fact, that such systems have too many degrees of freedom to get a mathematical solution for the training problem. Here the inclusion of a penalty term for model complexity into the training goal is of interest.

Radial basis function networks provide good modelling capabilities with smooth output, provided an appropriate centre point placement has been found. But the calculation effort grows exponentially with the input dimension. Thus for high input dimensions strategies have to be taken into account which select only appropriate basis functions φ_i for evaluation, based on a "nearest neighbours" search over the centre points $\vec{c}_i$. Networks with varying centre points and/or number of basis functions are difficult to control leading to unpredictable timing behaviour, which is not desirable for on-line application in control systems.

2.4. *Comprehensive Comparison of Modelling Strategies*

After presenting possible approaches to the problem of unstructered modelling, a final comparison (Table 2) can be done, of which the criteria are given in the following.

	CMAC	backprop nets	RBF nets	adaptive fuzzy systems	pure inter-polation	poly-nomial approxi-mation
training indication	+	–	–	–	+	–
access time	+	+/0	0	+	–	–
convergence	+	– –	+	+	+	–
calculation effort	+	–/0	–/0	+	0	– –
memory needs	–	+	0	0	–	0
noise filtering	+	+	+	+	–	+
adaptation on scattered data	–	–	–	–	+	–

Table 2: Qualitative comparison of methods for learning control; see text.

Training indication, i.e. an additional output telling something about the training situation (information for a considered input is available or not) is a mostly underestimated feature. The *access time* per iteration (model access, training step) should be constant and low, especially independent of the number of internal states. Learning should be of *fast convergence*. With *calculation effort* mainly the hardware effort to solve a given calculation task is meant. *Memory consumption* is a criterion depending on the application. Normally large memories are not a problem any more, at least, where superior process control hardware is in use. The requirement for small memory consumption can exclude progressive methods from being applied to control tasks at all. In real world applications *noise filtering* is important, because noise on the output of a process always has to be taken into account. Sometimes the *adaptation to scattered data* is of interest, especially for applications in high dimensional input spaces, where the data is expected to be sparsely scattered over the input space with large variations in density.

It may be concluded that in most categories CMAC-type memories are well suited for learning control and should especially be preferred to normal backpropagation networks, as [18] suggests. The requirement of adaptation on scattered data could prevent the use of CMAC, but many other modelling strategies will not be applicable from this point of view, either. Only pure interpolation memories are suitable for this case, because they interpolate even scattered data. The other disadvantage, large memory requirements, can be tolerated, if fast on-line operation is desired.

3. CMAC and Its Improvements

3.1. Introduction

There are different ways of interpreting the CMAC approach. Some people decided to call it an associative memory due to its input processing and memory addressing action. Others called it a neural network, a variant of the perceptron or an interpolating memory. In fact it has features justifying all these names.

Firstly, it is a neural network, more exactly a model of the cerebellum[19] (CMAC = Cerebellar Model Articulation Controller). In this function, it has been developed for the first time by Albus [20] and used for control of a (simulated) robot arm, following the basic structure of the real cerebellum, whose main task is to coordinate motion (sequences). A schematic model is shown in Figure 8.

It can be shown, that CMAC is a special case (a "localized" variant) of Rosenblatt's perceptron [21] which is said to be the first neural network applied to more or less "real" problems. Additionally, it is an interpolating memory, because each input point is internally represented as an *association area*, an approximately hypercubic environment around the input point. This means that each point (input vector with related output value) to be trained into CMAC affects not only this point itself in the input space, but also some area surrounding it.

It also has features of associative memories. The internal representation of an input point is used to address memory values, a kind of content driven addressing. Though CMAC avoids the sometimes enormous memory consumption of "real" associative memories by using hash functions in addressing.

In contrast to most other neural networks, where the inner states (weights) are

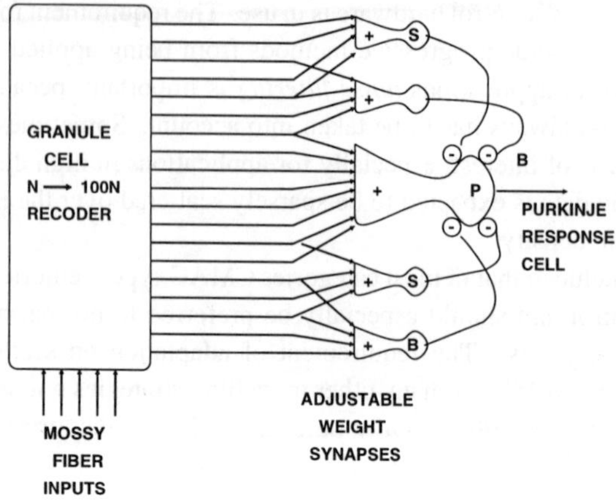

Figure 8: The cerebellum model of Albus.

processed in a non-linear way, the calculations following the input processing are carried out linearly. This results in a simple (linear) training algorithm with provable convergence behaviour.

Applications of CMAC to the problem of intelligent control can be found, e.g. in [22] [23] [24] [1]. Basics and an application are described in [25], while possible improvements in [26] [27].

3.2. *Input Processing*

3.2.1. Association

The input space for CMAC is discrete. For control purposes it is normally sufficient to consider an n-dimensional interval of integer values: $[0, 2^b] \times \ldots \times [0, 2^b]$, where b represents the (binary) accuracy, e.g. as given by the A/D converters of the process control hardware, and n is the input dimension typically ranging from 2 to 20. This input space is overlayed with coordinate grids of coarser resolution. The loss of resolution and the number of grid systems are equal. They have the well defined positive integer value ϱ, which is called the (amount of) *generalisation*. Figure 9 explains the method (the input vector is $\vec{s} = (s_1, \ldots, s_n)$):

$$a_{ji} = \begin{cases} \left\lfloor \frac{s_i}{\varrho} \right\rfloor, & \text{if } (s_i)_\varrho > j; \\ \left\lfloor \frac{s_i}{\varrho} \right\rfloor - 1, & \text{otherwise,} \end{cases} \qquad (15)$$

$$j = 0, \ldots, \varrho - 1, \quad i = 1, \ldots, n$$

($(x)_y$ is the rest (modulo) of the integer division $\lfloor x/y \rfloor$). Equation 15 calculates the hypercube (to be more exact: a code for its "lower left" coordinate) which meets the input vector from each coarser grid (each grid system is displaced relative to each other). The resulting vectors $\vec{a}_j$, $j = 0, \ldots, \varrho - 1$ represent the input vector exactly; Equation 15 defines a bijective relationship.

In the neural terminology: an input vector $\vec{s}$ activates ϱ overlapping receptors, which then "fire" with a constant rate of 1. The receptors are represented by coordinates identifying their "sensitivity" regions in the input space and (15) gives an effective computing rule for these coordinates.

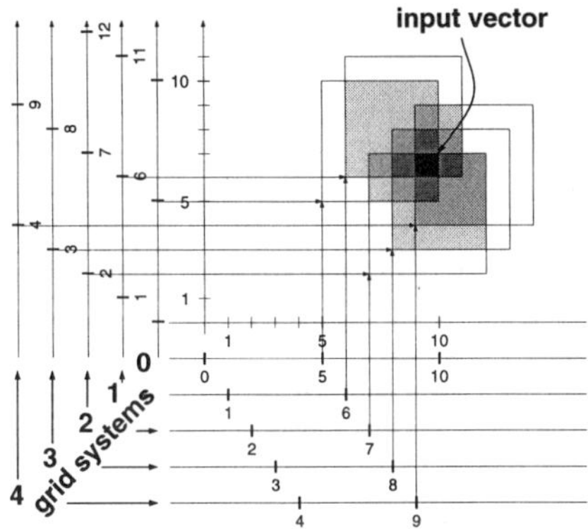

Figure 9: The Association in CMAC for a two-dimensional input vector and a generalization of $\varrho = 5$.

3.2.2. Hash Coding

Now, the task is to transform the receptive field coordinates $\vec{a}_j$ into memory addresses r_j to access the related synaptic weights effectively. In principle this can be done by simply constructing the number to the base $\sigma = \lfloor (2^b - 1)/\varrho \rfloor + 2$, which is one more than the value range resulting from the right side of (15): $-1, \ldots, \lfloor (2^b - 1)/\varrho \rfloor$.

$$r_j = \sum_{i=1}^{n} (a_{ji} + 1)\sigma^{i-1} + j\sigma^n, \quad j = 0, \ldots, \varrho - 1. \tag{16}$$

CMAC, due to its association algorithm, requires less memory than a pure tabulation of possible input/output values, but the number of different r_j can grow too large for normal memory resources. For control purposes it is normally not necessary to cover the whole input space S, but only small parts of it (trajectories of the process to be modelled/controlled). The classical solution for this problem, storing a relatively small amount of items out of a large input domain, is *hash coding*. This means a compression of the information in the $\vec{a}_j$ by a *hash function* μ, such that the result is of binary length $\log_2 R$, where R is the amount of memory (cells) supplied for storing the weights and additional information.

This compression does not harm the function of CMAC, provided:

(A) The concrete application needs only a small subset of the total input space.

(B) The compression function μ has pseudo-random properties, i.e. similar $\vec{a}_j$ should wherever possible result in rather different $r_j = \mu(\vec{a}_j)$ (called the *holographic* nature of the hash function).

(C) There is a possibility for detection and avoidance of *hash collisions*, i.e. the case, where

$$\vec{a}_j \neq \vec{a}_k \wedge \mu(\vec{a}_j) = \mu(\vec{a}_k) \tag{17}$$

can be detected and there is a strategy λ for resolving it: $\vec{a}_j \neq \vec{a}_k \Rightarrow \lambda(\mu(\vec{a}_j)) \neq \lambda(\mu(\vec{a}_k))$.

Assumption **(A)** is satisfied by the application, as already mentioned, while **(B)** depends on the calculation rule μ. Hash functions have been investigated for a long time [28], but the methods presented mostly use standard arithmetic (addition, multiplication) for the calculation of μ. In contrast, in [29] segmentation methods are given which are based on permutations of bits in the information to be "hashed". XOR operations are used to combine partial results from these permutation processes. Such calculations can be carried out in special purpose hardware very effectively (more effectively than the standard methods) and yielded good results in software simulations [25].

Assumption **(C)** can only be satisfied, if at the address $r_j = \mu(\vec{a}_j)$ not only the synaptic weight but also the *key* information (normally just the value $\vec{a}_j$) is stored. But this requires a flexible memory organization, which (especially in a hardware solution) is difficult to handle, because the length of information $\vec{a}_j$ depends on configuration parameters (input dimension n, generalization ϱ, input range of the components s_i of input vector $\vec{s}$). Therefore the information in $\vec{a}_j$ is compressed by a method ν, similar to the hash function μ:

$$\nu: \vec{a}_j \mapsto \nu(\vec{a}_j) \in [0, 2^{b_i}] \tag{18}$$

with some bit length b_i. In practice it has been shown, that $b_i = 16$ is sufficient. The $\nu(\vec{a}_j)$ is called an *identifier* for $\vec{a}_j$. If a hash collision (17) is detected, a popular method for collision avoidance (called λ above) is to add an increment δ, until a valid address has been found (stored identifier is equal to the calculated identifier) or a maximum of number of search steps, q_{max}, has been reached. The latter condition is introduced to limit the calculation time needed for the search process.

If the memory capacity is increased above about 70%, the output error grows slightly with the number of unavoidable collisions (strategy λ concludes without finding a valid memory address). This effect is known as *graceful degradation*.

3.3. *Weight Processing*

3.3.1. Output Calculation

The weights addressed by the above algorithm (*active weights*) are summed up to an output $\hat{p}$:

$$\hat{p} = \frac{\sum_{j=0}^{\varrho-1} \alpha_j w_j}{\sum_{j=0}^{\varrho-1} \alpha_j}. \tag{19}$$

If the association area is not or only in part trained, less than ϱ weights or even none will be found. The α_j are indicators whether a weight has been found ($\alpha_j = 1$) or not ($\alpha_j = 0$). As mentioned above, CMAC provides a training indicator. This is simply the ratio of the number of weights found to the maximum number of weights active at one access step, ϱ:

$$\tau = \frac{1}{\varrho} \sum_{j=0}^{\varrho-1} \alpha_j. \tag{20}$$

3.3.2. Training Procedures

Training in CMAC can be done like in backpropagation networks with one fundamental difference: after each training step the desired output value is represented exactly in the next access step. From another point of view this is an iterative LMS approach or the Kaczmarz algorithm of first order. Higher orders (i.e. $|S'| > 1$) are also possible, following earlier work in the field of modelling by linear equation systems [30].

Error Backpropagation Using the result $\hat{p}$ from an access step the active weights can be corrected with p being the desired output:

$$\begin{aligned} w_j^{(k+1)} &= w_j^{(k)} + p^{(k)} - \hat{p}^{(k)}, \\ \text{with: } w_j(k) &= 0, \text{ if } \alpha_j^{(k)} = 0, \\ & \quad j = 0, \ldots, \varrho - 1, \end{aligned} \tag{21}$$

where k, $k + 1$ are access step counters or discrete time values.

As a result of this local training CMAC surpasses other modelling strategies with regard to access time and convergence velocity. Only the ϱ active weights are adapted. This is in contrast to backpropagation networks where all available weights are processed.

A variation of this training method is the initialisation of previously untrained weights with the desired output p (instead of 0 as above). The output calculation is modified to:

$$\tilde{p}^{(k)} = \frac{1}{\varrho} \sum_{j=0}^{\varrho-1} \alpha_j^{(k)} w_j^{(k)} + (1 - \alpha_j^{(k)}) p^{(k)}. \tag{22}$$

Then the weight adaptation rule is:

$$\begin{aligned} w_j^{(k+1)} &= w_j^{(k)} + p^{(k)} - \tilde{p}^{(k)}, \\ \text{with: } w_j(k) &= p^{(k)}, \text{ if } \alpha_j^{(k)} = 0, \\ & \quad j = 0, \ldots, \varrho - 1. \end{aligned} \tag{23}$$

Stochastic Approximation Another training method can be used which needs access counters stored together with each weight. Given the ϱ active weights w_j and associated access counters z_j the training rule is:

$$w_j^{(k+1)} = \begin{cases} p, & \text{if } \alpha_j = 0, \\ \frac{z_j^{(k)} w_j^{(k)} + p}{z_j^{(k)} + 1}, & \text{otherwise,} \end{cases} \tag{24}$$
$$j = 0, \ldots, \varrho - 1.$$

Of course the access counters are updated: $z_j^{(k+1)} = z_j^{(k)} + 1$, $z_j = 0$, if $\alpha_j = 0$. Two features of this algorithm are worth noting:

1. It needs no output calculation before weight adaptation. Thus it can be realised in parallel (each weight update independently of each other).

2. The weighting of successive desired outputs p for one weight w_j is approximately $1/z_j$, which is a filter for noise with average 0.

3.4. Improvements

3.4.1. Modification of the Association

Militzer and Parks suggest a modification of the association algorithm (15) [31]. Figure 9 shows that the coarser grids are displaced along the main diagonal of the input space: grid system $j + 1$ is simply the translation of grid system j by the

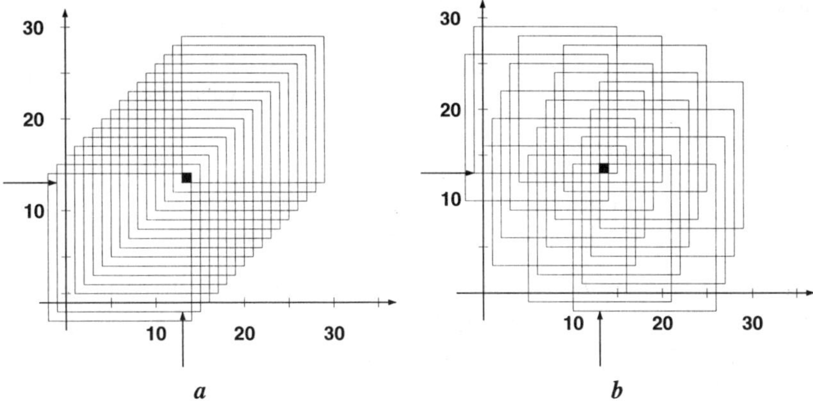

Figure 10: Modifying the association in CMAC; an example for $n = 2$ and $\varrho = 16$: ***a***: the association area is an extreme case for the old algorithm; ***b***: association area for the same input point following the new algorithm.

vector $(1, \ldots, 1)$. This sometimes leads to association areas not covering the input space in the desired (hypercubic) way. The idea of Militzer and Parks is to choose a different displacement of the grid systems in order to get a better distribution of the $\varrho \times \ldots \times \varrho$ hypercubes representing the association area. Figure 10 displays the old and the improved result in a 2-dimensional case with grid displacements of $(1, 1)$ and $(1, 3)$, respectively.

Even though the improvement of the modelling capability is not very significant (in or below the percent range), it has a positive effect on the methods presented below.

3.4.2. Continuous Receptive Fields

From a neuro-biological point of view it is not sufficient to consider only binary weight activation as given above (a weight is either in the active set or not) but also continuous activation. Provided the hypercubes forming the association area are seen as receptive fields, the activation of a weight should depend on the position of the input vector in the associated hypercube, e.g. by a radial basis function with the centre of the hypercube being the centre of the function; [32] gives more details. This method leads to a better convergence, because the underlying linear equation system for output processing is better conditioned.

Simplification leads to the following formula for an activation φ_j of weight w_i in the active set:

$$\varphi_j = \varphi(d_j) = e^{-(ad_j/d_{max})^2}, \quad d_{max} = \begin{cases} \frac{n\varrho}{2}, & \text{if } \varrho \text{ is even;} \\ \frac{n(\varrho-1)}{2}, & \text{otherwise} \end{cases} \tag{25}$$

with $d_j = \sum_{i=1}^{n} d_{ij}$ and

$$d_{ij} = \begin{cases} |(s_i)_\varrho - (j-1) - \lfloor \varrho/2 \rfloor|, & \text{if } (s_i)_\varrho > j-1; \\ |(s_i)_\varrho - (j-1) + \lfloor \varrho/2 \rfloor|, & \text{otherwise} \end{cases}$$

yielding in d_j being the L_1-norm of the distance between hyper-cube centre and input point. The used terms are already given by the association algorithm. For practical purposes it is sufficient to get the values of function (25) out of a table which has been calculated in advance.

The modifications leads to better convergence at high values ϱ (> 100).

3.4.3. Getting a Gradient's Direction out of CMAC

A probably new idea was motivated by the optimisation problem in LERNAS. In cases of high dimensionality the optimisation by simple search algorithms may be time consuming. The question is whether an estimate of gradient's direction for the actually modelled function can be calculated using CMAC's internal values.

The principle is to some extent derived from the already mentioned continuous receptive field approach. The components of the vector from the input point to the centre of the hypercubes in the association area are each weighted by the difference between the associated weight and the actual output $(w_j - \hat{p})$ and then summed up. If $\vec{d}_j$ are these vectors, a (scaled) estimate for the function's gradient can be expressed as follows:

$$\hat{p}_{s_i}(\vec{s}) = \sum_{j=0}^{\varrho-1} (w_j - \hat{p}) d_{ji}, \qquad i = 1, \ldots, n. \tag{26}$$

First tests with 2-dimensional functions show that the polar coordinate (angle) of the resulting gradient vector is rather good for practical applications (relative errors at or below 5%). But in regions, where the function surface is flat (i.e. gradient's norm is small), the error becomes very high. This is not surprising. The error contributed by $w_j - \hat{p}$ might then be disturbing.

4. Approach to a Hardware Solution for CMAC

CMAC is a good candidate for a digital hardware solution, for the following reasons:

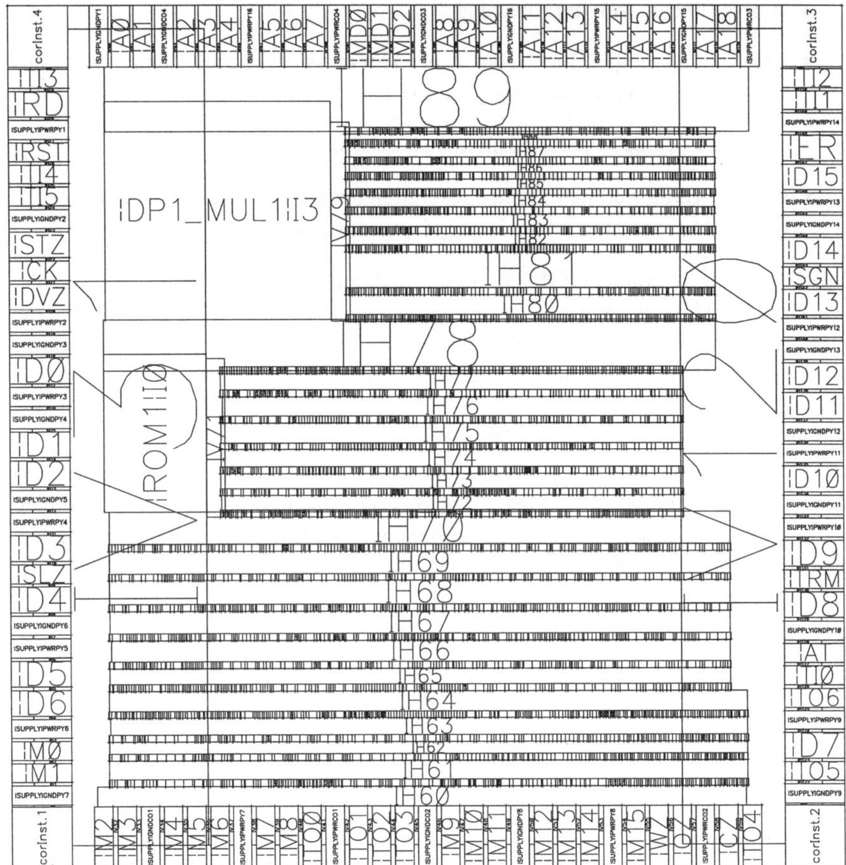

Figure 11: Schematic view of the CMAC chip surface. in the upper left edge is the multiplier, below it the ROM. All other logic circuitry is in the horizontal standard cell rows. The chip has an area of about 56mm^2. Nearly 58,000 transistors are implemented in 1μm double metal layer CMOS technology. Interfacing is done via 120 pads, 40 of them for power supply, 29 for a 16 Bit wide host interface, 13 for testing purposes, the remaining 38 pads are for the memory interface (16 Bit data, 19 Bit address). The chip runs with clock frequencies up to 28MHz. The operation speed depends on the access time of the external RAM. With 25ns-RAMs 16MHz operation is possible.

- non-trivial, yet simple access algorithm which can completely be carried out in integer arithmetic,

- simple (linear) weight processing,
- large memory requirements, but no need for massively parallel data access (thus standard memory devices can be used),
- moderate degree of parallelism in an access step,
- CMAC works well with limited resolution for weights and intermediate results of weight processing, so again only integer arithmetic is necessary.

Most of the above arguments do not hold for other neural networks.

The CMAC hardware has been carried out using synthesis.[g] The first stage of the investigation of CMAC was a software simulation with a lot of additional features as those described in Section 3. above. The second stage was the development of a hardware related software version in a high level language (C, in this case). It was followed by a third version in VHDL for synthesis.

The final abstract version was a gate level net list, the gates being instances from a digital standard cell library. This version is then put into the placement and routing process, which mainly is an automatic procedure concluding in a physical representation of the chip surface, which is shown in Figure 11.

4.1. *Some Principles of Hardware Design*

4.1.1. Association

If ϱ takes only powers of two, the division/modulo calculation of (15) becomes a simple shift and masking operation respectively. The VHDL formulation has only a few lines of code.

4.1.2. Address Calculation

The hash coding is processed by a hardwired multiplexed structure: something like "shaking" the bits of the association vectors $\vec{a}_j$ in order to get a first memory address which is not correlated to the information in the input vector. In addition the hardwired structure generates two other compressed versions of the input information. One is the identifier, and the other an increment for generating the new memory addresses in the memory search.

The method (memory read operation to get the key, key comparison, address addition) is carried out in a pipelined architecture operating independently of other parts of the chip.

---

[g] Automatic generation of a gate level representation from a description in a (mostly hardware orientated) high level language. Most popular for this purpose are *Verilog* and *VHDL*.

4.1.3. Weight Processing

The training algorithms require an arithmetic data path of moderate complexity. The largest part is the multiplier for the evaluation of Equation (25). The division in (25) has been replaced by the multiplication with the tabulated reciprocal value. Thus the expression for a weight adaptation step becomes:

$$\frac{\xi_j(z_j w_j + p + \Delta_j)}{2^{b_j + b_{\xi_j}}}, \quad \text{with:} \quad \xi_j \approx \left\lfloor \frac{2^{b_j + b_{\xi_j}}}{z_j + 1} \right\rfloor, \quad \Delta_j \approx (z_j + 1)/2, \tag{27}$$

where $b_j + b_{\xi_j}$ is a bit shift value to keep all values in the integer range. The variables ξ_j and Δ_j, which are functions of z_j, are calculated so that for a given resolution the maximum error of expression (27) is one bit compared to the rounded "exact" solution. (The error is actually less than one bit because it is only in the range of the maximum rounding error, 0.5).

If the resolutions of the weights w_j (16 bits) and the counters z_j (8 bits) are given, the bit lengths of ξ_j (17 bits) and Δ_j (8 bits) are determined. The latter two values are implemented on chip in a ROM of size 256×25. The value of b_j can simply be derived from z_j. It is the index of the most significant bit of z_j, which is '1' ($b_j = \lfloor \log_2 z_j \rfloor$).

5. Conclusion

The presented discussion of possible learning elements for control concluded that—among other candidates—CMAC is a successful and extensible method, especially for fast on-line applications. Further improvements of the algorithm are under investigation. Also for hardware implementation CMAC is well suited due to its fast and simple access algorithm. A first version of a CMAC-ASIC is available. This way a learning device satisfying extreme speed requirements can be applied to control problems.

6. Acknowledgements

This work was supported by the *Deutsche Forschungsgemeinschaft* (DFG), Bonn, Germany as part of the *Sonderforschungsbereich*[h] *241 IMES* "Integrated Mechanical-Electronic Systems" and the Zentrum für Neuroinformatik[i] GmbH, Bochum, Germany.

Additionally the author thanks the colleagues at the Microelectronic Systems

[h] special research initiative

[i] Centre for NeuroInformatics

Department of his university, especially Prof. Dr. M. Glesner and the member of his staff, Mr. P. Windirsch for their support.

7. References

1. W. T. Miller III, F. H. Glanz, and L. G. Kraft III, "Application of a general learning algorithm to the control of robotic manipulators," *The International Journal of Robotics and Control*, vol. 6, no. 2, pp. 84–98, 1987.

2. E. Ersü, "On the application of associative neural network models to technical control problems," in *Localization and Orientation in Biology and Engineering* (Varjú/Schnitzler, ed.), (Berlin, Heidelberg), pp. 90–93, Springer, 1984.

3. M. Kortmann and H. Unbehauen, "Ein neuer Algorithmus zur automatischen Selektion der optimalen Modellstruktur bei der Identifikation nichtlinearer Systeme," *Automatisierungstechnik*, vol. 35, no. 12, pp. 491–498, 1987. (in German).

4. H.-U. Flunkert, *Regelstrategien auf der Basis reduzierter nichtlinearer Modelle*. Fortschritt-Berichte VDI, Reihe 8, Nr. 319, VDI-Verlag, 1993. (in German).

5. D. E. Rumelhart, J. L. McClelland, and the PDP Research Group, *Parallel Distributed Processing. Explorations in the Microstructure of Cognition. Vol. I: Foundations*. Cambridge, Mass., London: MIT Press, 5th ed., 1987.

6. K. J. Hunt, D. Sbarbaro, R. Żbikowski, and P. J. Gawthrop, "Neural networks for control systems: A survey," *Automatica*, vol. 28, pp. 1083–1112, November 1992.

7. R. Żbikowski, K. J. Hunt, A. Dzieliński, R. Murray-Smith, and P. J. Gawthrop, "A review of advances in neural adaptive control systems," Technical Report of the ESPRIT NACT Project TP-1, Glasgow University and Daimler-Benz Research, 1994. (Available from FTP server `ftp.mech.gla.ac.uk` as PostScript file `/nact/nact_tp1.ps`).

8. A. Cichocki and R. Unbehauen, *Neural Networks for Optimization and Signal Processing*. Stuttgart/Chichester etc.: Teubner/Wiley, 1993.

9. R. Żbikowski and A. Dzieliński, "Neural approximation: A control perspective," in *Neural Network Engineering in Dynamic Control Systems* (K. J. Hunt, G. R. Irwin, and K. Warwick, eds.), Advances in Industrial Control, pp. 1–25, Berlin: Springer-Verlag, 1995.

10. G. P. Drago and S. Ridella, "SCAWI: an algorithm for weight initialization of a sigmoidal neural network," in *Artificial Neural Networks, 2* (I. Aleksander and J. Taylor, eds.), vol. II, pp. 983–986, Elsevier, 1992.

11. W. H. Schiffmann and H. W. Geffers, "Adaptive control of dynamic systems by backpropagation networks," *Neural Networks*, vol. 6, no. 4, pp. 517–524, 1993.

12. K. Hornik, M. Stinchcombe, and H. White, "Multilayer feedforward networks are universal approximators," *Neural Networks*, vol. 2, no. 5, pp. 359–366, 1989.

13. S. E. Fahlmann, "Faster learning variations on back-propagation: an empirical study," in *Connectionist Models Summer School* (D. Touretzky, G. Hinton, and T. Sejnowski, eds.), pp. 38–51, Carnegie Mellon University: Morgan Kaufmann, 1988.

14. R. Salomon and L. van Hemmen, "A new learning scheme for dynamic self-adaptation of learning-relevant parameters," in *Artificial Neural Networks, 2* (I. Aleksander and J. Taylor, eds.), vol. II, pp. 1047–1050, Elsevier, 1992.

15. D. H. McLain, "Drawing contours from arbitrary data points," *Computer Journal*, vol. 17, no. 4, pp. 318–324, 1974.

16. M. Schmitt, T. Ullrich, and H. Tolle, "Associative datafields in automotive control," in *3rd IEEE Conference on Control Applications*, (Glasgow, UK), pp. 1239–1244, 1994.

17. C. J. Harris, C. G. Moore, and M. Brown, *Intelligent Control—Aspects of Fuzzy Logic and Neural Nets*. World Scientific, 1993.

18. W. T. Miller, F. H. Glanz, and L. G. Kraft, "CMAC: An associative neural network alternative to backpropagation," *Proc. of the IEEE, Special Issue on Neural Networks, II*, vol. 78, pp. 1561–1567, 1990.

19. D. Marr, "A theory of cerebellar cortex," *Journal of Physiology*, vol. 202, pp. 437–470, 1969.

20. J. S. Albus, "A new approach to manipulator control: The cerebellar model articulation controller," *Transactions of the ASME. J. Dynamic Systems Measurement and Control*, vol. 63, no. 3, pp. 220–227, 1975.

21. F. Rosenblatt, *Principles of Neurodynamics: Perceptrons and the Theory of Brain Mechanism*. Washington DC: Spartan Books, 1961.

22. S. Gehlen and J. Kreuzig, "Learning by interpolating memories for modelling of fermentation processes," in *IFAC-Symposium on Advanced Control of Chemical Processes, October 14-16*, (Toulouse/France), 1991.

23. D. P. W. Graham and G. M. T. D'Eleuterio, "Robotic control using a modular architecture of cooperative artificial neural networks," in *Artificial Neural Networks, Vol. I* (T. Kohonen, K. Mäkisara, O. Simula, and J. Kangas, eds.), pp. 365–370, North-Holland: Elsevier, June 1991.

24. K. Tanaka, M. Shimizu, and K. Tsuchiya, "A solution to an inverse kinematics problem of a redundant manipulator using neural networks," in *Artificial Neural Networks, Vol I* (T. Kohonen, K. Mäkisara, O. Simula, and J. Kangas, eds.), pp. 327–332, North-Holland: Elsevier, 1991.

25. H. Tolle and E. Ersü, *Neurocontrol.* No. 172 in Lecture Notes in Control and Information Sciences, Springer-Verlag, 1992.

26. P. C. E. An, W. T. Miller, and P. C. Parks, "Design improvements in associative memories for cerebellar model articulation controllers (CMAC)," in *Artificial Neural Networks, Vol. II* (T. Kohonen, K. Mäkisara, O. Simula, and J. Kangas, eds.), vol. 2, pp. 1207–1210, Elsevier, 1991.

27. M. Hormel, "A self-organizing associative memory system for control applications," in *IEEE Conference on Neural Information Processing Systems - NIPS '89*, (Denver, Co., USA), November 1989.

28. D. Knuth, *The Art of Computer Programming, Vol 3.* Reading, Mass.: Addison-Wesley, 1968.

29. T. Kohonen, *Content Adressable Memories.* Berlin-Heidelberg-New York-Tokyo: Springer, 2 ed., 1987.

30. E. D. Aved'yan, "Modified Kaczmarz algorithms for estimating the parameters of linear plants," *Automation and Remote Control*, vol. 39, pp. 674–680, May 1978.

31. P. C. Parks and J. Militzer, "Improved allocation of weights for associative memory storage in learning control systems," in *Proc. 1st IFAC Symposium on Design Methods of Control Systems*, (Zürich/Switzerland), pp. 777–782, Pergamon Press, September 1991.

32. W. S. Mischo, "Receptive fields for CMAC. an efficient approach," in *Artificial Neural Networks, 2* (I. Aleksander and J. G. Taylor, eds.), vol. 1, pp. 595–598, Elsevier, 1992.

THE EQUIVALENCE OF SPLINE MODELS AND FUZZY LOGIC APPLIED TO MODEL CONSTRUCTION AND INTERPRETATION

Glenn Terje Lines and Tom Kavli
SINTEF Instrumentation, P.O.Box 124 Blindern
0314 Oslo, Norway
E-mail: Glenn.Lines@si.sintef.no
E-mail: Tom.Kavli@si.sintef.no

ABSTRACT

The ASMOD (Adaptive Spline Modelling of Observation Data) algorithm automatically identifies non-linear multivariate models from empirical data, using B-splines as the internal representation for the identified dependencies. The model structure is automatically adapted to the problem by an iterative search. The capability of ASMOD to find model structures which are both parsimonious and relevant for the problem have earlier been demonstrated for many real and simulated problems.

We here show that models of this type are equivalent to a certain type of fuzzy logic models. The B-spline basis functions in an ASMOD model correspond to the fuzzy membership functions of the input set in the equivalent fuzzy system, and each basis function corresponds to one rule in the fuzzy model. An automatically identified ASMOD model can thus be directly mapped into a set of fuzzy rules for manual interpretation and modification, or a fuzzy system set up from expert knowledge can be mapped into an ASMOD model structure for further refinement and adaptation to training data.

We discuss how this relation can be used to combine the strengths of both paradigms, and the potentials of this approach is demonstrated on a simulated problem where the task is to identify a state space model for a highly non-linear batch fermentation process.

1. Introduction

The objective of system modelling is to establish a mathematical description of the system behaviour, typically as functions describing dependencies among system variables. The construction of such models is central in many industrial engineering tasks, such as system analysis and development, system optimisation, control system design, and instrument calibration.

In developing such models one would want to be able to use as efficiently as possible all sources of system information, of which the most important are:

- mechanistic knowledge obtained from the first principles (physics and chemistry);
- empirical expert knowledge, often expressed as rules of thumb;
- empirical measurement data obtained during normal operation or from an experimental design programme.

Most of modelling techniques used today address only one of the above sources of information. When good mechanistic knowledge is available the first choice is to set up analytical differential equations based upon this knowledge, and to adapt the parameters of the model to measured data if such are available. Such models directly reflect the underlying phenomena in the system, and may thus give globally valid and parsimonious models.

On the other hand, such models rely on a correct understanding of the system, and if the system contains significant unmodelled phenomena, the models may be both inaccurate and misleading. Empirical modelling techniques, such as ARMAX models, artificial neural networks and ASMOD models aim at fitting models of a general structure to training data, and hope that the identified model represent the true and interesting properties of the system. Such models are often difficult to interpret and verify, and the possibilities to include any mechanistic or expert knowledge is generally very limited.

Expert knowledge is most easily entered into rule based systems, such as fuzzy logic models. These models are generally not aiming at problems with high precision requirements, but rather at situations where only a qualitative model is needed.

Different sources of information may be available for different operational domains of the system. Similarly, the accuracy requirements may vary for different domains. For instance, there may be plenty of observational data for the normal operational conditions of an industrial process, whereas there may be very few measurements from abnormal system conditions. The large amounts of data can then be used to obtain empirical models of high accuracy for the normal system operation, while a more globally valid but less accurate mechanistic model may be used when the process moves away from the domain where the empirical model is valid. Similarly, the expert knowledge may be a more appropriate source of information when the system is in a critical alarm condition or during startup and shutdown. Luckily, in these situations a very coarse model accuracy is often sufficient.

As can bessen from the above discussion, an integrated framework, where all sources of information can be utilised in the model construction, would be beneficial. We here suggest to use B-splines as the mathematical representation for both fuzzy models and empirical models. B-splines may also be used to interpolate smoothly between this kind of fuzzy/empirical models and local domain mechanistic models[10,11]. This makes it possible to include also such models when and where appropriate.

The equivalence between B-spline models and a certain type of fuzzy systems have been shown by several authors[4,17]. ASMOD is an algorithm for empirical modelling using B-splines for its internal model representation, and we here show the equivalence of ASMOD models and fuzzy systems. ASMOD and B-spline fuzzy systems thus represent a direct integration of empirical models and expert based models in the same mathematical representation. Johansen and Foss[10] have shown how fuzzy logic can be used as a basis for interpolating between local domain mechanistic models. Here we concentrate on the link between ASMOD models and fuzzy systems, and leave the possibility for also integrating mechanistic models to future developments.

This chapter is organised as follows. In Section 2 the ASMOD model structure is defined and the link between ASMOD models and fuzzy systems is established. Section 3 gives a description of the ASMOD search algorithm. Then, in Section 4, both methods are applied to the problem of identifying state space models for a simulated non-linear batch fermentation reactor. It is shown how expert knowledge can be used to construct an initial set of rules for the model, how the ASMOD algorithm can be used to automatically refine the initial model, and finally how the resulting ASMOD model can be interpreted and analysed in terms of fuzzy rules. Some further possibilities and potential problems are discussed in Section 5 before a short conclusion given in Section 6.

2. ASMOD Models and the Equivalence to Fuzzy Systems

2.1. The ASMOD Model Structure

ASMOD[12,13] (Adaptive Spline Modelling of Observation Data) is an algorithm for automatically identifying non-linear models for multivariate systems based on measurement data. The algorithm iteratively builds up the model by gradually increasing the model complexity, while always searching for the model structure with minimal complexity and optimal accuracy. The algorithm has proven to give sparse models of high accuracy for many real and simulated problems[5,3].

An ASMOD model is a mapping from an input variable $x \in \Omega \subset \mathcal{R}^n$ to an output variable $y \in \mathcal{R}$. For multidimensional input spaces the model is generally decomposed into a sum of several low dimensional submodels, each dependent only on a small subset of the input variables.

An ASMOD model $m(x)$ can thus be described as a sum of U lower dimensional functions denoted by $s_u(x)$, $u = 1, \ldots, U$. The sets of input variables to the submodels are mutually disjoint so that

$$m(x) = \sum_{u=1}^{U} s_u(x_u), \tag{1}$$

where $\{x_u\} \subseteq \{x\}$ and $\{x_i\} \cap \{x_j\} = \emptyset$ for $i \neq j$ ($\{x_u\}$ is the set of input variables present in the x_u vector). Figure 1 shows an example of an ASMOD decomposition into additive submodels.

Each submodel is a one-dimensional or a tensor product B-spline function. A B-spline function is constructed as a linear combination of B-spline basis functions.

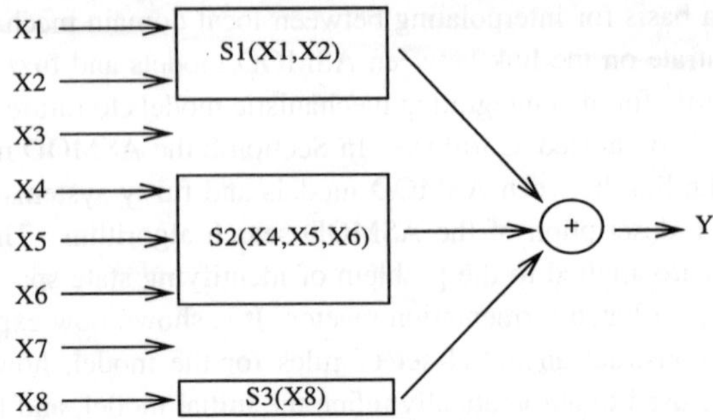

Figure 1: A possible ASMOD model structure. The model is decomposed into submodels where each submodel is a spline function representing an additive dependency on a few input variables. Some input variables may be unused in the model.

Hence we can write a submodel as

$$s_u(x) = \sum_{i=1}^{K_u} c_{u,i} b_{u,i}(x) = c_u^T b_u(x), \tag{2}$$

where K_u is the number of basis functions in $s_u(x)$, $b_u(x)$ is a vector of B-spline basis functions, and c_u is a coefficient vector.

Since the total model is a sum of the submodels, this can also be described as a combination of basis functions as

$$m(x) = \sum_{i=1}^{K} c_i b_i(x) = c^T b(x). \tag{3}$$

The set of B-spline basis function for a one-dimensional ASMOD submodel is defined by a selected polynomial degree of the spline, $p \in \{0, 1, \ldots\}$, and a knot vector $\tau \in \mathcal{R}^k$ with $k \geq 2p+2$ and $\tau_{i-1} \leq \tau_i$, $i = 2, \ldots, k$. The set of B-spline basis functions for a spline of degree p are defined recursively from the basis functions for the spline of degree $p-1$ as:

$$\begin{aligned} b_{i,p,\tau}(x) &= (x - \tau_i) q_{i,p-1,\tau}(x) + (\tau_{i+p+1} - x) q_{i+1,p-1,\tau}(x), \\ & \quad i = 1, 2, \ldots, k - p - 1, \end{aligned}$$

where

$$q_{i,p,\tau}(x) = \begin{cases} b_{i,p,\tau}(x)/(\tau_{i+p+1} - \tau_i), & \text{if } \tau_i < \tau_{i+p+1}; \\ 0, & \text{otherwise} \end{cases}$$

and the basis functions of degree zero are defined as

$$b_{i,0,\tau}(x) = \begin{cases} 1, & \text{if } \tau_i \leq x < \tau_{i+1}; \\ 0, & \text{otherwise,} \end{cases} \qquad i = 1, 2, \ldots, k-1.$$

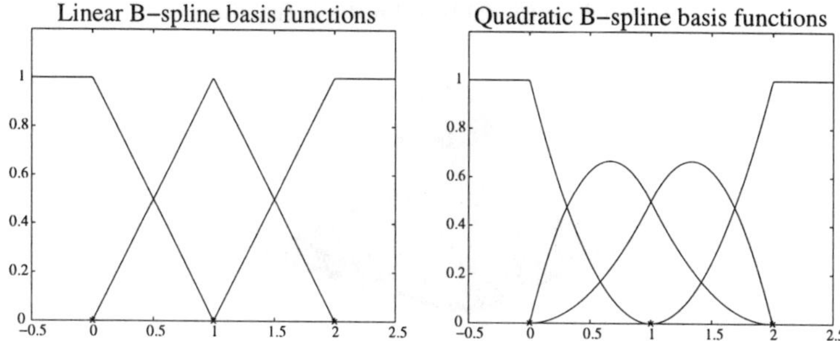

Figure 2: Examples of univariate B-spline basis functions of degree 1 and 2. The knot vectors are $\tau = (-\infty, 0, 1, 2, \infty)$ and $\tau = (-\infty, 0, 0, 1, 2, 2, \infty)$ respectively. The input domain is $[0, 2)$ in both cases. The boundary knots are placed at $\pm\infty$ to get constant extrapolation outside the input domains.

The number of basis functions in the submodel is $k - p - 1 \geq p + 1$ and its input domain is $[\tau_{p+1}, \tau_{k-p})$. The linear space spanned by all linear combinations of a set of B-spline basis functions is called a spline space, and an ASMOD submodel as given in (2) is thus one element in such a space. Examples of basis functions of degree one and two are given in Figure 2.

In order to model coupled dependencies of two or more input variables in an ASMOD submodel, multi-dimensional spline spaces are used. Multi-dimensional B-spline spaces are constructed as the tensor product of two or more univariate spline spaces. The set of basis functions for a two-dimensional space is thus constructed from the two sets of one-dimensional functions $b_{i_1,p_1,\tau_1}(x_1)$, $i_1 = 1, 2, \ldots, K_{u,1}$ and $b_{i_2,p_2,\tau_2}(x_2)$, $i_2 = 1, 2, \ldots, K_{u,2}$ as

$$b_{i_1,i_2,p_1,p_2,\tau_1,\tau_2}(x_1, x_2) = b_{i_1,p_1,\tau_1}(x_1) b_{i_2,p_2,\tau_2}(x_2),$$
$$i_1 = 1, 2, \ldots, K_{u,1}, \quad i_2 = 1, 2, \ldots, K_{u,2}.$$

Higher order tensor products are similarly constructed by repeatedly multiplying in new sets of one-dimensional functions. An n-dimensional B-spline basis function is thus constructed as

$$b_i(x) = \prod_{k=1}^{n} b_{i_k,p_k,\tau_k}(x_k). \tag{4}$$

To simplify the notation we have removed the indices indicating the spline degrees and the knot vectors and we have renumbered the new set of basis functions with a common index variable $i = 1, 2, \ldots, \prod_{k=1}^{n_u} K_{u,k}$, where n_u is the number of basis functions in submodel u. With this renumbering both one- and multi-dimensional submodels can be expressed in the form (2). Figure 3 shows an example of a two-dimensional basis function of polynomial degree two.

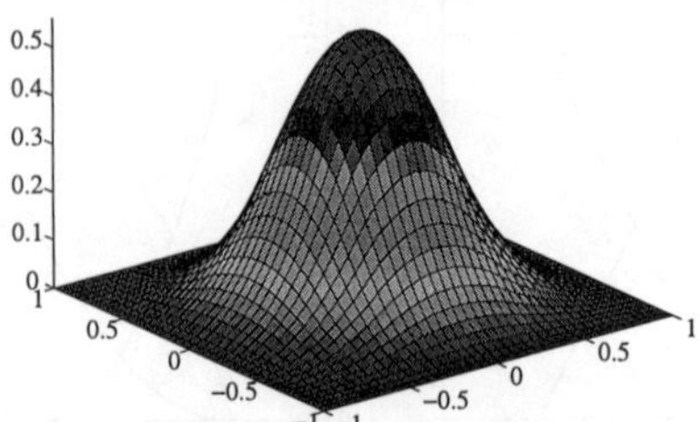

Figure 3: A bivariate B-spline basis function of degree 2 formed by the knot vector $\tau = (-1, -1/3, 1/3, 1)$ for both variables.

From the definition it can be shown that a B-spline is a piecewise polynomial, and the polynomial segments are given by the knot vector. Provided the knots are distinct, the splines and their derivatives, up to the $(p-1)$th derivative, are continuous. It also follows that the B-spline basis functions have local support, being non-zero only over $p+1$ knot intervals. Moreover the spline basis functions form a partitioning of unity (sum to one) at all points in the input domain. Finally, it can be shown that any polynomial of degree p or less can be written as a B-spline of degree p. For a more comprehensive presentation of spline theory see for instance Schumaker[14].

2.2. *The Relation to a Fuzzy Systems*

The relationship between B-spline models and fuzzy logic models has been pointed out by Brown *et al*[4], Wang *et al*[17] and others. We will here show that there is a direct equivalence between ASMOD models and a special class of fuzzy systems. We will for this purpose give only a very brief introduction to a small subset of the many varieties of fuzzy systems. For a more comprehensive presentation the reader is referred to any of the numerous texts about the subject[4,16,18].

Fuzzy systems in the general form operate on fuzzy input sets to produce fuzzy output sets. In industrial applications both the inputs and the outputs of the fuzzy system should normally be non-fuzzy (crisp) variables, so that the system can be integrated in an environment of sensors, control signals, etc. producing or requiring crisp variables. The fuzzy system will thus process the information in three steps: first the crisp inputs are fuzzified, then the fuzzified variables are given to the fuzzy model as inputs to create fuzzy output sets, and finally the fuzzy outputs are defuzzified to create crisp outputs to the environment. In such cases the fuzzy system (or fuzzy model) represents a mapping from a multi-dimensional input space to a multi-dimensional output space, both spaces of non-fuzzy variables.

Hence, the purpose of the fuzzy system and an ASMOD model is the same in this case: to represent general non-linear mappings. The main difference lies in the way the model is constructed, either by specifying a set of fuzzy rules from empirical system understanding, or by an automatic model identification algorithm taking its information from measured data. Here we will only consider systems with real valued crisp input and output variables. We will also limit ourselves to systems with only one output variable, since systems with multiple outputs can be obtained by using multiple models.

A fuzzy model is described by a set of rules of the form

$$\text{IF } (x_1 \text{ is } F_1 \text{ AND } \cdots \text{ AND } x_q \text{ is } F_q) \text{ THEN } (y \text{ is } G). \tag{5}$$

The premise of the rule measures to what extent the input variables x_1 to x_q are all members of the respective fuzzy input sets F_1 to F_q. The extent to which the premise is true determines to what extent the expression (y is G) is true. G is a fuzzy output set. The fuzzy sets F_i and G are generally assigned linguistic labels which describes a characteristic property of the set, like "small", "large", "cold" and "hot". This makes the rules easy to read and interpret.

The premise of rule (5) can be rewritten in multivariate form which gives

$$\text{IF } x \text{ is } (F_1 \text{ AND } \cdots \text{ AND } F_q) \text{ THEN } (y \text{ is } G). \tag{6}$$

Using vector notation, this can be simplified to

$$\text{IF } (x \text{ is } F) \text{ THEN } (y \text{ is } G) \tag{7}$$

where $F = (F_1 \text{ AND } \cdots \text{ AND } F_q)$.

To be able to describe the processing of the rules we need to describe several concepts of fuzzy logic. A fuzzy set S is a generalisation of ordinary sets by allowing its elements to belong more or less to the set. A fuzzy set S in a universe of discourse $\Omega \subset \mathcal{R}^n$ is characterised by a membership function $\mu_S(x) : \Omega \rightarrow [0, 1]$. The value of $\mu_S(x)$ says to what extend x belongs to S.

The membership functions often have a shape giving their elements a localised region of high membership. Triangles, trapezoids and normal distribution density functions are commonly used. The B-spline bases functions shown in Figure 2 are special cases which have several interesting properties described in Section 2.1. Among these is the partitioning of unity which will be used below.

To construct logical expressions we need an operator to compute the intersection of two fuzzy sets (AND) and another operator to compute the union of two sets (OR). Both operators work on the membership functions of the two operands to produce a membership function for the new fuzzy set. The intersection of the two sets F_1 and F_2, expressed by F_1 AND F_2, is implemented by a triangular norm of the membership functions. Common choices are the min operator and the algebraic product. With these norms the membership function of the intersection becomes

$$\begin{array}{ll} \text{min} & \mu_{F_A}(x_1, x_2) = \min(\mu_{F_1}(x_1), \mu_{F_2}(x_2)) \\ \text{product} & \mu_{F_A}(x_1, x_2) = \mu_{F_1}(x_1)\mu_{F_2}(x_2). \end{array}$$

The union, expressed by F_1 OR F_2, is implemented by a triangular conorm. Common choices here are the max operator and the bounded sum, giving the following implementation of the union

$$\begin{array}{ll} \text{max} & \mu_{F_O}(x_1, x_2) = \max(\mu_{F_1}(x_1), \mu_{F_2}(x_2)) \\ \text{sum} & \mu_{F_O}(x_1, x_2) = \min(1, \mu_{F_1}(x_1) + \mu_{F_2}(x_2)). \end{array}$$

If the two sets F_1 and F_2 are in the same space then the resulting set is also in the same space. But if they are in different spaces, the resulting set is in the union of the two spaces. This means that a premise of the form shown in (6) combines several one-dimensional membership functions into a multi-dimensional function. Assuming we use the algebraic product operator for implementing the fuzzy AND, we get a membership function for the expression $F = (F_1 \text{ AND } \cdots \text{ AND } F_q)$ given by

$$\mu_F(x) = \mu_F(x_1, \ldots, x_q) = \prod_{i=1}^{q} \mu_{F_i}(x_i). \tag{8}$$

We see that if B-spline basis functions are used for the one-dimensional membership functions on the right hand side of (8), then the multi-dimensional membership function will be a multivariate B-spline basis function as given in (4).

A fuzzy rule R defines a fuzzy set in the combined input and output spaces of F and G. If we choose the product-operation rule for fuzzy implication the membership function of this set is given by

$$\mu_R(x, y) = \mu_F(x)\mu_G(y).$$

For a given set A in the input space, the rule yields a set B in the output space, commonly defined as

$$\mu_B(y) = \sup_x[\mu_A(x)\mu_R(x, y)] = \sup_x[\mu_A(x)\mu_F(x)\mu_G(y)]. \tag{9}$$

In the systems we are considering the input x is a vector of crisp variables x'. This input needs to be fuzzified to get the input fuzzy set A. The most common choice is the singleton fuzzifier

$$\mu_A(x) = \begin{cases} 1, & \text{if } x = x'; \\ 0, & \text{otherwise.} \end{cases} \tag{10}$$

With this choice the expression (9) evaluates simply to

$$\mu_B(y) = \mu_F(x')\mu_G(y), \tag{11}$$

which is a scaling of the output set membership function $\mu_G(y)$.

A fuzzy model consist of a set of rules, $R = \{R^1, R^2, \ldots, R^M\}$, combined with the fuzzy OR operator

$$\begin{array}{ll} & R^1\text{: IF } (x \text{ is } F^1) \text{ THEN } (y \text{ is } G^1) \\ \text{OR} & R^2\text{: IF } (x \text{ is } F^2) \text{ THEN } (y \text{ is } G^2) \\ & \cdot \\ & \cdot \\ \text{OR} & R^M\text{: IF } (x \text{ is } F^M) \text{ THEN } (y \text{ is } G^M) \end{array} \tag{12}$$

Each rule produces a scaled output set $\mu_{F^l}(x')\mu_{G^l}(y)$, and if the bounded sum is used as the OR operator, then the total output from the model is a fuzzy set given by

$$\mu_R(y|x) = \min(1, \sum_{l=1}^{M} \mu_{F^l}(x)\mu_{G^l}(y)). \tag{13}$$

If we now choose to use singletons also for the output sets G^l, i.e.

$$\mu_{G^l}(y) = \begin{cases} 1, & \text{if } y = y^l; \\ 0, & \text{otherwise,} \end{cases} \tag{14}$$

then the membership function for the output set becomes a series of pulses at the locations y^l with heights $\mu_{F^l}(x)$. To be useful the fuzzy output set needs to be defuzzified. One of the most common choices for the defuzzification is to compute the crisp output as the centre of mass of the output membership function. Assuming that the upper clipping at one in (13) is not active for any value of y (which is always true for ASMOD models), this gives the output

$$y(x) = \frac{\sum_{l=1}^{M} y^l \mu_{F^l}(x)}{\sum_{l=1}^{M} \mu_{F^l}(x)}. \tag{15}$$

If we use B-spline membership functions we get

$$y(x) = \frac{\sum_{l=1}^{M} y^l b^l(x)}{\sum_{l=1}^{M} b^l(x)} = \sum_{l=1}^{M} y^l b^l(x). \tag{16}$$

The last equality holds because the basis functions in a B spline function (and an ASMOD submodel) add up to one for any value of x in the input domain. We see

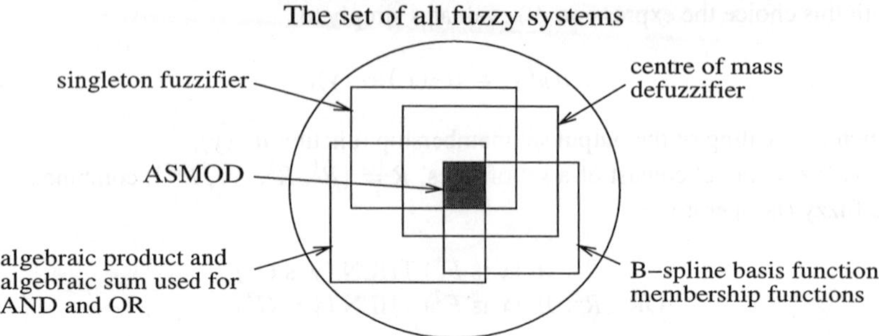

Figure 4: ASMOD models are equivalent to a small subset of all fuzzy systems, as illustrated here.

that (16) is equivalent to an ASMOD submodel as defined in (2). And we also see that an ASMOD submodel in this case is equivalent to a fuzzy system with $M = K_u$ rules, one rule for each basis function in the submodel. The B-spline basis function is used as the membership function of the input set F^l for the corresponding rule, and the spline coefficient correspond to the center location of the singleton output set G^l.

In the fuzzy logic theory there are many possibilities for fuzzy membership functions, logic operators, etc., and we have selected one of several options for each part of the fuzzy system. The equivalence of fuzzy logic and ASMOD models is thus valid only for a small subclass of all the possible implementations of fuzzy systems, as shown in Figure 4.

2.3. Interpretation of Fuzzy Rules Generated by the ASMOD Algorithm

As explained above, each basis function in the ASMOD model gives rise to one fuzzy rule in the equivalent fuzzy system as given in (12). Since the output sets are singletons determined by parameter optimisation, these can be arbitrarily placed on the real axis and will generally be distinct for each rule. Hence, it makes little sense to give these sets linguistic names, and we have chosen to just give the B-spline coefficient (singleton location) as the rule consequences. In some cases we have added a descriptive label like (S), (M) or (L) indicating that the coefficient is relatively small, medium or large. This label is not to be confused with fuzzy sets, also using the same labels.

The fuzzy input sets are either defined manually or they are automatically determined by the ASMOD algorithm. In either case, the input sets are totally defined by the knot vectors, the spline degrees and the tensor products formed. A knot vector specifies a partitioning of the input domain of the corresponding input variable. A minimal linear spline function will have two basis functions giving rise to two input sets which we label S (small) and L (large). For every knot inserted within the input domain a new basis functions is created. These are automatically named for instance as S, MS, M, ML, and L according to their center positions at 0, 1/4, 1/2, 3/4, and 1 relative to the input domain. With more knots a correspondingly higher resolution is used for the labeling. The input domains for each variable are normally scaled to their respective dynamic ranges. A fuzzy set labeled L may thus have a different support range when referring to different input variables, depending on the dynamic range of the variable, the knot vector and the spline degree. In a more formal setting these sets should thus have been given different labels.

A fuzzy rule generated by ASMOD may thus look like

If Cells is L and Glucose is S then $\Delta\chi$ is 0.041 (S)

This rule should be interpreted as follows. If the concentration of cell is large and the concentration of glucose is small compared to their variational range, then this influences the cell growth by the amount 0.041 (which is relatively **S**mall). Notice that we *should not* interpret the rule as: If the concentration of cells is large and the concentration of glucose is small compared to their variational range, then the cell growth *is* 0.041. This is because the rule only states how the cell growth is influenced by the two variables Cells and Glucose. There may still be other system variables, which may be included in other rules, that make the growth large. Such decoupled influences from different input variables are reflected in the ASMOD models by the decomposition into several low dimensional submodels.

Such a decomposition into additive submodels (or additive rule sets) is well known to give a dramatic reduction in the overall model complexity of high dimensional systems, an is in fact one of the major merits of the ASMOD modelling scheme. As a very simple example, let us say we have the following relationship: $y = ax_1 + bx_2, x_{1,2} \in [0, 1)$. In ASMOD it takes two basis functions to represent this relationship in two linear submodels. To represent this relation by standard fuzzy rules we would need three rules:

If x_1 is L and x_2 is S then y is a
If x_1 is S and x_2 is L then y is b
If x_1 is L and x_2 is L then y is $a + b$

However, if we interpret the consequences of the rules as independent influences on the output variable the relation can be represented by the two rules:

If x_1 is L then y is influenced by the amount a
If x_2 is L then y is influenced by the amount b

These are exactly the rules the ASMOD algorithm would generate, and is the correct interpretation of ASMOD rules.

3. The ASMOD Algorithm

The ASMOD algorithm represents a method for automatically identifying a relevant ASMOD model structures based upon data only. The idea is to start from an initially very simple model structure, or even an empty model (no submodels) and to gradually build up a more complex model. A stepwise model refinement procedure is repeatedly applied to the current model, every time including some new modelling capability which seems to be the most appropriate. Eventually, when no more improvements can be obtained, the model can be pruned back to obtain a more parsimonious model structure, and a correspondingly improved parameter estimation.

Any given model structure can be refined by applying one of the three possible refinement methods:

1. A new one-dimensional submodel can be inserted to model a dependency on an input variable not already included in any of the other submodels.

2. A new knot may be inserted in a knot vector of one of the current variables. This will locally introduce a greater ability to model a non-linear dependency of the corresponding input variable. To limit the number of refined candidate structures, the ASMOD algorithm only evaluates insertions of single knots at the midpoint between two existing knots.

3. Two of the current submodels may be combined and replaced by a new submodel constructed as the tensor product of the two. The new submodel comprises the modelling capability of the replaced submodels, and also allows for coupled dependencies of the input variables from the two original submodels.

Similarly, there are three methods for model pruning:

1. The spline degree may be decremented for splines with no knots in the interior of the input domain. Hence, a spline which was initially inserted as quadratic can in the pruning process be reduced to represent only a linear dependency. Splines (submodels) with degree zero are removed from the model.

2. A knot may be removed from a knot vector of one of the current input variables. A knot may typically be made superfluous when new knots are inserted next to it, and such knots will hence be removed.

3. A tensor product submodel may be split into two lower dimensional submodels. This will normally be done if the possibility for modelling a coupled dependency of the variables now being split did not give a significantly improved accuracy.

For every step in the model refinement or pruning sequence all possible candidates for refined or pruned model structures are evaluated. The candidate structures are evaluated by first optimising the model parameter vector (spline coefficients) and then by computing the optimisation criterion. The candidate having the best performance with respect to the chosen criterion smallest criterion is selected as the new model structure. The choice of optimisation criterion may influence the final model structure, and should thus be considered with care.

When fitting the model to the training data we want to determine the parameter vector c^* in (3) which minimises the expected value of the chosen non-negative criterion function $f(y - c^T b(x))$, i.e.

$$c^* = \arg\min_c f(y - c^T b(x)). \tag{17}$$

Here we will only consider the quadratic criterion given by

$$I(c) = E(y - c^T b(x))^2, \tag{18}$$

where E is the expectation operator. Other criteria may be chosen for more robustness with respect to non-normal noise distributions and outliers in the data.

Since the underlying probability distribution is unknown, we cannot evaluate the expected value of the criterion directly. Instead, we can compute the empirical value

$$I_{\text{emp}}(c) = \frac{1}{l}\sum_{i=1}^{l}(y^i - c^T b(x^i))^2 \tag{19}$$

from the training data $(x^1, y^1), \ldots, (x^l, y^l)$, and use this as an estimate for $I(c^*)$.

It is well known that if the criterion function is evaluated on the same data set as used for estimating the parameter vector, far too optimistic results may be obtained. If there is enough data it is better to split the data into independent training and test sets, where the test set is used to estimate the expectation error (18) for new independent data. Very often we do not have this much data, and other methods must be used.

Kavli and Weyer[13] have tested several possible criterion functions for the ASMOD algorithm. These all aim at estimating the expectation error $I(c^*)$ from the training data. The more successful criteria were Akaike's Information Theoretic[2] criterion, the Final Prediction Error[1] criterion, the Structural Risk Minimisation[15] criterion and full cross validation[7].

For all these criteria the vector c^* minimising the criterion is the same as the one minimising the empirical value I_{emp} (19). Minimising I_{emp} is a linear regression problem, and the solution is commonly known to be defined by the Moore-Penrose pseudo-inverse as

$$c^* = (B^T B)^{-1} B^T y.$$

Here B is the matrix of basis functions evaluated at the input vectors, i.e. $B_{i,j} = b_j(x^i)$, and y is the vector of observed output variables. Singular value decomposition or another numerically stable algorithm should be used for the matrix inversion. The backfitting algorithm can be used to decompose the fitting problem into several smaller problems of fitting individual submodels[7].

4. Modelling a Biological Batch Process

To demonstrate how the relationship between ASMOD models and fuzzy systems can be utilised in an interactive model building process, we have applied the methodologies for identification of a non-linear state space model for a batch fermentation process where the micro-organism *Pseudomonas ovalis* transforms glucose into gluconic acid. This process has been extensively studied[6], and an analytically derived state space model was used to generate simulated data for the tests[11]. This model uses the five state variables:

χ	Concentration of Cells
p	Concentration of Gluconic acid
l	Concentration of Gluconolactone
s	Concentration of Glucose
c	Concentration of Oxygen

and is given by:

$$\begin{aligned}
\dot{\chi} &= \mu_m \frac{sc}{k_s c + k_O s + sc}\chi \\
\dot{p} &= k_p l \\
\dot{l} &= \upsilon_l \frac{s}{k_l + s}\chi - 0.91 k_p l \\
\dot{s} &= -\frac{1}{Y_s}\mu_m \frac{sc}{k_s c + k_O s + sc}\chi - 1.011\upsilon_l \frac{s}{k_l + s}\chi \\
\dot{c} &= k_l a(c^* - c) - \frac{1}{Y_s}\mu_m \frac{sc}{k_s c + k_O s + sc}\chi - 0.09\upsilon_l \frac{s}{k_l + s}\chi .
\end{aligned}$$

The principal reaction chain in the process is given by

$$\text{Glucose} + O_2 \xrightarrow{\text{Cells}} \text{Gluconolactone} + H_2O \rightarrow \text{Gluconic Acid}. \qquad (20)$$

Glucose and oxygen are consumed by the cells for the cell growth, while producing gluconolactone as a by-product. Gluconolactone then reacts with water to form the final product gluconic acid. Oxygen is continuously added to the system during fermentation trying to keep the oxygen concentration at a steady level.

The aim of our fuzzy and ASMOD models was to predict the change in the state variables half an hour ahead in time, based on the currently measured or predicted state variables. The models were then used for ballistic prediction of 10 hour batches, based upon measured initial conditions.

Johansen and Foss generated simulation data sets used for model identification and testing by integrating the equations over batch durations of 10 hours, starting from randomised initial states. The five state variables were sampled every half hour, giving 20 samples from each batch. The sampled variables were scaled to a similar range and added random normally distributed noise. The data set used for identification consisted of 10 batches, and the test data had 30 batches. For details on the parameters and simulations see Johansen and Foss[9,11].

4.1. Fuzzy State Space Models

Initially, a very simple model was constructed based upon the general system understanding as described above. The different state variables were modelled independently by the following set of rules:

Table 1: Initial model based on *a priori* fuzzy knowledge.

χ: Cells, s: Glucose, c: Oxygen			
Cells	If χ is L and s is L and c is L	then $\Delta\chi$ is	O^1
Gluconic acid	If l is L	then Δp is	O^2
Gluconolactone	If χ is L and s is L	then Δl is	O^3
	If l is L	then Δl is	O^4
Glucose	If χ is L and s is L and c is L	then Δs is	O^5
Oxygen	If c is L	then Δc is	O^6
	If χ is L and s is L	then Δc is	O^7

The consequence part of the rule describes the expected change in the state variable one half hour ahead in time. The fuzzy input set L(arge) was triangular (linear B-spline) covering the entire dynamic range of the input variables. The fuzzy output sets O^i were singletons determined by the least squares fitting of the model to the training data. With only one input set per variable this model was only able to represent linear dependencies. Initial testing showed that this rule base was too small to make any decent predictions, cf. the first row of table 7.

Although the rule set above is not optimal, it does represent the fundamental dynamics of the system. We used this representation as a basis for a more accurate fuzzy description. Using only linear dependencies is a strong restriction, so we inserted a knot where we believed more flexibility was needed. By doing this we were able to operate with the fuzzy input set Medium, in addition to Large and Small. This is useful in, e.g. the cell growth description, where medium amount of cells and medium amount of glucose represent the state which has the largest cell growth. Only

Table 2: Models for Cells

	Manual			
R^1	If Cells is L and Glucose is S	then $\Delta\chi$ is	0.041	(S)
R^2	If Cells is M and Glucose is M	then $\Delta\chi$ is	0.152	(L)
R^3	If Cells is S and Glucose is L	then $\Delta\chi$ is	0.015	(S)
	Refined Manual			
	χ: Cells, s: Glucose, c: Oxygen			
R^1	If χ is S and s is S and c is S	then $\Delta\chi$ is	0.134	(L)
R^2	If χ is M and s is S and c is S	then $\Delta\chi$ is	0.104	(M)
R^3	If χ is S and s is M and c is S	then $\Delta\chi$ is	0.137	(L)
R^4	If χ is M and s is M and c is S	then $\Delta\chi$ is	-0.040	(-S)
R^5	If χ is M and s is S and c is L	then $\Delta\chi$ is	0.020	(S)
R^6	If χ is L and s is S and c is L	then $\Delta\chi$ is	-0.089	(-M)
R^7	If χ is S and s is M and c is L	then $\Delta\chi$ is	-0.033	(-S)
R^8	If χ is M and s is M and c is L	then $\Delta\chi$ is	0.079	(M)
R^9	If χ is S and s is L and c is L	then $\Delta\chi$ is	0.044	(S)
	Automatic			
R^1	If Cells is S	then $\Delta\chi$ is	0.026	(M)
R^2	If Cells is MS	then $\Delta\chi$ is	0.018	(S)
R^3	If Cells is ML	then $\Delta\chi$ is	0.046	(L)
R^4	If Cells is L	then $\Delta\chi$ is	-0.060	(-L)
R^5	If Oxygen is S	then $\Delta\chi$ is	0.019	(S)
R^6	If Oxygen is M	then $\Delta\chi$ is	0.037	(M)
R^7	If Oxygen is L	then $\Delta\chi$ is	-0.026	(-M)

rules corresponding to physically plausible situations were used, so a rule describing the output at large cell levels and large glucose levels, was not considered since this situation is never encountered in the process.

With these guidelines, new models were set up where we made use of both process understanding and the knowledge we have gained from the empirical data. The resulting models are shown in table 2 through 6 under the label *Manual*. The linguistic labels (**S**mall, **M**edium etc.) on the consequences have been added to facilitate the reading.

We have also used the ASMOD algorithm to identify state space models automatically. Two different settings have been used. For the models labeled *Refined Manual*, the manually derived models were used as starting points for the ASMOD algorithm, which was run without supervision until no further improvements could be obtained.

Table 3: Models for Gluconic Acid.

	Manual			
R^1	If Gluconolactone is S	then Δp is	0.002	(S)
R^2	If Gluconolactone is L	then Δp is	0.142	(L)
	Refined Manual			
R^1	If Gluconolactone is S	then Δp is	-0.026	(-S)
R^2	If Gluconolactone is L	then Δp is	0.092	(L)
R^3	If Oxygen is S	then Δp is	0.041	(M)
R^4	If Oxygen is L	then Δp is	0.026	(S)
	Automatic			
R^1	If Gluconolactone is S	then Δp is	0.002	(S)
R^2	If Gluconolactone is L	then Δp is	0.142	(L)

For the models labeled *Automatic* the ASMOD algorithm was initialised with empty models and was then run without human interaction. The Structural Risk Minimisation criterion and linear B-splines were used for all runs. The purpose of these runs was to study the interpretability of the automatically generated models, and how the incorporation of *a priori* knowledge affected the models, both with respect to performance and interpretability.

The total root mean square errors (RMS) for ballistic prediction of 30 test batches (which were independent of the training data) are given in table 7. The ballistic predictions were obtained by repeatedly applying the models to the estimated state vectors, starting from the noisy measured initial state for each batch. The given RMS values are mean values for all 30 test batches, and all 20 measurements from each batch. Figure 5 shows ballistic predictions of the first test batch by the three different model types.

4.2. Discussion of Results

We were mainly concerned with two properties of the derived models: interpretability and accuracy. The manually derived models had for obvious reasons input sets and rules with high interpretability. The output sets (the B-spline coefficients) were automatically identified by the least squares fitting of the model to the training data. Also, these coefficients reasonably agreed with our understanding of the system.

The automatically derived models differed from manual models both with respect to structure and coefficients. We see that the ASMOD algorithm when applied to the manual models tend to increase the model complexity, either by introducing new variables, by inserting more knots, or by making new tensor products. By these

Table 4: Models for Gluconolactone.

	Manual			
R^1	If Cells is L and Glucose is S	then Δl is	-0.171	(-M)
R^2	If Cells is M and Glucose is M	then Δl is	0.380	(L)
R^3	If Cells is S and Glucose is L	then Δl is	0.069	(S)
R^4	If Gluconolactone is S	then Δl is	0.000	(S)
R^5	If Gluconolactone is L	then Δl is	-0.223	(-L)
	Refined Manual			
R^1	If Gluconolactone is S	then Δl is	0.322	(L)
R^2	If Gluconolactone is L	then Δl is	-0.018	(-S)
R^3	If Cells is M and Glucose is S	then Δl is	-0.300	(-L)
R^4	If Cells is L and Glucose is S	then Δl is	-0.450	(-L)
R^5	If Cells is S and Glucose is M	then Δl is	-0.302	(-L)
R^6	If Cells is M and Glucose is M	then Δl is	-0.031	(-S)
R^7	If Cells is S and Glucose is L	then Δl is	-0.338	(-L)
R^8	If Cells is M and Glucose is L	then Δl is	0.259	(M)
	Automatic			
R^1	If Gluconolactone is S	then Δl is	0.271	(L)
R^2	If Gluconolactone is M	then Δl is	0.100	(S)
R^3	If Glucose is MS and Cells is S	then Δl is	-0.248	(-L)
R^4	If Glucose is ML and Cells is S	then Δl is	-0.275	(-L)
R^5	If Glucose is S and Cells is MS	then Δl is	-0.248	(-L)
R^6	If Glucose is MS and Cells is MS	then Δl is	0.020	(S)
R^7	If Glucose is ML and Cells is MS	then Δl is	0.213	(L)
R^8	If Glucose is S and Cells is ML	then Δl is	-0.323	(-L)

Table 5: Models for Glucose.

	Manual			
R^1	If Cells is L and Glucose is S	then Δs is	-0.016	(-S)
R^2	If Cells is M and Glucose is M	then Δs is	-0.216	(-L)
R^3	If Cells is S and Glucose is L	then Δs is	-0.012	(-S)
	Refined Manual			
R^1	If Cells is S and Glucose is S	then Δs is	-0.103	(-M)
R^2	If Cells is L and Glucose is S	then Δs is	0.065	(S)
R^3	If Cells is S and Glucose is M	then Δs is	0.024	(S)
R^4	If Cells is L and Glucose is M	then Δs is	-0.302	(-L)
R^5	If Cells is S and Glucose is L	then Δs is	-0.011	(-S)
R^6	If Cells is L and Glucose is L	then Δs is	-0.548	(-L)
	Automatic			
R^1	If Glucose is XS	then Δs is	0.099	(L)
R^2	If Glucose is S	then Δs is	0.011	(S)
R^3	If Glucose is M	then Δs is	-0.039	(-M)
R^4	If Glucose is L	then Δs is	-0.075	(-L)
R^5	If Glucose is XL	then Δs is	-0.091	(-L)
R^6	If Cells is S	then Δs is	0.081	(M)
R^7	If Cells is M	then Δs is	-0.109	(-L)
R^8	If Cells is L	then Δs is	-0.067	(-M)

refinements the modelling error was reduced to less than half the error of the manual models.

Since we had already selected what we thought were the most relevant input variables when ASMOD was started, it was very unlikely that the algorithm should replace these variables with other variables, something that also did not happen. The inclusion of oxygen as a new variable in the model for cell growth sounds reasonable, since we know that oxygen is a requirement for cell growth. But, the additional dependency on oxygen in the model for gluconic acid is totally unreasonable and must be a result of overfitting to the random noise. By supervising the selection of candidates in the model refinement process, this obvious error could have been avoided, and a simple and more intuitive model could have been obtained. All the other refined manual models had structures which corresponded to the original manual models, and were thus in compliance with our *a priori* understanding.

The models which were automatically derived from empty initial models also had structures which were comparable to the manual and refined manual models. Except for the cell growth model, the same input variables were selected by the ASMOD

Table 6: Models for Oxygen.

	Manual			
R^1	If Cells is L and Glucose is S	then Δc is	0.159	(M)
R^2	If Cells is M and Glucose is M	then Δc is	-0.580	(-L)
R^3	If Cells is S and Glucose is L	then Δc is	-0.111	(-M)
R^4	If Oxygen is S	then Δc is	0.245	(L)
R^5	If Oxygen is L	then Δc is	-0.010	(-S)
	Refined Manual			
	χ: Cells, s: Glucose, c: Oxygen			
R^1	If χ is MS and s is MS and c is S	then Δc is	-0.011	(-S)
R^2	If χ is ML and s is MS and c is S	then Δc is	0.068	(M)
R^3	If χ is ML and s is S and c is MS	then Δc is	0.273	(L)
R^4	If χ is MS and s is MS and c is MS	then Δc is	-0.010	(-S)
R^5	If χ is ML and s is MS and c is MS	then Δc is	-0.115	(-L)
R^6	If χ is MS and s is ML and c is MS	then Δc is	-0.061	(-M)
R^7	If χ is ML and s is S and c is ML	then Δc is	0.271	(L)
R^8	If χ is S and s is ML and c is ML	then Δc is	-0.039	(-M)
R^9	If χ is MS and s is ML and c is ML	then Δc is	-0.074	(-M)
R^{10}	If χ is S and s is L and c is ML	then Δc is	-0.088	(-M)
R^{11}	If χ is ML and s is S and c is L	then Δc is	-0.011	(-S)
R^{12}	If χ is L and s is S and c is L	then Δc is	0.015	(S)
R^{13}	If χ is S and s is ML and c is L	then Δc is	0.004	(S)
	Automatic			
R^1	If Glucose is MS and Oxygen is S	then Δc is	-0.065	(-S)
R^2	If Glucose is ML and Oxygen is S	then Δc is	-0.079	(-M)
R^3	If Glucose is S and Oxygen is MS	then Δc is	0.174	(L)
R^4	If Glucose is MS and Oxygen is MS	then Δc is	-0.043	(-S)
R^5	If Glucose is ML and Oxygen is MS	then Δc is	-0.094	(-M)
R^6	If Glucose is L and Oxygen is MS	then Δc is	-0.151	(-L)
R^7	If Glucose is S and Oxygen is ML	then Δc is	0.122	(L)
R^8	If Glucose is ML and Oxygen is ML	then Δc is	-0.122	(-L)
R^9	If Glucose is L and Oxygen is ML	then Δc is	-0.138	(-L)
R^{10}	If Glucose is S and Oxygen is L	then Δc is	-0.121	(-L)
R^{11}	If Glucose is ML and Oxygen is L	then Δc is	-0.144	(-L)
R^{12}	If Glucose is L and Oxygen is L	then Δc is	-0.058	(-S)
R^{13}	If Cells is S	then Δc is	0.098	(M)
R^{14}	If Cells is M	then Δc is	0.060	(S)
R^{15}	If Cells is L	then Δc is	0.165	(L)

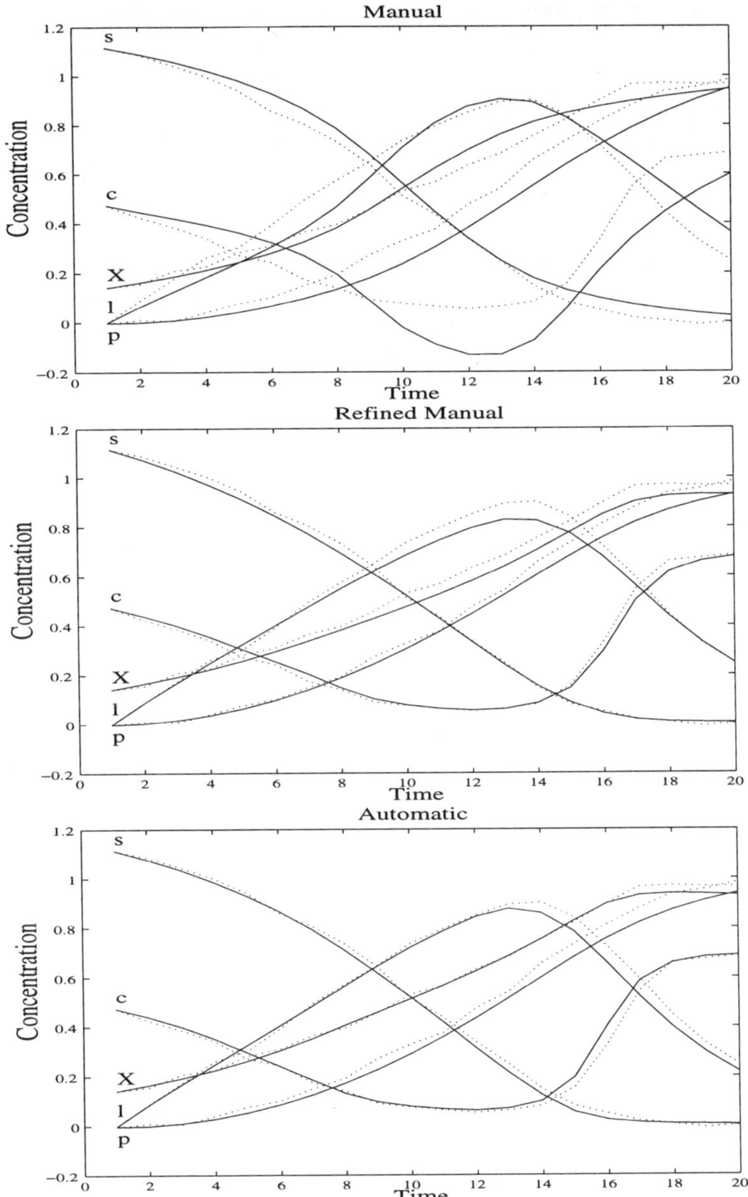

Figure 5: Ballistic prediction using the three different models sets. The predictions are shown as solid lines.

Table 7: The RMS error for ballistic prediction of 30 test batches.

	$\Delta\chi$	Δp	Δl	Δs	Δc	Total
Initial	0.254	0.131	0.334	0.349	0.571	0.328
Manual	0.059	0.039	0.062	0.037	0.089	0.057
Refined Manual	0.027	0.020	0.031	0.022	0.027	0.025
Automatic	0.031	0.025	0.037	0.025	0.038	0.031

algorithm as was selected in the manual model construction. For the cell growth model, cell and oxygen concentrations were used instead of the manually selected cells and glucose. But, as we noticed for the refined manual model, all three variables are relevant. For the other models, the differences between the manual and the fully automatic models were the selection of tensor products formed and the number of functions used for individual input variables.

When analysing the rule consequences, we discovered that they were not always in accordance with our system understanding. We must keep in mind that when we set up the manual model we try to capture the general dependencies among the system variables, whereas the automatic modelling aims at describing the special situations captured in the batch process data. A batch always starts with high concentrations of glucose, low concentrations of cells, gluconolactone and gluconic acid and some medium concentration of oxygen. And the batch ends up with low concentrations of glucose, and high concentrations of cells and gluconic acid and again some medium concentration of oxygen (see figure 5). The process thus runs through similar trajectories in the state space every time, covering only a small fraction of the five dimensional hypercube defining the model input domain. Moreover, there is a high degree of correlation between how the different state variables develops over the batch duration. This makes it difficult for the automatic algorithm to determine which dependencies are causal, and which are not.

If we for instance look at the automatically generated model for cell growth, it does not show the monotonic increasing dependency in cell concentration as we might expect from the differential equations. We rather see an abrupt decline in cell growth when the amount of cells is large. This, of course, is a result of that the growth stops when there is no more glucose left for the fermentation. This dependency on glucose is implicitly modelled as a dependency on cell concentration, avoiding the extra cost of including glucose as a new variable to the model. Since the oxygen concentration also has a repeated development in every batch, oxygen can be used to compensate errors in the model based on cell concentration only, thus giving a complex interplay between the two variables. Hence, the different rules can not be interpreted independently, and the overall model may become difficult to interpret.

As discussed above, the structure of the automatically generated models was in this case reasonable. But the algorithm searches in the data for any relationship that can be used for estimating the target variable. With the high degree of correlation among the variables, as found in this data set, the relations chosen by the algorithm does not need to correspond to causal dependencies. In fact, ASMOD has been run several times on these data with other parameter settings, giving different and often not so reasonable model structures. However, the model accuracy measured as ballistic predictions were comparable in all cases. This shows that there exist many different model structures which gives similar and good results on this problem, but not all are interpretable.

5. Discussion

There are several ways of incorporating fuzzy knowledge into an ASMOD model. The simplest is to initialise the model in accordance with *a priori* knowledge. This means including the variables one believes are important for the output and to form the proper tensor products. Also, the degree of non-linearity can be set by choosing the spline order and the knot resolution.

A more detailed control can be achieved by supervising, and if necessary override, the automatic selection of the refinement candidates. This gives a nice interaction between the algorithm and the user. The algorithm may suggest a set of possible model refinements together with estimated improvements in accuracy. The user can then manually choose among the candidates and select the one which best matches his *a priori* knowledge. Candidates that are obviously wrong can be eliminated.

The fuzzy input sets corresponding to an ASMOD submodel always fill up a grid that partitions the input domain. But subsets of the full rule set can in effect be obtained by forcing the coefficients of some basis functions to zero, thus removing the influence of the corresponding rules. From the practical point of view this can be obtained by simply removing the basis functions from the fit. But, for the equivalence between fuzzy systems and ASMOD to hold (by maintaining the partitioning of unity by the basis functions), we must interpret this as if the basis functions were present with coefficients set to zero.

Removal of basis functions has been used in our experiment by removing all functions which have little support in the training data set. A basis function is said to have little support if the sum of its evaluation over all training points is less than a threshold, in our case set to 5.

In this Chapter we have used training data to compute the consequence part of the rules. We could also have set the consequences manually based on our system knowledge. This approach may be appropriate if no measurement data is available, or if the data material is too sparse in some parts of the input domain to perform parameter estimation.

Ideally one would like to use both sources of information simultaneously. This can be achieved by a weighed least square fitting of the model to training data, weighing between the *a priori* coefficients set up by an expert and the coefficients determined by the data. Different methods can be used to locally weigh the two sources of information, putting more emphasis on the data where the data density is high relative to the density of membership functions, and putting the emphasis on the expert knowledge where data are lacking or sparse.

Further investigation of these possibilities remains a topic for future research, as does the possibility to include mechanistic models as the consequence of some rules.

6. Conclusion

In this Chapter we have established an equivalence between ASMOD models and a certain type of fuzzy models. The ASMOD algorithm thus represents an automatic scheme to generate a fuzzy rule description of a system. We examined how this could be utilised in the construction of non-linear state space models for a batch fermentation process. Models based purely on fuzzy knowledge were inferior to models automatically identified. However, this does not mean that fuzzy knowledge should not be incorporated into the ASMOD scheme. Models which were manually initialised based on expert knowledge and then automatically refined with the ASMOD algorithm gave the most accurate results.

Interpreting ASMOD models in terms of the corresponding fuzzy rules is another application of the link to fuzzy logic. This linguistic representation may be easier to comprehend than inspecting the model parameters directly. Fuzzy interpretation can be an aid to better understand the model and can also be used for interactive control of the ASMOD model refinement process.

Analysis of the fuzzy rules obtained with the ASMOD algorithm for the fermentation process showed that great care must be taken when trying to interpret such models. Due to the repeated runs of similar batches, there were strong internal dependencies between the five state variables of the system. This resulted in models where features in the target variables were explained by noncausal features in the input vectors. A direct interpretation of the rules as reflecting properties of the fermentation process could thus lead to erroneous conclusions.

The problems encountered here are not specific to ASMOD, but are common for all empirically based modelling schemes. The more expert that knowledge can be built into the model, by manual model initialisation or by manual supervision and control of the model construction process, the more interpretable the final model will be.

7. References

1. H. Akaike. Fitting autoregressive models for prediction. Annals of the Institute of Statistical Mathematics, **21**:425–439, 1969.
2. H. Akaike. A new look at the statistical model identification. IEEE Transactions on Automatic Control, **19**:716–723, 1974.
3. K.M. Bossley, D.J. Mills, M. Brown and C.J. Harris. Construction and Design of Parsimonious Neurofuzzy Systems. In *Neural Networks Engineering in Dynamic Control Systems*, eds. K.J. Hunt G.R. Irwin and K. Warwick, Advances in Industrial Control. Springer 1995.
4. M. Brown and C. Harris. *Neurofuzzy adaptive modelling and control.* Prentice-Hall International (UK) Limited, 1994.
5. M. Carlin, T. Kavli, and B. Lillekjendlie. A comparison of four methods for non-linear data modelling. Chemometrics and Intelligent Laboratory Systems, **23**:163–177, 1994.
6. T.K. Ghose and P. Ghosh. Kinetic analysis of gluconic acid production by pseudomonas ovalis. Journal of Applied Chemical Biotechnology, **26**:768–777, 1976.
7. T.J. Hastie and R.J. Tibshirani. *Generalized Additive Models.* Chapman and Hall, London, 1990.
8. D.R Hush and B.G. Horne. Progress in supervised neural networks. What is new since Lippman. IEEE Signal Processing Magazine, 8–39, Jan 1993.
9. T.A. Johansen. *Operating Regime based Process Modeling and Identification.* Dr. Ing. thesis, Department of Engineering Cybernetics, Norwegian Institute of Technology, Norway 1994
10. T.A. Johansen and B.A. Foss. Fuzzy model based control: Stability, robustness, and performance issues. IEEE Transactions on Fuzzy Systems **2**, 221–234, 1994
11. T.A. Johansen and B.A. Foss. Semi-empirical modelling of non-linear dynamic systems through identification of operating regimes and local models. In *Neural Networks Engineering in Dynamic Control Systems*, K.J. Hunt, G.R. Irwin and K. Warwick (Eds.), pages 105–126, Springer Series on Advances in Industrial Control, Springer Verlag, 1995.
12. T. Kavli. ASMOD—An algorithm for adaptive spline modelling of observation data. International Journal of Control, **58(4)**:947–967, 1993.
13. T. Kavli and E. Weyer. On ASMOD—An algorithm for empirical modelling using spline functions. In *Neural Networks Engineering in Dynamic Control*

Systems, K.J. Hunt, G.R. Irwin and K. Warwick (Eds.), pages 83–104, Springer Series on Advances in Industrial Control, Springer Verlag, 1995.
14. L.L. Schumaker. *Spline Functions: Basic Theory*. Wiley, New York, 1981.
15. V. Vapnik. *Estimation of Dependencies Based on Empirical Data*. Springer-Verlag, 1982.
16. L.X. Wang. *Adaptive fuzzy systems and control, design and stability analysis*, Prentice Hall, 1994
17. C.H. Wang, W.Y. Wang, T.T. Lee, P.S. Tseng. Fuzzy B-Spline Membership Functions (BMF) and Its Applications in Fuzzy-Neural Control. IEEE Transactions on Systems, Man, and Cybernetics, **25**, No. 5, May 1995.
18. L. A. Zadeh. Outline of a new approach to the analysis of complex systems and decision processes. IEEE Transactions on Systems, Man, and Cybernetics, **3** 1973.

INDEX